Introduction to
Digital
Astrophotography

Imaging the Universe with a Digital Camera

Robert Reeves

Willmann-Bell, Inc.

P.O. Box 35025 • Richmond, VA 23235 • **TOLL FREE 1 (800) 825-7827** • (804) 320-7016 • FAX (804) 272-5920
www.willbell.com

Published by Willmann-Bell, Inc., P.O. Box 35025, Richmond, Virginia 23235
First Printing May 2005
Second Printing April 2006

Printed in the United States of America

Library of Congress Cataloging in Publication Data
Reeves, Robert, 1946-
  Introduction to digital astrophotography : imaging the Universe with a digital camera / by Robert Reeves
    p. cm.
  Includes bibliographical references and index
  ISBN 0-943396-83-2
  1. Astronomical photography. I. Title,

QB121.R446 2005
522'.63--dc22

                                                        2004063706

06 07 08 09 10 9 8 7 6 5 4 3 2

Table of Contents

Preface
Why Do Astrophotography?

It wasn't until I wrote the book *Wide-Field Astrophotography*, published by Willmann-Bell in 2000, that I gave much thought to why we amateur astronomers photograph the sky. Having done just that for over 40 years using various film cameras, astrophotography was simply an ingrained part of who I was, and I did not question it. But having to explain the process of astrophotography to various audiences forced the question: why do it?

Everyone has a different motivation for aiming their camera at the sky. For me, the roots of it date back to 1957 when *National Geographic* published an article about the ongoing Palomar Observatory Sky Survey (POSS) using the big 48-inch Schmidt telescope. The images of the sky in that article were spellbinding. I was amazed at the detail and depth of space. What I had previously perceived as a desolate void was actually filled with stars and swirling galaxies, each as unique as a snowflake. Though I had no interest in photography in general, seeing the Palomar images flipped a switch in my mind; taking images like these was something I wanted to do myself.

In that pre-space age time, there weren't nearly the resources for the astrophotographer that exist today. There were fewer telescope companies, few publications offering advice, no industry devoted to adapting photographic devices to astronomical instruments, and virtually no contact between fellow amateur astronomers with similar interests except those who belonged to the same astronomy club. The individual was basically alone and had to work out the art of astrophotography through trial and error. My initial star voyage consisted of taking the family's WW II-vintage Voightlander 120-format camera to the roof of our south-Texas garage. Using a block of wood to brace it in place, I snapped a one-second image of the brightest object in the sky. I later learned that I had targeted Jupiter. My initial success encouraged me to lengthen my exposures to a half hour. The images revealed hundreds of stars and showed the patterns of the constellations circling the north celestial pole.

In the pre-light pollution era of the 1950s, the Milky Way was easily visible from the suburbs of large cities like my native San Antonio. The sight of our galaxy was inspiring and exciting. It quickly became a magnet for my celestial photography, and remains so today. In the introduction to his classic *Atlas of Selected Regions of the Milky Way* (1927), Edward E. Barnard summed up the attraction it holds for me:

> ...the Milky Way reveals all its wonderful structure, which is so magnificent in photographs made with the portrait lens. The observer with the more powerful telescopes, and necessarily the more restricted field-of-view, has many things to compensate him for his small field-of-view, but loses essentially all the wonders of the Milky Way....It was these views of the great structures in the Sagittarius region of the Milky Way that inspired me with the desire to photograph these extraordinary features, and one of the greatest pleasures of my life was when this was successfully done at Lick Observatory in the summer of 1889.

Today, I have come to view the astrophotographer as both an explorer and someone who loves nature. It is easy to appreciate the natural beauties that surround us; the colors of different species of birds and butterflies that visit our back yard, the crystalline structure and varied hues of different minerals, even the shape and flow of clouds in our dynamic atmosphere. However, an unfortunate side-effect of our increasingly urbanized society is that many people have a "geocentric" view of nature; that is, they see only the Earth-based portions of it. The heavens are ignored by most people, partly because they simply can't see them due to man-made obstacles such as light pollution and urban apartment living that prevents us from simply going out in the back yard and looking up at night. The astronomer and astrophotographer, however, have a broader vision of nature. They know what is up in the night sky and have a curiosity about it.

I believe that one of the reasons why astrophotographers work to capture the beauties of the night sky is because their handiwork communicates astounding celestial majesty, and they want to share it with others. Also, photography extends the range of human vision and gives us the power to see things that our ancestors never knew existed. For instance, under dark skies the average person can see stars down to about 6th magnitude—that is, about 6000 stars with a brightness range of perhaps 250 to 1. A one-inch aperture camera lens extends this to 5 million stars with a brightness range of about 60,000 to 1. Indeed, a digital SLR camera can easily image all the stars plotted in *Sky Atlas 2000.0* with simple tripod-mounted, unguided, 20-second exposures, and show them in color. What seems like a familiar sky view becomes a thrilling new area of discovery when the camera reveals far more than the eye can see.

A telescope is not needed to record many of the beauties in the sky. This combination of celestial and atmospheric targets was recorded with a handheld Canon 10D at ISO 400 and a 70-200-mm zoom lens at f/9.5 at a 1/350-sec. exposure. Photo by Jeff Ball.

The "near instant" results achieved with digital cameras also bring much excitement to astrophotography. There is no more waiting to get the film developed. Indeed, the thrill of instant results was a turning point in my own astrophotography adventures. I am used to shooting astronomical images with my Nikon film cameras and sometimes waiting several months before all the pictures on the film have been exposed. There have been times when I forgot what was on the roll of film by the time it is processed. While this can lead to pleasant surprises, it also takes out some of the fun because I know it may be a long while before I see the results.

In 2001 I entered the realm of digital photography with the purchase of a 3 megapixel Olympus 3020. The camera was a joy to work with for snapshot photography, and I quickly applied it to the Moon with my trusty 8-inch Schmidt-Cassegrain telescope. I was astonished to see that I could take better Moon pictures by accident with the digital camera than I could on purpose with the old film cameras, and get the results immediately. Several years later I acquired a Canon 10D. My initial imaging used a 200-mm telephoto lens which gave approximately the same field-of-view as my venerable 8-inch Schmidt camera. As the first 10D image appeared on my computer screen just seconds after the shutter closed, I was absolutely fascinated. There was an image of the star cluster M13 swimming in front of a starry backdrop, and in color! I was hooked! The old thrill I first felt years ago when photographically prowling the skies was back and stronger than ever. The pure fun of seeing the unseen is so strong that I cannot wait to get back under dark skies with a camera. I haven't met another astrophotographer who didn't feel the same way after seeing their first digital image.

The challenge of tinkering with technical apparatus and the desire to master a new skill are also motivators to those who pursue astrophotography. But there is artistic satisfaction as well. In spite of the technicalities involved, astrophotography can go beyond optics, electronics, and computers and be aesthetically pleasing. There is a great deal of pride in displaying a fine celestial photograph and knowing, "I did this; it is my picture." Imaging the sky offers a universe beyond description and imagination to those who are willing to seek it out.

I do astrophotography to satisfy one person—me! I want to capture the natural beauty of the constellations, to chart the river of stars that makes up the Milky Way, to see the unseen, and record these vistas to enjoy again and again. If I am satisfied with what I do, there is a trickle-down effect. Others around me are exposed to new ideas and new worlds through my enthusiasm for astrophotography.

To succeed at astrophotography, my philosophy dictates that I pursue

A beautiful sunrise over the ocean is punctuated by the transit of Venus in June, 2004. This image was taken with a Canon 10D at ISO 200 using a 70-200-mm zoom lens and a Canon 2X converter at 1/760-second at f/5.6. Photo by Jeff Ball.

projects that I can actually accomplish within the range of my equipment's capability. This does not preclude experimenting to see what is possible; indeed, I often push the limits of my instruments. But realistic goals must be maintained. Biting off more than you can chew is a sure way to become discouraged. New skills take time to master. Simply spending money on fancy equipment does not automatically guarantee success.

Critical analysis of my work is also important. If something failed, I have to figure out why in order to achieve success on another try. Bad results can be very instructive if carefully analyzed. Was it an equipment problem, a procedural error, or was it simply beyond the capacity of my equipment? Recognizing that I can make mistakes is important—simply blaming the camera or telescope will only perpetuate an error. It is essential to persevere when the inevitable failure appears. Good astrophotography is a learned skill requiring time, patience and practice. Tutorials and publications can guide your efforts, but practical experience is the real teacher.

I believe that an exceptional view of the sky should be shared with others. It should be shown to audiences, posted on a web page, or submitted to a magazine. I encourage others to do the same and not worry about having their images "ripped off." That is not likely to happen because the only realistic venue for publication is the relatively small amateur astronomy magazine and book market where economics dictate very little in the way of monetary reward.

Good astrophotography is also a natural ambassador for astronomy and science in general. I have yet to see someone who was not moved by a beautiful color portrait of the Milky Way. My efforts are thus geared toward capturing views of our universe and to share with others.

Robert Reeves

Foreword

Author Aldous Huxley provided an exceptionally versatile entry in the lexicon of modern phrases when he penned the title to his 1932 novel *Brave New World.* There comes a time during almost every generation and in every avenue of human endeavor when it seems appropriate to invoke Huxley's words. For the hobby of astrophotography, that time is now.

After more than a century of dominance, emulsion-based astronomical photography has not only passed its Golden Age, but is in such rapid decline that I suspect we'll see someone writing its definitive history before the end of the present decade. Rolls of film and the alchemic interiors of darkrooms have given way to silicon chips and personal computers.

In the world of conventional photography this transition happened so suddenly that even pundits were taken by surprise. But those of us in the world of astrophotography saw it coming for years. Indeed, by the beginning of the 1990s, long before most people ever heard the term "digital photography," amateur astronomers were prowling the heavens with specialized CCD cameras. Many of these individuals were either well heeled or irresistibly drawn to the cutting edge (or both), for back then CCD imaging was expensive, and there were few places where one could turn for advice.

The rewards of digital imaging, however, more than offset its challenges. With CCDs, backyard telescopes in suburban neighborhoods could probe the universe to depths only seen by the largest mountaintop telescopes in the days of emulsion-base photography. Furthermore, digital imaging changed many of astrophotography's long-standing ground rules. Whereas bigger telescopes traditionally produced better pictures with film, with digital imaging you could compensate for a small aperture by simply taking longer exposures. And amateurs leveraged this advantage with the freedom of their observing agendas to create not only images that competed with those made at professional observatories, but in many cases surpassed them. And amateur images grow better with each passing day.

As profound as these changes are, they are not the reasons why I see astrophotography today as a "brave new world." For those we must turn to traditional consumer photography where the digital revolution began changing the landscape just a few short years ago. A new photographer today is more likely to purchase a digital camera than a film camera. And since most astrophotographers ascend from the ranks of conventional photographers, today's budding astrophotographer is likely to be starting out with a digital camera. Until now the wealth of astrophotography literature came from the age of film, and as such leaves gaping holes for the beginner. Because even such basics as focusing a lens and making a time exposure are so vastly different between film and digital cameras, beginners are indeed facing a "brave new world."

With *Digital Astrophotography,* Robert Reeves has prepared a superb new roadmap that will help not only beginners navigate astrophotography's digital landscape, but will also aid those of us who have crossed the territory with film cameras and are now switching to digital. His years of experience have taught him the pitfalls of this transition, so he knows what points bear special emphasis when teaching anyone how to get started with digital astrophotography.

Reeves has also carved himself a fascinating niche in the history of amateur astronomical photography. In my opinion, his previous book, *Wide-Field Astrophotography,* was the last significant work to appear on the subject when film still ruled. This new volume is the first significant one to appear in the brave new world of digital photography.

Dennis di Cicco, Senior Editor
Sky and Telescope Magazine

Acknowledgements

It is challenging today for a single individual to write a book about astrophotography, a subject that embraces many branches of technology and astronomy. At best, an individual can be a good astrophotographer and communicate ideas about how he or she achieves success. But when it comes down to the technical nuts and bolts of the equipment and software needed to fully encompass the topic of celestial photography, that individual also must be a good reporter. For a reporter to write a good story, he or she must have good information sources and credit those sources in their story. That is what I wish to do here. After almost 50 years behind a camera under the stars, I feel I am reasonably competent at astrophotography, but still rely on support and advice from my peers in order to learn more about the topic. The following are those in the worldwide astronomy imaging community that helped make this a well-rounded book about digital astrophotography.

First, this book wouldn't exist without the love and encouragement of Mary Reeves, my wife and companion for the past 25 years. Mary not only supported my work in writing this book, but also went out of her way to provide a home environment conducive to my being able to devote a lot of time to this project when I needed to. I feel blessed to have her support.

Others who helped were: Becky Ramotowski who read early versions of the manuscript, Gregory Hallock Smith who helped me understand camera optics and the various forms of image noise and Jordan Blessing from Scopetronix, Richard Shell from Stellar Technologies International, and Chris Venter from DSLRFocus who generously allowed me access to their astrophotography products. I also gratefully mention computer technician Jay Rubio who saved my bacon not once, but twice, when hardware problems threatened progress on this book.

An astrophotography book would be dull without any astrophotos, so I am very appreciative of the following individuals who generously provided illustrations for this book: Jeff Ball, Jay Ballauer, Fred Bruenjes, Dennis di Cicco, Antonio Fernandez, R. A. Greiner, Hap Griffin, Mark Hanson, Phil Hart, Paul Hyndman, Dale Ireland, Ted Ishikawa, Dave and Jean Kodama, Rick Krejci, Peter Langsford, Terry Lovejoy, Rod Mollise,

Becky Ramotowski, Johannes Schedler, Mark Schmidt, Thomas Schumm, and José Suro.

The public affairs departments of the following corporations also contributed images of various cameras and products: Canon Cameras, Fuji Photo Film USA, Hap's Astrocables, Hoodman Corporation, Hutech Corporation, Konica-Minolta, Olympus America, Nikon, Pentax Imaging Company, Sigma Corporation of America, and Xantrex.

Ideas and concepts are the mainstay of a book like *Digital Astrophotography* and those presented in the following pages come from sources and individuals too numerous to attempt individual acknowledgement. I will however point out that Jordan Blessing and Becky Ramotowski provided the initial concept of core digital astrophotography techniques discussed in this book. Another important source was the technical discussion and advice freely given by the many members of Ginger Mayfield's wonderful *digital.astro* Internet mailing list, a gathering place for 6000 digital astrophotographers from all over the world.

Also Richard Berry, James Burnell, Mike Unsold, and Chris Venter were very generous with their time in helping me understand some of the many software packages that make digital astrophotography the thrilling endeavor that it is today.

This book is dedicated to the memory of Dennis Harrell Smith, a friend who, after showing me the meaning of love for family and fellow man, has taken his place among the imperishable stars.

Chapter 1
Entering the Digital Realm

1.1 Introduction

In 1999 when I finished the book *Wide-Field Astrophotography* the field of astrophotography was on the threshold of change. The century-old technique of processing film images and printing them in a home darkroom was moving toward electronic film scanners and photo editing using computer software. Trays full of foul-smelling chemicals for print making were in the process of being replaced by a desktop covered with inexpensive, yet sophisticated, ink jet printers and home computers. While this was happening, consumer digital cameras that bypassed film development and directly produced images in electronic form, were gaining market acceptance. These early cameras were expensive and produced pixilated low-resolution images, but there were a few professional cameras that heralded the future direction of the market that we now see today.

Although I have spent half a century following amazing advances in space technology and electronics, I was not prepared for the lightning speed at which digital photography overtook film-based imaging. It is true that some film-based astrophotography will remain with us for the foreseeable future because it is a complementary tool for the astroimager. But there is also no denying that digital imaging has leapfrogged film into mainstream astrophotography. Today's larger and higher resolution sensors with lower electronic noise that enable longer exposures, along with powerful image processing techniques, have established digital imaging as a widely accepted and powerful tool in astronomy.

The popularity of digital imaging is readily apparent from the large numbers who discuss it on various Internet forums. Film astrophotography attracts about 500 adherents to the *Astrophoto* mailing list.[1] But in a few short years the *digital_astro* mailing list has grown to include thousands of subscribers. Indeed, as Figure 1.1 demonstrates, amateurs using off-the-

---

[1] In this book I will, from time to time, mention web sites. Since the web is a dynamic medium, specific addresses are listed in Appendix A. Further, an updated list will be available at my publisher's web site (www.willbell.com) in the section devoted to this book.

Fig. 1.1 *Today's digital single lens reflex cameras have unprecedented sensitivity to faint light. This image of the Rossette Nebula was taken with a Canon 10D at ISO 1600 through an Orion 80ED refractor with an f/6.3 focal reducer, and shows faint hydrogen-alpha emission extending to the lower left that is not recorded by the author's 8-inch f/1.5 Schmidt camera. The image is a combination of 42 separate three-minute exposures, 11 four-minute exposures, and 7 five-minute exposures for a total of 3 hours and 35 minutes combined exposure. Photo by Michael Howell.*

shelf consumer digital cameras can achieve results on par with, or better than, they used to with film cameras. In no uncertain terms, digital astrophotography is now an established branch of astronomy. I believe that it has evolved to the point that it deserves its own formal book on the subject. The field has grown beyond what can be done by merely updating an existing astrophotography book, so this one was designed from the start to address that need and acquaint both the novice and the experienced film astrophotographer with the new world of digital astrophotography.

Electronic imaging using cooled charged-coupled device (CCD) cameras has been in widespread use for several decades and still reigns as the highest resolution and most sensitive medium for recording faint celestial objects. However, a scientific grade CCD camera lacks the familiarity of conventional photography equipment. There is a big jump in user-friendliness between standard cameras and specialized CCDs. However, the advent of high-quality sub-$1000 digital single-lens reflex (DSLR) cameras now presents the astroimager with a more familiar alternative. Up to this point, digital cameras were primarily fixed-lens point-and-shoot models that had limited long-exposure capability. But now, a high-performance digital camera can be used to do wide-field astrophotography with its own lenses, or image directly through a telescope just like the standard 35-mm SLR cameras we are familiar with. Since the introduction of consumer digital cameras, it has not been a question of "if" but "how soon" digital photography will mature to a state where it is a viable replacement for film astrophotography. The answer to that question is ... *NOW!*

1.2 Is Digital Good Enough for Astrophotography?

The introduction of high-quality sub-$1000 DSLR cameras marked a true turning point in the acceptability of digital astrophotography. Prior to their appearance, the debate about the viability of digital cameras in astrophotography was a polarizing subject with believers of film astrophotography in one camp and digital devotees in the other. The Canon EOS DSLR models—beginning with the 30D, 60D, 10D, 20D — have convincingly proven that consumer digital cameras do have the sensitivity to be effective astroimagers. The acceptance has accelerated as prices have dropped.

Jay Ballauer took up the digital challenge and compared the performance of the Canon Digital Rebel to the output of a top-of-the-line Santa Barbara Instrument Group ST-10XME CCD astrocamera. Both instruments were coupled to his Takahashi FSQ-106, that operated at $f/5$. Jay shot a triple exposure of the Orion Nebula, M42, with the ST10XME for 20 minutes each through a red, blue and green filter, then combined them into a color composite with a total exposure time of 60 minutes. He then imaged M42 with a Digital Rebel, taking 18 separate 5-minute exposures, and then digitally combine them into a single 90-minute exposure. The resulting images are shown in Figure 1.2.

Did the ST-10XME outperform the Digital Rebel in Jay's comparison? Of course it did. The cooled CCD chip in the ST-10 camera, using three exposures each through a different filter (tri-color imaging), has far more dynamic range than the Digital Rebel's single exposure complementary metal oxide semiconductor (CMOS) sensor that images all colors at once. The ST10XME's CCD was also far more sensitive to the important hydrogen-alpha wavelength present in emission nebulae than the Digital Rebel which performed more like a film camera using a red insensitive "four-layer" color film.

But look closely at the two images. Consider that the ST10XME image was obtained with a camera that is dedicated only to telescopic astrophotography, while the Digital Rebel image was obtained by a camera that

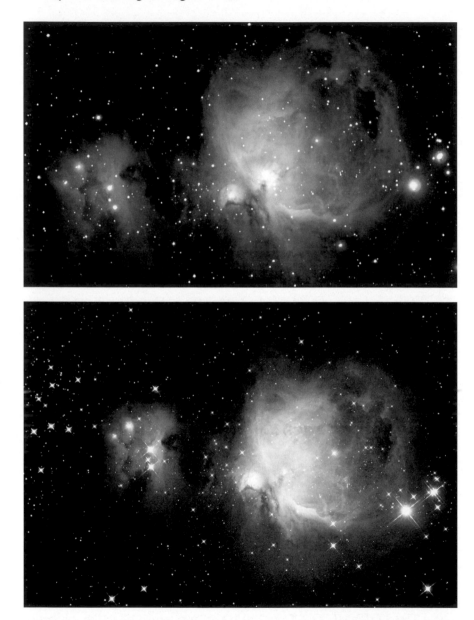

Fig. 1.2 *Jay Ballauer took these comparison photos of M42 to demonstrate the amazing image quality possible with today's DSLR cameras. The upper image was taken with a Santa Barbara Instrument Group ST10XME specialized astronomical CCD camera. The lower one was taken on the same telescope with a Canon Digital Rebel camera. The diffraction spikes around bright stellar images in the lower photograph were added for aesthetic appeal by placing two crossed strings in front of the refractor objective lens. Photos by Jay Ballauer.*

is versatile enough to also use for family vacation pictures. The Digital Rebel's CMOS sensor will never replace a CCD like that used in the ST-10, but the results are still amazing. It is a fact that consumer digital cameras are less sensitive and more prone to image noise than specialized CCD cameras, but Ballauer's comparison shows they do offer a simple way to begin astrophotography adventures. There is no doubt that digital astronomical imaging has arrived for the masses.

1.3 Is Digital Astrophotography Hard to Master?

We have all admired the great astrophotography displayed in magazines like *Sky and Telescope* and *Astronomy*. We have also known that producing such images requires the three "Ps"; that is, practice, patience and persistence. Celestial imaging is not like taking ordinary snapshots and digital astrophotography differs in some ways from film astrophotography. In this endeavor we are chasing dim objects, often invisible to the naked eye, that are literally millions of light years away. To image these elusive objects, we have to rely on basic knowledge of astronomy and telescopes to navigate our way to the target, acquire the image using established principles of technical photography, and then finally apply computer image processing techniques to "develop" the image for display. For some, this might require a lot of expensive new equipment.

Fortunately, most amateur astronomers already have a majority of the high-priced items needed to begin digital astrophotography. The fact that you are reading this book means it's likely that you already have a telescope. You probably have a digital camera, or plan to get one soon. It is also likely that you have a home computer that has some sort of image processing software already in it. If this is the case, you are essentially ready; there are only small details like adapters to mount the camera to the telescope to deal with.

We have all heard the story of an astrophotographic novice who simply hand-held a digital camera to the eyepiece of a telescope and snapped a fantastic close-up image of the Moon. But what about more advanced and challenging deep-sky imaging through a telescope? Is that hard to do? Peter Langsford found that it was not if he properly chose an object that is almost guaranteed to be a success—a bright globular cluster. Using his alt-azimuth mounted Meade 8-inch LX-90 *f*/6.3 and his Canon 300D, Peter took the stunning image of Omega Centauri shown in Figure 1.3 on his first deep-sky astrophotography attempt.

Langsford's image of Omega Centauri would be the envy of any film astrophotographer—indeed, one of my fondest memories is spending an

Fig. 1.3 *Peter Langsford's first astrophoto attempt resulted in this excellent Omega Centauri image. It is a combination of 4 nineteen-second exposures at ISO 800 using a Canon 300D and an 8-inch f/6.3 Schmidt-Cassegrain telescope. Peter achieved this success by applying techniques that are only possible with digital astrophotography. Photo by Peter Langsford.*

Fig. 1.4 *Peter Langsford's Meade alt-az mounted LX-90 is typical of most beginner's astrophotography set up. His telescope features some handmade wooden accessories such as a laptop stand, refractor mount, and counterweight bar. Photo by Peter Langsford.*

evening with Jack Newton, a premier expert in celestial imaging, as he photographed Omega Centauri at the 1984 Texas Star Party. I can attest that Jack, a master at film astrophotography, worked hard with advanced film camera equipment to capture an image of Omega that was no better than the one Langsford digitally captured from his own backyard.

So how did Langsford capture Omega Centauri? The key was a digital camera and the rapid feedback it provides. Here is how he went about it: Using an inexpensive but very useful software program called DSLRFocus, he took a rapid series of images with his 300D set at ISO 800 that were downloaded via a cable connected to his laptop computer. The software sequentially displayed the images on his laptop as he adjusted the focus of the telescope. Once proper focused was achieved, Peter shot a sequence of four 19-second images of his target. Longer exposures would have been possible with an equatorially-mounted telescope, but Langsford's LX-90 was driven only in altitude and azimuth. Thus the image seen through it experienced field rotation. By using a series of short exposures, he eliminated visible field rotation in each separate image. Next, using one of several astronomical image processing programs perfected for digital photography (ImagesPlus), he slightly rotated and aligned the successive images to counter the effects of field rotation. Then, he digitally combined all four images into a single one that simulates just over a minute's exposure. Other standard image processing techniques such as contrast and sharpness enhancement were applied to create the final result: a stunning deep-sky image on his first attempt.

Langsford did nothing extraordinary to achieve his fine image of Omega Centauri. In fact, he used the same tools available to most amateur astronomers and quickly produced an image that anyone would be proud to display. The digital camera, software, and techniques used are available to anyone who can afford moderately-priced equipment, and is willing to learn some new twists to the old art of astrophotography.

However, I do not want you to conclude from this description that imaging deep-sky objects with a digital camera shooting through a telescope has become point-and-shoot imaging. Peter chose a target that exploited the strengths of his camera. He took a series of images that were short enough to overcome his tracking problems and then combined them digitally using an image processing program. Langsford's results represent, in a sense, a "best case scenario" for novice digital imaging success. But what are the typical beginner's results more likely to be like? To find out, Richard Berry purchased a Nikon D70 DSLR and took a series of photographs to determine what the novice will see with a single, relatively short expo-

Fig. 1.5 *Richard Berry writes: This is the first astro-picture I took with my new Nikon D70 camera. Since it was late in the season, the Lagoon (M8) was nearly in the trees when I started this 60-second exposure, and the sky was not yet fully dark. I used an ISO setting of 1600 with noise reduction on. Both tracking and focus are poor, but the nebula is clearly visible. For processing, I loaded the image into Astronomical Image Processing for Windows (AIP4Win), split it into RGB color channels, applied AIP's Rolling Pin Tool to fix vignetting, and reassembled the channels using the Join Colors Tool.*

sure. Here is what he had to say:

> The common theme is that these are "first light" images with the Nikon D70. I didn't bother guiding or anything fancy—it was bad enough trying to find the little buttons, focus on dim stars, and start and stop exposures. I used a speed of ISO 1600 with noise reduction and normal JPG compression for every exposure. While the Byers 812 mount normally tracks well if it's properly polar aligned and you're in the right part of the sky I did not fiddle with it. It was also a chilly night and everything got wet.

Figures 1.5 through 1.8 show Richard's pictures, and the captions provide background material on how they were shot and processed. These are the kind of exposures you are likely to try your first night out. It is quite possible to be disappointed if you are not forewarned.[2] While imaging the Sun, Moon and planets is relatively easy because of the abundance of light, deep-sky objects like gaseous nebulae, galaxies and planetary nebulae

---

[2] However, if you are an old timer who tried film based astro-imaging in the past and were frustrated with the first results which were usually rather poor, you probably will be amazed and pleased.

Fig. 1.6 *Richard Berry writes: This is a five-minute exposure of M31 with my 6-inch f/5 Newtonian on the Byers 812 mount, at an ISO setting of 1600 with noise reduction on, saved as a JPG. The clock drive was running, but I did not guide the telescope, so the stars are trailed. This image is straight out of the camera with no processing, and the corners are dark from vignetting. This is the real McCoy—a completely honest example of "first light" with my new DSLR.*

Fig. 1.7 *Richard Berry writes: The last-quarter Moon was shining brightly, and the eastern sky was starting to brighten when M42 finally cleared the trees, so this is the last picture I took the first night I had my Nikon D70. The camera was on my 6-inch f/5 Newtonian on a Byers 812 mount. I focused on Betelgeuse, so I had a nice bright star that was easy to see in the viewfinder. The ISO was 1600.*

Fig. 1.8 *Richard Berry writes: This is a 60-second exposure of M35 with my 6-inch f/5 Newtonian, shot about an hour before dawn. There was no guiding of any kind. I processed the image by splitting it into RGB color channels, which I used to remove the vignetting, and then I reassembled the image using AIP4Win's Join Colors Tool. The Nikon does well with star clusters.*

have never been "easy" targets whether you use a film camera, astronomical-grade CCD, or a consumer digital camera. One of the facts of life with digital camera imaging is that the hardware limits exposure times, and to build up a nice image you will have to take a number of short exposures and combine them. With the proper processing, you will get very pleasing results that will compare very favorably with film. At least for now, the laws of physics prevent digital cameras from head-to-head competition with dedicated astronomical CCD cameras. As I wrote at the beginning of this section, successful imaging takes practice, patience and persistence. If you like the pictures you see in this book and are willing to purchase a good mount and spend the time, you too can successfully image "faint fuzzies."

1.4 The Digital Advantage

As someone who has exposed countless rolls of film over 50 years, I think. "What's not to love about digital imaging?" Digital imaging is quick, giving new meaning to the term "instant gratification." You see the results immediately on the camera's own image-viewing screen and know immediately if you have a "keeper." Digital images are versatile—they can be printed, archived to storage media, transmitted via E-mail or displayed

Fig. 1.9 *After the initial investment, image storage is practically free and you never have to keep a bad exposure. This inexpensive 256-megabyte CompactFlash card (foreground), when used with a 6-megapixel camera in RAW format mode, will hold about the same number of images as a single roll of 35-mm film. On the other hand, a 4-gigabyte card will hold the same number of RAW format exposures as 16 36-exposure rolls of 35-mm film. Photo by Robert Reeves.*

on a web site. Digital is inexpensive (after the initial equipment investment). With 35-mm film, you may get one (if that!) good high-resolution lunar image out of 36 exposures on a roll. Counting the cost of the film and developing, followed by the time to scan the negatives in search of the best image, that one roll can cost up to $10 and consume an entire evening of time, possibly without yielding a good image. With digital, that same imaging session costs literally pennies—just the price of the electricity to recharge the camera's batteries and run the computer to view the images. Expenses only accumulate with successful images that deserve to be archived on a recording medium and/or printed. The sub-par images go away with a tap of the computer's delete key. Digital imaging also provides the ability to instantly correct problems. If an image is out of focus, underexposed, suffering from camera shake or any other correctable problem, you will know it now instead of after wasting an entire imaging session. Digital techniques open up a whole new spectrum of possibilities for the astrophotographer. After all, there is no photo lab to send you back a dark, overexposed print because they didn't know what an astrophoto was. When you move to digital you can be in complete control of the image from shutter snap to finished print.

Experienced film imagers know that digital is not ready to entirely replace film photography. There are certain applications where the latter is still superior, such as ultra-high resolution wide-field Schmidt camera work. But digital has certainly matured to the point where it takes its place alongside film in a majority of celestial applications. Indeed, digital is now superior to film in telescopic lunar and planetary work, and is equal to film in deep-sky imaging and wide-field work in all areas except recording emission nebulae that shine in the far red end of the spectrum. However, as we shall see in later chapters, astrophotographers are finding ways around this inherent limited sensitivity to red light by performing modifications to their digital cameras.

1.5 Comparing Digital and Film Cameras

Digital cameras are also electronic devices, and in some cases this leads to an unconventional appearance. They all have a lens in front and a view-finder with which to frame the image. Some digital models have fixed lenses that cannot be removed while other models are of the conventional single-lens reflex design and have interchangeable lenses. Additionally, digitals have a liquid crystal display (LCD) screen on the back of the camera that can act as a viewfinder with fixed-lens cameras or be used to preview any of the images stored in the camera's memory. Some camera models like the Olympus Camedia series strongly resemble small conventional 35-mm cameras. Others, such as older Nikon Coolpix models, have a two-piece body where the camera body and LCD display can swivel at an angle to the lens and imaging head, giving it a decidedly futuristic look. What separates digitals from film cameras is the fact that instead of opening the back of the camera and threading a roll of film onto the take-up spool, you insert a reusable memory card to record your images.

Since digital cameras use no film, you do not face one of the major practical problems in film astrophotography—which film to use? Another digital advantage is that all images can be in color, although some cameras have an option for recording images in pseudo-black and white. These black-and-white images appear to be monochrome, but are recorded as color images whose red, blue and green components combined to create all whites, blacks and grays. But the default selection of only a single imaging sensor has drawbacks for certain types of astrophotography. Most digital camera sensors are covered with an infrared filter that makes them insensitive to the far-red region of the spectrum where emission nebulae shine in the red hydrogen-alpha wavelength.

With film cameras, the type of film used governs the exposure's sensitivity to light and the extent to which colors can be perceived. A fine-grained high-resolution film may be used, but at the expense of low sensi-

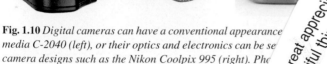

Fig. 1.10 *Digital cameras can have a conventional appearance* ~~~~~~~~ *del Olympus Ca-media C-2040 (left), or their optics and electronics can be se* ~~~~~~~ *to radical-looking camera designs such as the Nikon Coolpix 995 (right). Ph* ~~~~~~ *lympus and Nikon.*

tivity to light. A faster film more sensitive ~~~~~~~ may be used, but at the disadvantage of having coarser-graine ~~~~~~~ olution images. Film speeds are gauged by their ISO speed ra ~~~~~~ ne doubling of the ISO number indicating a doubling of the fi' ~~~~~~ nt sensitivity to a given amount of light. 200-speed film is tw: ~~~~~~ s 100-speed, 400-speed is twice as fast as 200-speed, and so forth. r.. cal film speeds range from ISO 100 to ISO 1600. Digital cameras carry over the concept of ISO speed with their imaging sensors, but the same sensor is adjustable between ISO 100 and 400, and even up to ISO 6400 on some, by making electronic adjustments with the camera's controls.

Under light-polluted skies, digital cameras have an advantage: the skyglow can be subtracted through image processing. If the sky is too bright, the color image can be converted to a grayscale black-and-white image to allow deeper processing that will reveal faint objects that would ordinarily be lost in the glow of light pollution. Under dark skies digital cameras show another great advantage: their system of electronic imaging records light linearly; that is, there is no "reciprocity law failure" as there is with chemical-based film. Doubling the exposure produces an image that is twice as bright.

The end result of the photographic process is similar with the two types of cameras—an image printed on paper, or a "picture" is produced. With a film camera, photosensitive chemicals on the film react to incoming photons during the exposure. The exposed film is developed in a succession of chemicals to produce either a negative image or a positive transparency. This is in turn printed onto photographic paper with an enlarger and developed chemically, or it is scanned electronically for computer processing and printing with an ink jet or other appropriate printer. With a digital camera, incoming photons strike an electronic detector that converts the

Fig. 1.11 *Heavily light-polluted sky inhibits backyard astrophotography of diffuse nebulae, but bright star clusters will shine through light pollution. Convert the image to grayscale to eliminate excess color shifts caused by light pollution, and image starry objects like globular cluster M3 in black and white. This image is composed of 4 stacked thirty-second exposures at ISO 800 with a Canon 10D and a 200-mm lens stopped to f/5.6. Photo by Robert Reeves.*

image into electronic signals. These signals are then processed and record-ed electronically on the memory card, then downloaded to a computer or printer in order to produce the final picture. More detail on how a digital camera works is presented in Chapter 3.

1.6 Expectations

The digital beginner would naturally ask, "How good can my astronomical images be?" Unfortunately, there is no clear-cut answer. Many factors en-ter the photo quality equation, and most of them are beyond the immediate control of the photographer. Different cameras vary greatly in capabilities, usually in direct proportion to their price. The type of astrophotography be-ing done and the size of the telescope used also set limits on what can be achieved. Of course, environmental factors play a big role. Long-exposure deep-sky imaging is very difficult from light-polluted areas, and high-res-olution telescopic images are equally difficult to achieve if wind is shaking the telescope. But digital cameras by their very nature offer the user an array of capabilities that in some cases allows them to achieve images that would not be possible with conventional film cameras.

Among the factors governing final image quality, perhaps the most important is the camera itself. Almost any digital camera is capable of taking breathtaking images of the Sun, Moon and planets. Indeed, you will find that a digital model can often take better lunar images than a film camera. But most of point-and-shoot digital cameras have limited long-exposure capability usually between to 16 to 30 seconds—due to the camera's internal operating system; and digital processing must be used for imaging dim deep-sky objects. More advanced interchangeable lens DSLR cameras are capable of longer exposures, but they are also more expensive.

The resolution, or number of pixels, used to form the image in a particular camera, also sets a limit on how much detail can be recorded. A threshold past which a digital camera has enough pixels to produce 4 by 5-inch, "film-like" photographic prints is usually considered to be 3 megapixels. However, today even entry-level point-and-shoot models now easily exceed this threshold. We will discuss this in detail in Chapter 3.

A 35-mm film camera has a fixed imaging size: 24 by 36 mm. We have grown accustomed to gauging our field-of-view with a particular lens or calculating the size of the target's image on the focal plane using these numbers. However, digital cameras usually have a much smaller imaging surface. This presents us with puzzling lens focal length "equivalency factor" charts in camera advertisements. The smaller imaging detector means that the field-of-view with a given lens focal length is smaller with a digital camera. On fixed-lens models this is no problem as they come already equipped with shorter focal length lenses, sometimes of less than 10-mm focus, to provide normal or wide-angle views on their smaller imaging surfaces. Interchangeable lens DSLR cameras, such as the Canon 10D, 20D, 300D, or Nikon D70, use lenses originally designed for 35-mm film cameras and one quickly finds that the smaller imaging chip on the DSLR changes the expected image scale by quite a bit. For the Canon models above, the ratio is 1.6; that is any lens previously used on 35-mm film cameras now seems to have 1.6 times greater focal length when used on digital ones. In some cases this is a bonus. Our moderate 200-mm focal length *f*/4 lens used on a 35-mm camera now behaves like a whopping 320-mm focal length lens with the same *f*/ratio. But there is a down side. Our existing 35-mm camera wide-angle lenses are now merely "normal" lenses with the DSLR. We will have to purchase a 20-mm focal length lens to get the same digital field-of-view we used to have with 35-mm focal length lenses on film cameras.

Another problem with the smaller imaging sensors in digital cameras is that even if yours has a sufficient number of pixels to produce detailed images that does not mean it will necessarily work well for long exposures. As the size of the sensor gets smaller, or more pixels are packed onto the same size sensor, the chip's full-well capacity is decreased. Since the full-well capacity is smaller, the readout noise becomes a more significant part of the overall signal, because the smaller pixel cannot be exposed as long as a larger pixel before it saturates. This is the digital camera's electronic corollary to the dreaded reciprocity law failure that makes certain 35-mm films less efficient with increasing exposure length. This phenomenon creates speckles, or hot pixels, that dot the image like snow and grow worse as exposure time increases. It also worsens at higher temperatures, doubling in amount for every seven-degree Centigrade increase. However, again, the digital world has ways to minimize the effects of noise in long-exposure images as will be discussed in Chapter 14.

Even the very best digital camera will not ensure good astrophotos if the telescope is not up to the task. Sharp optics are a must for high-resolution lunar and planetary imaging and a large aperture is necessary for capturing dim, deep-sky objects. Good polar alignment and accurate tracking by the telescope's mount are is also needed for successful long-exposure imaging.[3] The telescope and camera mounting and how the camera is adapted to the telescope also play a critical role in the resulting image quality.

Since a majority of digital cameras have a fixed lens, telescopic photography with these models must be done by the afocal-projection method, which often introduces the problem of vignetting of the image. Vignetting is the effect where the edges of the field-of-view are darkened because the camera's sensor is not viewing the entire cone of light exiting the eyepiece. Fortunately, there are methods for controlling vignetting with the creative use of a fixed-lens digital camera's built-in zoom features and the careful selection of eyepieces. We will discuss this in more detail in Section 4.6.

Image processing is the vital link in the chain of events that ultimately produces the finished print. Just as computer image processing became an integral part of film astrophotography after the popularization of film scanners, digital astrophotography—through its electronic character—naturally lends itself to image enhancement with various image-processing techniques. But again, the details of these concepts will be explored in Chapter 14.

---

[3] But even here, digital astrophotography can compensate for less than perfect alignment and tracking to a surprising degree if you are willing to take many short exposures and combine them.

Fig. 1.12 *The most common mistake made by novices and veterans alike during afocal lunar photography is forgetting to turn off the camera's built-in flash. Photo by Becky Ramotowski.*

1.7 For Instant Gratification, Shoot the Moon

Before we dive into the specific details of digital astrophotography, as a confidence building exercise, let's jump right in and actually do some imaging of an object that is easily viewed from almost anywhere—the Moon. Large amounts of skyglow have no affect on the Moon and bright planets. These targets can be imaged from almost any location so long as there is not a light source shining directly into the telescope that would reduce image contrast.

Over a period of four decades, I have exposed thousands of frames of Plus-X, Panatomic-X and Technical Pan film through various telescopes seeking the ultimate lunar image. Often I achieved reasonably good results, but at great expense in both time and money. Now, by utilizing the inherent traits of lightweight digital cameras—high sensitivity, autoexposure, autofocus, reusable memory cards, the ability to quickly detect and delete substandard images, and digital-image processing software—the success ratio is near 100%. I am not exaggerating when I say that I can now take better Moon pictures by accident with a simple digital camera than I previously did on purpose with top-of–the-line film cameras and exacting darkroom work. Let's see how you can start doing this right now.

To image the Moon, one only needs tripod-mounted binoculars or a telescope and a digital camera. A clock drive to track the Moon is not required. Low-magnification scenes of the entire lunar globe can be photographed by simply holding the camera to the eyepiece. The exposure times are so short, just hundredths of a second, that movement from unsteady hands, and the orbital movement of the Earth and Moon, will not be noticeable. Only when higher-magnification views are attempted through a telescope will a clock drive or means of holding the camera steady be required.

Since a majority of digital cameras have a fixed lens, we will be using the afocal-projection method whereby the telescope eyepiece projects the Moon's image directly into the camera's infinity-focused lens. Choose a low-power eyepiece with the widest apparent field-of-view. The wide field-of-view is necessary to prevent the vignetting problem previously mentioned. Focus the telescope carefully on the Moon using corrected vision—that is, if you wear glasses, use them while focusing the telescope. Now, just for fun, simply hold the camera to the eyepiece and aim using the LCD image preview screen. If the Moon fills a majority of the field-of-view, let the camera's autoexposure and autofocus functions control the exposure. Make sure the camera lens is square with, and perpendicular to, the eyepiece. Use the camera's optical zoom to help frame the image or reduce vignetting. Now press the shutter release.

If you are typical, and I also made the same mistake, you just had your first lesson in digital astrophotography with a point-and-shoot camera … turn off the flash! Now try it again. Aim squarely into the eyepiece and press the shutter. Congratulations! Now the LCD viewfinder is displaying your first lunar photograph, and it is probably better than the vast majority of amateur film-based Moon pictures from decades past. Now continue to take images through the eyepiece until you fill the camera's entire memory card. There is strength in numbers; take a long series of images, because a few will always be sharper than others due to variables such as camera movement and atmospheric turbulence. Remember, you can later delete all but the best ones.

Higher-magnification views of the Moon will require a steadier camera mounting to allow the longer exposures necessary with higher-power eyepieces. The easiest way to do this without first purchasing special adapters to mount the camera onto the telescope is to use an ordinary camera tripod. Now that the camera is stable and not in our hands, we have more control over the image. Using the LCD preview monitor on the back of the camera, we can detect focus, composition, and exposure errors and correct them on the spot.

1.8 Improving Your Moon Shots

Now that you see celestial imaging is possible with your camera, let's look

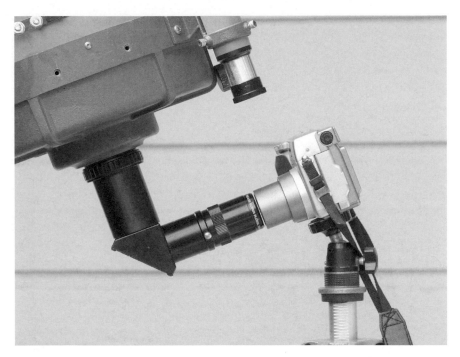

Fig. 1.13 *An option for camera stability during afocal astrophotography is to mount the camera on a separate tripod. This eliminates telescope shake during the exposure, but it is inconvenient to constantly shift the camera to keep up with the moving eyepiece as the telescope tracks the sky. Photo by Robert Reeves.*

at some basic techniques to improve on what you just did. The following is a list of pointers to insure digital success on your digital lunar voyage.

1. Be sure the camera's battery is fully charged before an imaging session. Keep a spare set of batteries handy in case it becomes depleted. A digital camera with a dead battery is little more than an expensive paperweight.

2. Set the camera to the full-manual control setting that allows you to select the aperture and shutter speed. If the camera has an auto-off option, be sure to disable it so the camera will not shut down after a period of inactivity while you are working with the telescope. Shoot with the lens wide open. If the camera is allowed to automatically stop down the lens, it will vignette the image at smaller *f*/stops. Find the proper shutter speed by experimenting and viewing the results on the camera's LCD monitor. Once the proper exposure is found, delete the misexposed test shots.

3. If the camera is attached to the telescope with an adapter, make sure it will not strike any portion of the mount when the telescope is slewed from one part of the sky to another. This is especially important with motorized computer-controlled Go-To mounts.

Fig. 1.14 *At low magnification it is possible to take great Moon pictures by simply holding the camera to the eyepiece. Photo by Robert Reeves.*

4. Focus the telescope by eye, with corrective eyeglasses if needed, then let the camera's autofocus option refine the focus.

5. Turn off the flash!

6. When shooting the full disk of the Moon at low magnification, such as with a 32-mm Plössl through a common 8-inch $f/10$ SCT, the camera can actually be hand-held because the shutter speed will be in the $\frac{1}{500}$- to $\frac{1}{800}$-sec. range at ISO 200 or 400 settings, (see Figure 1.14).

7. At slightly higher magnifications—say, using a 25-mm eyepiece on the same 8-inch SCT—the camera can still be hand-held, but only half the

shots will be acceptable because of camera shake. With still higher magnification, a tripod, or some means to attach the camera to the telescope is necessary as the shutter speeds drop to between about ¼₀– and ⅛–sec. At slower shutter speeds, it may be necessary to use the camera's self-timer option to trigger the shutter without vibration if the camera does not have remote-control options.

8. When aiming the camera into the eyepiece, either by hand or on a tripod, make sure the lens is square with the eyepiece. If it is not, the edges of the photo will be out of focus. If you are holding the camera by hand, remember that not all models fire the shutter as soon as the shutter button is pushed. Some have up to a two-second delay while the autofocus routine completes its operation before exposing.

9. If using a tripod, position it so that the clock drive will move the eyepiece either directly away from, or toward, the camera, *not* across the camera's field-of-view. This allows about two minutes of shooting before the camera needs to be nudged either closer or further away. Preferably, place the tripod so the eyepiece will be moved away from the camera. However, depending on where the telescope is aimed, this may not be practical. If the eyepiece must approach the camera, monitor it closely to avoid a slow collision.

10. Cameras with a fixed lens assembly that does not automatically extend outward when the camera is turned on can be attached directly to the telescope eyepiece using various adapters that screw into the lens' filter thread. WARNING: If your camera lens automatically extends when the camera is turned on, you cannot attach the camera with the threads at the end of the lens. You must use a camera-to-eyepiece mounting system (see Section 8.5 for examples). This adapter threads into the fixed barrel surrounding the extendable portion of the lens assembly. These are the same threads that the camera's lens cap attaches to. The reason it is critical to use this type of camera-to-telescope adapter is that the motor-driven extendable lens elements are retained with small plastic tabs that can break very easily when a side load is applied to them.

11. Do not trigger the shutter on a scope-mounted camera by hand. This will induce vibration that is magnified by the telescope and will blur the image. If a remote trigger, either electric cable operated or wireless, is not available, start the exposure using the selftimer option. By the time the camera shutter trips after the 10-second delay, any telescope vibrations will have damped out.

12. Use the widest field-of-view eyepieces possible to avoid vignetting. My Meade Series 4000 32-mm Super Plössl covers the camera's entire field-of-view, but my Brandon eyepieces cause progressively greater vignetting with each decrease in focal length. Use the camera's optical zoom to increase detail or eliminate vignetting. Avoid using digital zoom because

Fig. 1.15 *To avoid vignetting when using afocal projection, choose a low-power wide-field eyepiece and use the camera's zoom lens to increase image magnification. Photo by Robert Reeves.*

it merely enlarges the image by resampling the central pixels in the field-of-view and enlarges that portion of the image over more pixels. There is no increase in detail with digital zoom.

13. If you do not have an external power source to run the camera during extended sessions at the telescope, use nickel metal hydride (NiMH) rechargeable batteries in the camera. These last several times longer than

regular alkaline batteries, which will allow you to use the camera's LCD monitor screen as needed to monitor the telescope's aim, without worrying about battery depletion.

14. Use the highest quality JPG setting with the largest pixel-size image available. Although the image counter will show you have a certain number of exposures for a given size memory card, you will get more images per megabyte on the Moon and planets than with normal daylight imaging. My Olympus 3020Z indicates I should get 34 maximum resolution JPG images on a 64-Mb memory card. But in practice, I get about 60 lunar, and over 100 planetary, images because the JPG algorithm in the camera compresses the black background more efficiently than for normal fully-illuminated daylight snapshots.

15. Take as many shots as the camera's memory card will allow—there is strength in numbers. Out of a series of 50 repetitive lunar images, at least one will be great even if atmospheric turbulence is blurring most of them. It is possible to take over 100 lunar images in a half-hour imaging session. Just delete the substandard ones.

16. When processing the JPG images saved from a photo session, first convert them to TIF format to prevent losing any image quality as happens with repeated "saves" of JPGs that have been processed and changed. Once lunar images are processed and adjusted to your liking, you can convert them to grayscale (black and white) to reduce the file size by two-thirds. If no further changes in the image are expected, the TIF file can be converted back to a JPG to further reduce the file size for archiving.

17. Depending upon what type of telescope you use, and whether you use a star diagonal or not, the orientation of the resulting image may be normal, reversed, or both upside down and reversed. The image will have to be rectified in an image processing program before it can be viewed and printed normally.

18. Advanced image processing programs such as Photoshop are not required to process your lunar images. The necessary conversion from JPG to TIF format, the inversion or reversal of the image, contrast manipulation, conversion to grayscale, sharpening, and printing the image can all be done with the freeware program Irfanview.

19. A useful accessory to have is a memory-card reader to download images into the computer. This allows downloading the contents of the memory card without having to remove the camera from the telescope. Once the card is erased, it can go back to the telescope for more imaging. A second memory card will also prove useful as lunar and planetary imaging can produce a massive amount of exposures in a short time.

In a nutshell, it is fun to be able to do astrophotography from the driveway at home with a digital camera. Try your hand at this the next time the Moon is available and follow the tips listed above to get your first digital astrophoto experience. As you gain confidence in your ability and in your camera's capability, follow the tips and advice in the following chapters and join the legions of new astrophotographers who are imaging the universe with a digital camera.

Chapter 2
Digital Imaging Using a Webcam

2.1 Introduction

As we shall see in the coming chapters, digital cameras are capable of taking excellent photos of the Moon and planets and are now challenging film cameras in the photography of deep-sky objects. But there is one area of astrophotography where a simple and inexpensive device that requires virtually no photographic experience to operate can, in many respects, outperform both film and digital cameras: solar system photography with webcams.

Since this is a book about digital astrophotography, it may seem odd that I am jumping into webcams, especially since the last chapter concluded with a discussion of how easy it is to image the Moon with today's digital cameras. My reasons for this are twofold. First, webcams are inexpensive, and image processing software for them is free. Therefore, they offer an easy way for a newcomer to quickly enter the world of digital astrophotography. Second, working with a webcam to image the Moon has turned out to be the most fun I have had in astronomy for the past 20 years. Anything that turns on a jaded imager like me has to be good! I would be remiss in my duties as an astrophotography popularizer if I didn't encourage others to also try this fun way to image the solar system.

Unlike digital cameras, webcams require a computer at the telescope in order to work, so a laptop or a desktop rolled to the telescope on a cart, is a necessity. However, since today's webcams are powered through the computer's USB connecting cable, you will not need another power supply. And, if you already have a suitable computer, a webcam is more than just a perfect low-cost lunar and planetary camera; it is now the best way to image the Moon and brighter planets.

Many brands of webcams are available. Some examples are:

- The Philips ToUcam (~$125 with adapters) and Logitech QuickCam Pro 4000 (~$75). Both of these use the ¼-inch Sony ICX098 640x480 pixel

CCD sensor with 5.6-micron pixels. The ToUcam Pro II PCVC 840 is the latest version of Philips cameras and the one that I use. These models are a little more expensive than others available from computer stores, but they have a proven record of providing excellent astronomical results. The small X-10 cameras commonly marketed through Internet popup ads use a CMOS sensor and are not as suitable for astronomical uses, but their wireless operation is intriguing.

- For about $125 another option is the Meade Instruments Lunar and Planetary Imager (LPI). The LPI differs from the average webcam in that it uses a CMOS rather than a CCD chip, and it can couple directly to the back of a Schmidt-Cassegrain telescope. One plus is that it's capable of exposures of up to 15 seconds without the modifications that a webcam requires for long-exposure capability. It can be operated in either manual or automatic exposure mode. It uses a ⅓-inch sensor that provides high resolution when used with a Barlow on smaller telescopes. If you want to shoot big lunar vistas and don't want to do a mosaic, an *f*/3.3 focal reducer is a must for the SCT owner.

- The Astrovid StellaCam II is designed for long exposures of deep-sky objects, and at more than $700, is overkill for lunar and planetary imagers. Another is the SAC7B camera by SAC CCD Imaging Systems which is a Logitech QuickCam Pro 400 that has been modified for cooled long-exposure photography of deep-sky objects. This one also works well for short exposures of solar-system objects, but its price is between $400 and $500.

Longtime astrophotographers who have spent countless hours imaging the Moon and planets with conventional cameras may find it hard to believe that a tiny television camera can outshoot their top-of-the-line equipment, but it is true. Webcam images are now commonly displayed in magazines and on Internet sites that show that these simple little plug-and-play cameras are capable of recording better detail on the Moon and planets than one could ever hope for with a film camera. The secret to achieving such high image detail with a simple webcam is the use of a powerful freeware program called RegiStax. RegiStax sorts through all the individual frames in an AVI video sequence, selects the best images, and then aligns and combines them into a single frame that contains more detail than any similar view with other types of cameras.

Although capable of amazing high-resolution results, webcams have two major drawbacks. First, they have maximum exposure times of only a fraction of a second and are thus insensitive to very dim light. Unmodified webcams cannot image even the brightest deep-sky objects. Second, their small imaging sensors result in an inherently small field-of-view. This restricted field is not a problem with the planets due to their small angular

Fig. 2.1 *A typical "turn-key" webcam astrophoto system is the Philips ToUcam Pro II 840 camera (left) seen here with a Scopetronix 1¼-inch eyepiece-holder adapter. The eyepiece adapter (right) can screw so deeply into the camera that it may strike the CCD circuit board so a spacer (in this case two ½-inch flat washers) is needed to keep the adapter from contacting it. Photos by Robert Reeves.*

size, but imaging broad areas of the Moon requires mosaic-building techniques to combine multiple images into a single wide-field one.

As a long-time Moon imager with film, I now recall with some amusement how difficult it was to capture just a smudge of the small craterlets on the dark floor of the crater Plato, or achieve even a hint of the rille that runs down the Alpine Valley. Today, these features are considered "quality checks" for webcam images, and those that do not show them are considered to be inferior.

2.2 Basic Equipment for Webcam Imaging

Until fairly recently, to use a webcam for celestial photography required that it be dismantle and modified so it could be attached to the focuser of a telescope. Today, all of this has become much simpler. "Turn-key" webcam astrophoto systems can be purchased for less than $200 that include the 640x480 pixel Philips ToUcam Pro II PCVC 840K camera, a 1¼-inch eyepiece-holder adapter, and an infrared (IR) blocking filter that threads into the eyepiece adapter. The camera comes with all the needed software to operate it using any Pentium II 233 Mhz or higher computer running Windows 98 through XP with a USB 1.1 connection.

In addition to the turn-key package, you need only two other things: a 6-foot USB extension cable to conveniently connect to the camera (the three-foot cable that comes with the camera is too short for proper computer placement), and a spindle of CDs to archive the torrent of data the

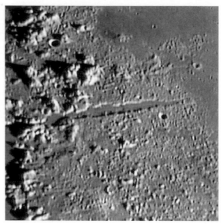

Fig. 2.2 *Webcams have so changed the criterion of what is a "high-resolution" lunar image that previously "impossible" targets are now considered image quality checks. If features like the tiny craterlets on the floor of the crater Plato (left) and the rille on the floor of the Alpine Valley (right) are not visible in webcam images, they are considered inferior. Both images were taken with a SAC7B on an 8-inch f/10 SCT with a 2x Barlow. Photos by Rod Mollise.*

webcam will generate.

At 100 grams the Philips ToUcam Pro II is so light that even with its adapter and filter it weighs less than many visual eyepieces so no telescope re-balancing is required. This camera uses a CCD instead of a CMOS sensor and is capable of resolutions between 320x240 and 640x480 pixels. Unfortunately, the 320x240 resolution is not a partial scan of the center of the field-of-view that would be valuable for planetary imaging, but a reduced scale view of the entire 640x480 view. The camera is thus best used at its full VGA resolution of 640x480. When used as a "terrestrial" webcam, the ToUcam Pro II is capable of producing excellent interpolated 1280x960 resolution images that rival those from standard video camcorders; but for astronomical applications, we need the full uninterpolated data from the camera, and thus the 640x480 is best. The camera is capable of operating in light levels as low as one lux—fully adequate for imaging the Moon and planets, but not for deep-sky objects. It also allows adjustment of white balance between 2500K and 7500K, or the option of black and white imaging. With the Moon, the white balance can be left on "auto," but small, colorful targets like Jupiter may need manual adjustment of the white balance. The normal 5700K-daylight setting will work well. Since this camera has a built-in microphone, your verbal notes about the imaging sequence can be recorded right along with the video.

The ToUcam Pro II comes with a 6-mm *f/2* lens for "normal" use.

This will be unscrewed from the camera body and a 1¼-inch eyepiece-holder adapter threaded into its place, so the camera can be used at a telescope's prime focus.

It is a good idea to use a computer with at least a 20-gigabyte hard drive with a majority of that capacity available to capture webcam video files. Depending on the camera settings and length of the video capture, typical imaging runs to create a single finished webcam image will produce video files in the range of 100- to 250-megabytes. Use no video compression when imaging, as the full amount of video data will be needed for processing the final image.

Once the hardware is purchased, the next step is to download from the web the software you will use. There are two possibilities:

1. RegiStax is freeware authored by Cor Berrevoets and can be found at his website (see Appendix B). Do not let the freeware status of RegiStax fool you into thinking this is not a powerful program. RegiStax is capable of analyzing and grading images for quality, rejecting those that are subpar, aligning and stacking the good images into a single master frame, and then allowing advanced wavelet-image processing and noise reduction that brings out stunning amounts of detail from what started as a stream of images from a relatively low-resolution video camera.

2. AstroStack is another program that performs functions similar to RegiStax.

2.3 Setting Up for a Webcam Imaging Session

Let's assume you have just taken delivery of a new ToUcam Pro II with IR-blocking filter and eyepiece-holder adapter and go through the steps needed to begin a lunar-imaging session. First, load and run the camera software installation CD on your computer. Like most applications, the software will attempt to load more options than you will need for astronomical imaging. If storage space is an issue, or you do not wish to load options for online video conferencing, etc., select only the Philips Video Lounge option. This will load the drivers needed to check camera compatibility with your computer, operate the camera, and change the camera exposure parameters.

As with any high-magnification astrophotography, the telescope mount must be polar aligned. Fortunately, the nature of webcam imaging allows some leeway in alignment accuracy. Since the RegiStax program can "track" image details as they slowly drift from frame to frame, exacting polar alignment is not a necessity.[1] However, good tracking will make it much easier to keep the target in the field-of-view long enough to capture

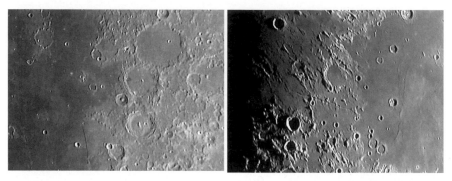

Fig. 2.3 *Mare Nubium to the left of the three large crater Ptolemaeus, Alphonsus, and Arzachel (left) and the area around the Ariadaeus Rille (right) were imaged at the prime focus of an 8-inch f/10 SCT using a Philips ToUcam Pro II 840. (These single-frame images bear a striking resemblance to those returned by NASA's early Ranger lunar probes in the 1960s.) Photos by Robert Reeves.*

the needed video sequence. From my driveway a large oak tree blocks Polaris thus preventing exact alignment with the polar scope built into my mount. In practice, I find that pointing my mount generally north, and setting its polar axis elevation to my latitude is sufficient to allow successful webcam imaging. If by chance the polar alignment is not good and the image slowly drifts during the video capture, I simply place the computer cursor on a prominent feature of the target as displayed on the image preview and use the telescope's slow motion controls to guide, keeping the image aligned with the cursor during the exposure sequence.

Next, insure that the finder scope is accurately aligned with the main one using a high-power eyepiece. The tiny CCD sensor on a webcam provides about the same field-of-view as does a 4-mm eyepiece. This is equivalent to about 500 power with a common 8-inch f/10 SCT so a well-aligned finder will be essential to aim the webcam accurately at such extreme magnification.

Place your computer on a cart or table close to the telescope. Now prepare the camera for installation onto the telescope. This involves unscrewing the lens assembly from the camera body and substituting the eyepiece-holder adapter that threads into the assembly opening. At this point, be careful—the adapter that came with my webcam has sufficient threads to allow it to bottom-out onto the camera's circuit board, possibly damaging the CCD. I solved this problem by making a spacer from two ½-inch I.D. metal flat washers that prevents the adapter from bottoming out. A similar spacer can be made from plastic or wood as long as it is about ¼-inch in

---

[1] The webcam does not cause the telescope to track the object however, so it must be somewhere in the image frame.

 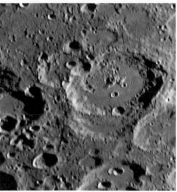

Fig. 2.4 *These high-resolution close-ups of the Moon—the area around the craters Theophilus and Cyrillus (left) and Maurolycus (right) —are typical of webcam results. Both were taken through 8-inch f /10 SCT telescopes with webcams using the same Sony ICX098 640x480 pixel CCD sensor. The left image was taken at prime focus while that on the right used a 2x Barlow. Photos by Robert Reeves (left) and Rod Mollise (right).*

thickness. Keep the time the camera is without a lens or filter-covered adapter to a minimum. Any dust that gets on the tiny CCD imaging sensor will cause a big spot on the video image.

Once the eyepiece-holder adapter is safely installed on the camera, thread the IR blocking filter into the open end of the adapter. This filter is absolutely vital if imaging with a refractor and remains helpful with Newtonian and Schmidt-Cassegrain optical systems. The filter is required because the webcam's CCD is sensitive to infrared as well as visible wavelengths. Most refractors will not bring the long infrared wavelengths to the same focus as visual rays, causing a loss of sharpness in the image. Reflectors are not as affected by this problem, but do suffer from a phenomenon known as "IR bleed" where the infrared wavelengths are interpreted as magenta by the camera, which renders images with a violet tint if an IR-blocking filter is not used.

2.4 Using a Webcam for Astrophotography

Now you are ready to image the Moon or a planet. Since the Moon can easily be imaged in daylight, do your first trial runs using a daylight Moon so you don't have the added problem of navigation in the dark. For serious high-resolution "keeper" images, be sure to set up your telescope early so it can reach thermal stability before you begin imaging. Insert the camera into the eyepiece-holder and bottom it against the shoulder on the adapter. Lock it in place with the set-screw. At this point it will become apparent just how short the 3-foot USB cable that came with the camera is and why

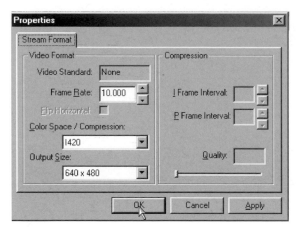

Fig. 2.5 *For best astronomical results with a webcam, select the highest uninterpolated resolution (640x480) and no more than 10 frames per second for a video capture sequence. Image by Robert Reeves*

I recommend that you install the 6-foot USB extension cable. Secure the cable to the telescope body at a pivot point on the telescope mount so it will remain in a fixed position as the telescope moves and not swing loose in the dark. Do not use a cable that extends the USB connection more than 10 feet because some USB devices have trouble communicating over greater lengths.

You are now ready to begin imaging. To start the camera, simply plug it into the computer's USB port, and it will power up. Click on the Philip's VRecord program from the Windows Start menu, and a preview screen will appear. At the top of the screen will be four drop-down menus: File, Devices, Options, and Capture. Click on Options → Preview, and a live low-resolution output from the camera will be displayed. Click on Options → Video Format. Here you set the frame rate and image output size. Although the ToUcam is capable of between 5 and 60 frames per second, it is best to use no more than 10 frames per second. This setting tends to produce better images when the seeing is poor. Set the output size to 640x480 and leave the Color Space/Compression option at the default setting. Again, no compression is desired, as this will degrade the image.

At this point the camera will be operating in full "auto" mode, so any bright object will be recorded.

Point the telescope toward the Moon and watch the computer monitor for signs of a lunar image. Most likely, the first view will be so out of focus that the Moon will be unrecognizable. Turn the telescope focus control until the image sharpens. At this point you will be amazed at the highly magnified image of the Moon on your computer monitor. Take a moment to enjoy the sight.

Now it is time to align the camera head so the image is square with the R. A. and Dec. axes of the telescope. This will place lunar north at the

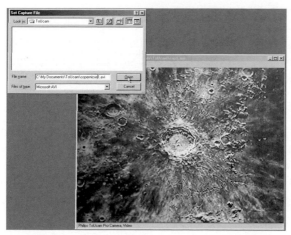

Fig. 2.6 *Before a video capture can be written to the computer hard drive, it must be named under the* Set Capture File *option (top). The easiest way to capture a video sequence is to set a maximum time limit for the exposure using the* Set Time Limit *option. Photos by Robert Reeves.*

top of the image and also simplify aligning any mosaic segments. Perform the alignment by placing a lunar feature along one edge of the field-of-view and moving the telescope north or south. The lunar feature will move accurately parallel to the edge of the field-of-view when the camera is properly positioned, so adjust it accordingly.

After aligning the camera, finish focusing the telescope. The image may already look focused; but to get the full benefits of high-resolution imaging, the focus must be exact. This is a process that has to be done with painstaking care. It is likely that the live webcam image will be gently rippling due to atmospheric distortion making focusing a challenge. Time spent insuring the best focus will be rewarded with outstanding images. Although the live image may look wavy and blurred by poor seeing, post-imaging processing with RegiStax will cull the best images grabbed during brief moments of good seeing. If the telescope is not properly focused, none of the images will show the detail potentially available when seeing allows the capture of undistorted views. As the image becomes clearer, a soft focusing touch is needed to prevent shaking the telescope at such great magnification.

When poor seeing makes focusing difficult, slew to a nearby bright star, manually set the shutter speed to $\frac{1}{5000}$- or $\frac{1}{10,000}$- sec. to reduce blooming from overexposure and focus the star until it is as small as possible. The point of exact focus is easier to determine for a star—which has a pinpoint image—than for extended objects like the Moon and planets. (Schmidt-Cassegrain users should continuously monitor focus during the imaging session because the main mirror may shift on its focusing collar due to gravity, which will cause a change in focus.)

An electric focuser is best because it minimizes telescope shake.

Since the webcam image will have about the same scale as that of a 4-mm eyepiece, just touching the focus knob will make the image shake, making it very difficult to locate the exact point of best focus. If you do not have an electric one, you can make a "Jiffy" focuser that will make fine manual focusing easier. A Jiffy focuser is made from the lid of a wide-topped mayonnaise or peanut butter jar that has been attached a telescope's smaller focus knob. Bill Arnett originated the idea and termed it a Jiffy focuser when he used the lid of a Jif Peanut Butter jar. Drill or grind a hole in the center of the lid to fit snugly when pressed over the standard focus knob. When finished, you will find that turning a focus knob of much larger radius makes it easier to keep a stable image than the standard one.

Once best focus is achieved, it is time to do some "housekeeping" functions. The first thing to do before beginning video imaging is to designate the file name for the video capture. This must be done prior to all video captures or the program will simply write a new capture file over the previous one. This is done by selecting File then Set Capture File. A window will appear allowing you to create a file name for the next video capture. Once the capture file name is entered, a second window will appear showing the available hard drive space and ask for the desired capture file size. This is one of a number of ways to set the capture file size. This option can be used to limit the capture to a given number of megabytes. If you wish to use other options to manually stop the capture when desired, or to set the capture length to a finite time limit, click Cancel for this option. The other option to limit file size can be found in the Capture menu. Start Capture will manually start the capture sequence and enable the Stop Capture option. The Set Time Limit option allows designating a fixed time for the video capture to record. When this option displays, be sure to check the Use Time Limit box above the window where the time is entered. I personally prefer to use the time limit to automatically stop the capture after a set time.

Now that the housekeeping functions have been handled, it is time to set the camera controls for the actual imaging session. Return to the Options menu then click Video Properties. Here three options display: Image Controls, Camera Controls, and Audio Controls. First, go to Image Controls and deselect the full auto camera control option and look at the frame rate option. If the frame rate does not show the previously selected 10 frames per second, go ahead and reset that option. If the same frame rate is going to be used for the entire imaging session, it is more convenient to lock the frame rate at 10 per second using the Set Frame rate option under the Capture menu. Be sure to check the Frame rate box that enables this option. Next, jump to the Camera Controls section. Leave the White balance

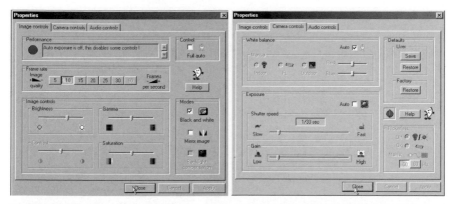

Fig. 2.7 *The control program that comes with the Philips ToUcam Pro II 840 allows full manual control of exposure parameters. In* Image Controls *(left) uncheck the* Full auto *box to allow manually setting the* Frame rate, Brightness, Gamma, *and* Black and white *option. In* Camera Controls *(right) the* Shutter speed, Gain, *and* White balance *can be adjusted. Images by Robert Reeves.*

on the Auto and deselect the Auto option on the exposure controls. Watch the image of the Moon on the computer monitor and increase or decrease the Exposure control as needed to properly expose the Moon. Leave the Gain control as low as possible to prevent electronic noise from speckling the image.

Now, return to the Image Controls window. Note the option to record in black and white—planets do display color so use the Black and White option only on the Moon. The default exposure setting is Auto where autoexposure controls the brightness and color balance. Individual images showing large areas of the Moon may expose well with the Auto setting, but can usually be improved with manual exposure setting. The camera controls will have settings different from those seen in conventional photography, for example: gain, gamma, and brightness.

- Gain can be thought of as the ISO speed setting for the webcam; the higher the gain, the more sensitive it will be, but at the expense of a noisier and grainier image.
- Gamma is similar to the contrast setting.
- Brightness controls how bright an image will appear.

For lunar imaging, I have found that setting the gain and gamma settings low and the brightness level high allows the selection of a higher shutter speed that is beneficial. Similar settings work on the planets, but you are encouraged to try different camera control settings since imaging the Moon or a planet is very subjective.

There is no one standard exposure selection because every new view

Fig. 2.8 *Jupiter's cloud belts and Saturn's ring system are always inviting targets for webcam imaging. Jupiter was imaged at f/25 through a Celestron 11 while Saturn was imaged at f/20 with an 8-inch SCT. Photos by Rod Mollise (left) and Robert Reeves (right).*

through the webcam will require some adjustment of its controls to look its best. Only when shooting a lunar mosaic should the camera be set once and then left alone for the entire imaging session. This is necessary to maintain consistent image density between mosaic frames. For mosaics, the camera controls should be adjusted for the brightest area you will image. This often means that detail along the dark terminator will be lost, but the overall image will look better than if the bright highlights are overexposed and washed out.

The Audio Controls option under the Video Properties menu allows you to take advantage of the ToUcam's built-in microphone to make verbal notes about the video capture while it is recording.

Once the camera is aligned, focused, and the exposure set, it is time to capture some video. When the chosen lunar scene is properly framed, try both the Manual and Timed options under the Capture menu. Click on Start Capture then click OK on the Ready to Capture subwindow. At this point the camera is rolling! The status bar at the bottom of the preview screen will show the number of frames captured, the elapsed time, and the frame rate per second. I find good results are achieved with a video sequence of about 500 frames. If the controls are set to manually end the video capture, click Stop Capture and the sequence will be saved to the file name you designated at the beginning of the session. If using a timed capture, the sequence will stop by itself. At this point, return to the Set Capture File option and enter a new file name for the next capture sequence. If this isn't done now, you run the risk of having the next capture sequence over-

Fig. 2.10 *Locating a planet in a webcam's small field-of-view can be a challenge. A flip-mirror system such as this unit from Meade Instruments enables the user to preview a wider field-of-view making it easier to center the target in the camera's field-of-view. Photos by Rod Mollise.*

writing the last one.

The capture-control option I prefer is Set Time Limit. To capture 500 frames at 10 frames per second, simply set the time limit to 50 seconds. Be sure to check the Use Time Limit box. Click OK then return to the Capture menu to begin recording using the Start Capture option. Once the capture starts it will automatically shut off after the designated time.

To achieve high detail in the webcam's narrow field under steady non-turbulent seeing conditions nothing will beat a large telescope. But large telescopes often do poorly because of atmospheric turbulence. An 8-inch instrument will often produce a sharper image than a larger telescope because the smaller one will more often be able to view through individual cells of turbulent atmosphere while the larger telescope may be viewing through several cells at once, each of which introduces its own aberration to the image.

As was discussed earlier, a webcam can image the Moon in daylight under clear blue skies. The visual contrast of the Moon on the computer monitor will not be as high as it is at night, but the image can be tweaked during processing to increase the contrast.

In my experience, using the Philips ToUcam Pro II with an aging laptop, the camera-control program becomes unresponsive to manual commands during a video capture sequence after about a half-dozen videos have been recorded. The control program works well when the camera is programmed to take a timed video sequence; but if the video sequences are manually stopped with the mouse controls, the program eventually stops responding. The solution is to restart RegiStax and then continue processing. Of course, your video resolution settings and any custom exposure settings will have to be reset.

Once you have some experience with a webcam, you may want to

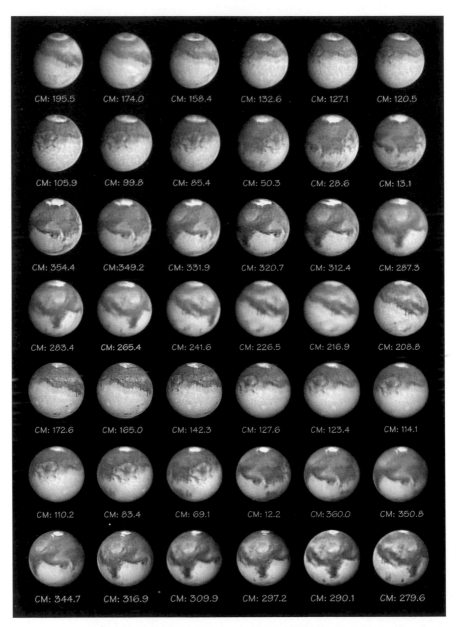

Fig. 2.9 *The high-resolution results attainable by stacking many planetary images can be seen in this montage of the 2003 Mars opposition. Between 8 and 90 separate frames were combined to produce each of these Mars images taken between July 9 and September 10, 2003. They were taken at f /40 through a Celestron 14 SCT and show almost two complete rotations of the planet. Photos by Mark Schmidt.*

expand its capabilities. Accessories such as a telescope focal reducer will allow wider-field imaging of lunar scenes. Conversely, using a 2x or 3x Barlow, under really good seeing conditions, will increase detail on planets. At normal prime focus, a well-aligned finder will allow aiming the webcam adequately, but the narrower field when using a Barlow will make finding a planet a challenge. The solution is to use a flip-mirror accessory like that used with CCD imaging. Meade Instruments makes a unit that will attach to Schmidt-Cassegrain telescopes. With this device, the target is centered using a low-power eyepiece that views the image via the mirror, then the mirror is flipped up allowing the image to pass through to the Barlow and webcam. Another useful accessory is a T-mount adapter to couple the camera to a variety of optical components.

The camera-control program that comes with the ToUcam Pro II will be adequate to perform image capture of celestial objects, but you should try other control programs such as K3CCD Tools by Peter Katreniak.

While imaging the Moon, one thing you will probably notice is the amazing number of birds and/or bats that fly in front of the Moon. From my urban backyard I capture some crossing the face of the Moon in two or three videos from each observing run. Considering that the field-of-view through my telescope is roughly $\frac{1}{16}$ of a degree, one can infer that there are a lot of nocturnal insect-feeders in the night sky. Since the processed image will be composed of many stacked frames, they are not a problem. They will appear as an out-of-focus blob, but it is fascinating to watch the flapping wings as they traverse the lunar scene.

I have noticed that if I can see a star in the 6x30 finder of my Schmidt-Cassegrain, the ToUcam Pro II will usually be able to display an image of it on the monitor screen. Exposure, gain, contrast, and brightness controls may have to be set to maximum limits to show dimmer stars. This offers the possibility of using the webcam to visually guide a telescope carrying a wide-field piggyback camera. This option is only useful if the imaging scale of the wide-field camera is forgiving of slight telescope aiming adjustments in order to find a star bright enough to display with the webcam. If a bright star is found, simply place the cursor over the star and guide with the telescope's slow-motion controls to keep the star within an inch of the cursor. With the webcam's very narrow field-of-view, this one-inch radius around the cursor equates to keeping the guide star within the guiding box of a 12-mm dual-crosshair guiding eyepiece used on an 8-inch $f/10$ telescope.

2.5 Image Processing Using RegiStax

Many observers use an older laptop for video image acquisition, but when they turn to processing, find that it is too slow. I transfer captured video from my slow laptop to a much faster desktop computer for processing. However, the transfer of gigabytes of video data takes a long time. My solution is patience. I simply transfer the video from laptop to desktop over a wireless home network connection, a process that can take several hours after a productive evening of webcam imaging. To make this less painful, I perform the transfer at night while I sleep so the next day all the video files are residing on the faster desktop where they can be processed with RegiStax, and the best of them archived to CD.

A faster but less convenient transfer method is to manually shuttle each video file between computers using a digital camera's USB card reader as an external drive. By installing a high-capacity digital-camera memory card in the reader, you can transfer a video file from the laptop to the card reader, then plug the reader into the desktop and download the file. This process is tedious with a large number of video files, but it will work, and enables one to quickly process a video sequence as a spot-check of how the imaging session went.

Once the video files are in your image processing computer, it is time to start RegiStax and watch the magic. Remember that even on a turbulent night a webcam and RegiStax will achieve better results than film cameras could on good nights. Being able to use a $\frac{1}{100}$-sec shutter even at high magnifications works wonders for freezing lunar detail. RegiStax will discard images ruined by bad seeing. Be sure to take sufficiently long video sequences to insure plenty of "keeper" frames because the fewer images that are stacked, the grainier the result will be.

To begin processing a video sequence with RegiStax, click on the program's Select Input box, then choose the AVI video you wish to process. The first frame of the video will appear on the preview screen. If color video of the planets is being processed, check the Color processing box. If lunar images are being processed, the color option can be left off. Within the Processing Area selection box, click on the 1024 option so the program will process the entire area of the frame. Below that in the Alignment box (pixels) selection, choose 32. Now, while the cursor is over the image, a small box 32 pixels square with a crosshair in it will appear. This will select the image feature that will be the "anchor" for aligning all subsequent images

The best frame in the video sequence should be used to select the alignment "anchor" point. To find the best image, use the slider at the bottom of the preview screen to scroll through the sequence of frames. Once

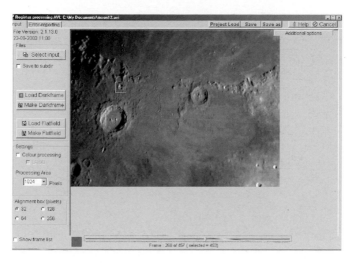

Fig. 2.11 *From the RegiStax main window, select the video input to be processed, then select the processing area and alignment box size. Use the slider at the bottom to preview the video frames to select the best one as a master alignment frame. Image by Robert Reeves.*

a good one is found, place the cursor over a bright lunar crater or the ball of a planet. The cursor will then turn into the 32-pixel square selection box. Left-click it when it is centered on the alignment feature. Next, two more windows will appear that display the brightness properties of the alignment feature selected and the image quality parameters for the video frames being analyzed. Click on the Align & Stack button; the program will begin scanning each frame in the video sequence, grade each image for quality, select the best images and automatically align and stack them. The program will, by default, select and stack all frames with a quality of at least 80 percent that of the first reference image selected. This is why it is important to initially select the best image as a reference. A higher-quality percentage can be chosen, such as 90 percent or 95 percent, but dramatically fewer images will "make the cut" for the final stacking process. While this will improve the sharpness and focus of the images that are stacked, there is a trade-off in that when fewer images are stacked, the resulting combined image is noisier and "grainier." Because of this, it is best to leave the quality setting at the default 80 percent. Some of the images may not be as sharp as we wish, but the final image will be smoother and better looking, even if it is slightly less sharp. The process of aligning and stacking will take several minutes even with a fast computer—a status display shows the progress of the operation.

The RegiStax default settings for grading and combining images produce such excellent results that I have not bothered to try alternative set-

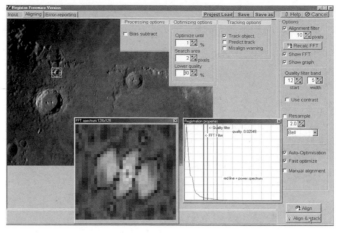

Fig. 2.12 *During the alignment phase, RegiStax displays two windows showing the centroid of the selected alignment feature and the FFT (Fast Fourier Transform) Spectrum that shows the alignment quality between the reference image and the frames that were combined. Photo by Robert Reeves.*

tings. However, the alignment, quality, and optimization filters are fully configurable; and those inclined to experiment with different settings have many options from which to choose.

Once the grading, aligning, and stacking are complete, wavelet image processing can be performed. At this point, RegiStax presents the stacked image in a window that contains six sliders that control different layers of detail with the wavelet image processing parameters. These sliders are selectively applying unsharp masking techniques to the image. This has the ability of recovering weak signals from the background electronic noise. Each slider offers settings from –5 through 100, and sliders 1 through 6 progressively process larger details. In practice, I have found that sliders 1, 2, and 3 offer the most improvement in image quality and generally leave sliders 4, 5, and 6 alone. For the image scale my telescope provides at prime focus with seeing conditions typical of my locale, I find slider 1 can usually be pushed to "15," slider 2 to about half that value at "7" or "8," and slider 3 to about half that of slider 2, or about "3.5." Once the wavelet image processing is complete, click on the Noise Reduction box, and a popup window will display more processing controls for all six layers and offer the option of selecting one of four default methods for noise reduction within each layer. Although the noise reduction routines can be customized with the use a slider for each layer, I have found the results provided by the selection of the "canned" noise reduction options to be adequate. Option B seems to work best with my particular telescope setup.

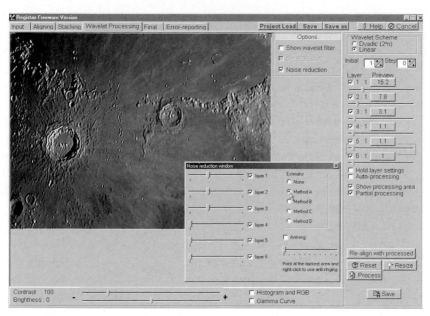

Fig. 2.13 *Extensive wavelet image-processing and noise-control options are available with RegiStax, as well as contrast, brightness, and color correction tools to tweak images to their best appearance. Photo by Robert Reeves.*

The settings of the wavelet sliders can be highly subjective. It is best to experiment and see what looks right. On planetary images, slider 3 can be placed as high as "15" or more. With a stack of many relatively noise-free frames, the sliders can be set higher; but if the image is composed of only a few noisy frames taken through poor seeing, the image will dissolve into a noisy mess if pushed too far. Be careful not to overprocess an image. Compare the results of each processing step to the previous image. Signs of overprocessing are crater rims that become bloated into donuts, and detail in rugged lunar highland areas that has become bright and washed-out. Watch subtle detail and make sure it does not become very contrasty and unnatural looking. Remember that webcams operate at 8-bits and thus have a limited dynamic range of 256 brightness shades, making it is easy to overprocess the images. A good point to remember is that most people tend to overprocess their images without realizing it. Process and sharpen an image until it looks good, then back the adjustments off slightly. I have found that identical image processing settings can usually be applied to all video captures taken under similar exposure, lighting, and atmospheric seeing conditions.

Once image processing is complete, RegiStax offers you options that can adjust the contrast, brightness, histogram, and gamma curve of the image. The final RegiStax window allows control of the hue, saturation, and lightness of color images. To complete the processing, the image can be rotated and resized before being saved in a number of different image formats.

Because of telescope tracking errors or atmospheric turbulence, the image will move around several pixels between the start and finish frames in a video sequence. Sometimes the atmosphere can be so turbulent that small lunar craters appear to wink off and on like a light. RegiStax will reject up to 95 percent of the images if the atmosphere is very turbulent, but the results will still be pleasing. When the frames are stacked, this will create a white border around the finished stack because several images do not perfectly overlap. This border will have to be cropped either for aesthetic reasons or before it is inserted into a mosaic.

Sometimes, in spite of the processing area being set correctly, RegiStax will not fully process one corner of the image. Simply click on the unprocessed area, and the program will perform more processing itinerations and "catch up" in the unprocessed portion of the image.

Running RegiStax with large AVI files takes a while. Fortunately, the program will run nicely in the background while you do other things like reading email or surfing the web.

Occasionally a glitch happens when recording an AVI video capture file, and the file becomes corrupted. I have found that once RegiStax reads a corrupt AVI file, it will not respond when it reads the next AVI file. The solution is to reboot RegiStax and then continue processing.

Once the AVI files have been processed into single images, be sure to archive the best videos on CD. Future versions of webcam-processing programs may contain features that will bring out greater image detail, or the operator may acquire a greater understanding of how the program works and wish to reprocess an old video to bring out better image detail.

The bottom line is that for less than the price of a premium eyepiece, a webcam, coupled with a powerful processing program like RegiStax, can introduce the user to the world of high-resolution lunar and planetary imaging; he or she can even create NASA-like images of the Moon, and achieve planetary images that surpass those taken by professional astronomers just a decade ago.

The small scale, but high-resolution, lunar images produced by a webcam opens up the possibilities for wide-area lunar mosaics. While DSLR cameras can image large areas of the Moon with a single shot because of their larger sensors webcams can produce sharper images, even with a much smaller array, because their pixels are smaller than the DSLR

Fig. 2.14 *Assembling high-resolution lunar webcam output into a mosaic can produce some de-cidedly NASA-like images. This mosaic is incomplete because the limited hard drive capacity of the author's laptop prevented further imaging. This problem was addressed by installing a high-capacity hard drive to handle the huge amount of data produced by a webcam imaging session. Photo by Robert Reeves.*

Fig. 2.15 *After increasing the hard drive capacity of my laptop computer (See Fig 2.14), I was able to image the entire lunar disk. This composite of the waxing crescent moon was created in Photoshop 7.0 using 20 separate images captured through a Celestron-8 at prime focus using a Philips ToUcam Pro II 840k. Photo by Robert Reeves.*

pixels. While the field of view will be much smaller and it will take up to two dozen overlapped webcam images to equal the same area imaged through a DSLR, the webcam mosaic will possess superior resolution.

When shooting a lunar mosaic, it is very important to align the webcam so its field-of-view is square with cardinal directions of the telescope mount (see Section 2.4). This will aid in the alignment of the mosaic's overlapping frames. It is tempting to start imaging along the curved limb of the Moon, but this will quickly become a navigational nightmare. It is better to start along the terminator and work your way north and south in sequential rows of images that progress toward the curved limb.

Individual webcam images can be manually assembled into a mosaic using the "layers" option in Photoshop, or they can be stitched together using the panorama software that comes with most digital cameras. The panorama software bundled with Nikon cameras, Panorama Maker by ArcSoft, usually does a good job. If you do not own a Nikon camera, this reasonably priced software can be purchased from ArcSoft. Be aware that the default image-stitching options may overlap lunar images in such a way that lunar landmarks are not spaced properly. Panorama Maker allows the assignment of registration points on images, so any misalignment can be refined.

The key to creating lunar mosaics is to preplan and properly space the individual lunar scenes so they overlap by a generous amount—about 25 percent works well. To avoid having a gap between images, use a good lunar chart to plot the mosaic. Also, use the same exposure settings for all segments of the mosaic so adjacent portions will match in density.

The high-resolution astrophotography possible with a webcam will happily occupy many hours at the telescope while you observe the ever-changing lunar surface in high detail on the computer monitor. The satisfaction of seeing the space probe-like images produced by your own telescope will be further enhanced by the fact that such amazing image quality was achieved with simple, modestly-priced webcam-imaging equipment.

Chapter 3
The Differences Between Film and Digital Cameras

3.1 Introduction

The term "digital" refers to the binary representation of data as bits and bytes. A bit, in turn, is the smallest unit of information that a computer can store and process. It represents an electrical state of either "on" or "off" with a value of either zero or one. The zero and one are used to create binary numbers, usually with eight bits associated together to form what is called a byte.

The digital photography term "pixel" is a contraction of PICture Element, but spelled pixel. It is the smallest point of a bitmapped image that can be assigned color and intensity. A bitmap is the digital representation of a picture where pixels are arranged in a grid and assigned a specific color and intensity.

Digital imaging has progressed amazingly fast especially when compared to the plodding history of film photography. The basic methods of capturing a film image evolved very slowly. It was nearly 40 years after the perfection of the Daguerreotype process (1839), with its complex and toxic chemical process, before the dry-plate photographic process was created (1877). It was almost another half a century before the photographic industry selected a common film format size (35 mm) upon which to build consumer photographic systems (Leica 1925). It was more than 75 years ago that the 35-mm SLR camera (Exakta 1933) was introduced and went on to become a worldwide standard for photography. But since the 1960s, film photography has remained basically unchanged. Lenses have become sharper with special glasses, coatings and computer-aided design. There have been improvements in film speed, resolution and color rendition. "Bells and whistles" have been added to cameras with items like automated exposure and focus settings. But overall, film photography operates today in a manner very similar to the way it did when the parents of the readers of this book were in their childhood. Digital photography on the other hand took only about a decade to evolve from very expensive cameras and low-quality images to become an affordable "film-like quality" replacement for film-based photography.

Today's digital cameras are computers that contain imaging, memory and processing chips in addition to the optical components needed to create an image. The camera's software can handle megabytes of data within a fraction of a second and record a fully-processed image within several seconds. The image is then moved from the imaging sensor to an EPROM (Erasable PROgrammable Memory) chip for processing, and finally transferred to a storage card. Most digital cameras can also network with a computer to download the images for display and printing.

3.2 How Digital Cameras Work

The image sensor in a digital camera is a gridwork of individual photodiodes that convert light into an electronic digital signal. More specifically, photons strike the photodiode and are converted into electrons. This is somewhat analogous to the film process where photons strike the individual film grains causing a chemical reaction that becomes visible when the film is developed in a chemical bath.

One way to visualize how a digital-camera imaging sensor works is to compare the photodiode array that makes up the sensor to an array of millions of very small solar cells. As the camera lens focuses an image, with varying light intensities, upon the different cells across the array, they produce a mosaic of electrical charges. The number of electrons produced by each solar cell is proportional to the brightness of the image immediately above each cell. We can use the electricity produced by each solar cell to illuminate an equal-sized array of light bulbs to reproduce the image since the brightness of each bulb on the display array will be proportional to the electrical output of the corresponding solar cell on the camera array.

In practice, the electrical charge contained in each photodiode is amplified and then converted by an analog-to-digital converter into a binary number. The camera's software creates colors using filters. (The color conversion process will be discussed in Section 3.6.) The image is then transferred to the camera's memory storage card.

The process of saving an image for most consumer cameras is relatively long even if a fast-writing memory card is used. This problem is receiving considerable developmental effort so that while the overall image size has steadily increased, recording times are less than with earlier lower-resolution cameras.

To speed things up, most cameras have random access memory

Fig. 3.1 *This appealing portrait of the constellation Cassiopeia is a single two-minute exposure taken with a Canon D60 at ISO 200. The view is cropped from the field of a 20-mm f/2.8 lens. Photo by Johannes Schedler.*

(RAM) to buffer a rapid sequence of images. The capacity of the image buffer varies from camera to camera, and after the buffer is full, the camera must wait for all images to process and write to the memory card before another image can be taken.

Fortunately, since digital cameras are geared for general use one does not have to know a lot about computers to take pictures. However, when they're applied to celestial photography you will have to use advanced image processing software to process the resulting photographs. The programs, though, are not hard to master and call for no more actual computer skill than that required to operate a good word-processing program.

3.3 Noise

In low-light applications, such as deep-sky astrophotography, you would like the highest percentage possible of the available photons to make a contribution to the image. CCD and CMOS image sensors, like those used in digital cameras, have a great sensitivity advantage over film and older electronic detectors. This is because CCD and CMOS sensors have high quantum efficiencies. Quantum efficiency is the probability (between 0 and 100%) that an incident photon will interact with the sensor to produce a detectable photon event.

The quantum efficiency of photographic film at its wavelength of peak sensitivity is rarely greater than about one percent, and it is often much less. Most of the photons simply pass through the emulsion without causing photon events. Certain photocathode-type electronic detectors, such as photomultipliers and image intensifiers, have values as high as 10 to 15 percent. However, solid state sensors like CCDs have quantum efficiencies as high as 50 to 80 percent, depending on whether they are front-illuminated or back-illuminated. This is one reason why astronomers have embraced these new detectors with such enthusiasm.

Another measure of sensitivity is detective quantum efficiency (DQE). DQE is defined as the signal-to-noise ratio of the recorded output divided by the signal-to-noise ratio of the input as determined by the incident photons, all quantity-squared. It can be shown that this really does give a measure that is an effective quantum efficiency. Simply stated, DQE is the efficiency of recording the information contained in the incident photons. Thus noise is basic to detector sensitivity.

In addition to being less than 100% sensitive, all photon detectors introduce additional noise of their own. This is why DQE is always somewhat less than the sensor's underlying quantum efficiency. Throughout this and subsequent chapters we will repeatedly deal with the effects of

Fig. 3.2 *This 30-second exposure using a five-megapixel Sony Cybershot demonstrates the noise encountered with the small pixels common to compact fixed-lens digital cameras. Photo by Jean Kodama.*

noise. Noise results in lower contrast and more washed-out images. The noise in a recorded image can be subdivided into two general types: random and non-random (fixed-pattern) noise.

3.3.1 Random Noise

Random noise varies with time and is statistical in nature. It is thus different in every image. It comes primarily from four sources: photon noise, detection noise, dark current noise, and readout noise. All contribute to the recorded output signal-to-noise ratio.

- **Photon noise** is independent of the sensor itself. It is the noise in the input light and is the result of the inherent statistical variation in the arrival rate of photons incident on the sensor. When calculating DQE, this gives your input signal-to-noise ratio. The total number of photons arriving at a pixel during an exposure is the signal. The expected statistical variation in this number is the noise and is equal to the square root of the number of photons. Thus the photon signal-to-noise ratio is equal to the square root of the signal. For instance, if there were only 100 photons, the expectation is there would be a variation of about 10 photons, or about 10 percent variation. If there were 1000 photons, the expected variation would be 32 photons more or less, or about three percent variation. This demonstrates that when the light level is low, such as in astrophotography, the variations get larger and cause more random noise.

- **Detection noise** is caused by the sensor's imperfect quantum efficiency. Some of the incident photons do not interact to produce photoelectrons and are lost. The number of photon events in a sensor is always less than the number of photons. Thus the number of effective photons is reduced through the statistical detection process. Because the recorded noise is determined by the number of interacting photons and its statistical variation, the recorded noise is correspondingly greater than the underlying photon noise, that is, DQE is less than 100%.

- **Dark current noise** arises from electrons that are thermally generated within the sensor's silicon structure at each pixel. These thermal electrons, often called dark current, mimic the photoelectrons generated by the incoming photons, except they contain no information. As with photon noise, dark noise varies statistically. The number of thermal electrons generated at a given pixel has an expected variation equal to the square root of the number itself. Although the basic dark background can later be subtracted out, the random noise in these thermal electrons cannot be removed. Thus dark current adds more noise to reduce the overall output signal-to-noise ratio and make DQE still lower.

- **Readout noise** is caused by the inability of the sensor electronics to perfectly detect the exact number of electrons stored in each pixel. The process involves converting the stored charges into voltages, one pixel after another, and then converting these analog voltages into digital values by an analog-to-digital (A/D) converter. Each of these steps can introduce more noise into the signal to further degrade the signal-to-noise ratio, which reduces DQE even more.

Random noise limits the ability of a sensor to reach its full imaging potential. The makers of CCD and CMOS sensors strive to minimize their detectors' random noise over and above the photon noise. Users of these detectors should similarly strive to introduce as little extra noise as possible. Users should understand that:

The full-well capacity (saturation) of a sensor represents the maximum number of electrons that can be stored in each pixel, and thus establishes the maximum amount of signal available for readout. The ratio of each pixel's full-well value to the number of read noise electrons characterizes the ability of the sensor to record a full range of image contrast.

Dark current accumulation in a pixel's potential well limits the exposure time over which a useful signal can be accumulated.

The total available signal relative to noise governs the dynamic range of a sensor and the maximum number of gray-level steps the camera's analog-to-digital converter can discriminate.

Fig. 3.3 *With long exposures, all electronic imaging sensors will display noise. This dark frame is a one-minute exposure with a Nikon Coolpix 995 at ISO 800 with internal noise-reduction mode turned on. Photo by Paul Hyndman.*

The rate at which thermal electrons are created in a sensor is very temperature-dependent and the cooler the sensor is, the less dark current it will produce. Cooling the sensor reduces the dark current dramatically. In general, good sensors display half as much dark current for every 5 to 9 degrees Centigrade they are cooled below room temperature. This value is called the doubling temperature, and the benefits of cooling remain steady until the temperature reaches about 5 to 10 degrees below zero Centigrade. Beyond this point, the rate of dark current reduction diminishes rapidly. Since random noise increases with temperature, the most direct way to reduce noise is to reduce the sensor's temperature. This is the technique used in dedicated astronomical CCD cameras, but it is impractical to add heavy, power-consuming cooling equipment to a consumer model. You cannot control the ambient conditions outdoors, but there are a number of ways to minimize the sensor's temperature during a long exposure. The most effective is simply to leave the heat-producing LCD preview screen turned off as much as possible.

To a minor degree, some noise is caused by cosmic rays. When these energetic particles strike a photodiode in the imaging sensor, their high energy levels can react like incoming photons and create a bright pixel in an ordinarily dark background. Fortunately, the Earth's atmosphere shields us from most cosmic rays. Their effects are minor compared to other sources of noise in an imaging sensor, and it is impractical to shield the camera with lead.

3.3.2 Non-Random Noise

Variations in the sensitivity of individual pixels can cause a non-random noise that repeats to some degree among images. This is sometimes called fixed-pattern noise. Ideally, all pixels on a sensor should be uniformly sensitive, but minor deviations in the manufacturing process can lead to some statistical distribution of pixels with varying sensitivity. A pixel which has a sensitivity far from the mean of most pixels on the sensor will always appear brighter or darker than equally illuminated surrounding pixels. When this condition becomes severe, the sensor will have hot pixels (always brighter than normal), stuck pixels (always on maximum brightness), or dead pixels (always dark). Such non-random noise is essentially built into the camera's sensor, but many cameras now have internal pixel-mapping software that will detect hot or dead pixels and eliminate them by interpolating surrounding pixel brightness to fill in the spot caused by the defective pixel. Hot pixels are discussed in greater detail in Section 4.5.

Another kind of non-random noise is caused by light reaching the lens from the side of the camera rather than only through the eyepiece with an afocal coupling. Any stray light getting in this way can cause noise and is noticeable if it is not uniform. Shielding the exposed camera lens barrel will eliminate this type of noise.

Higher ISO speed settings will also increase noise. This is because a higher sensitivity is achieved by increasing the electronic gain on the sensor's photodiodes. This process amplifies the signal created by incoming light, but also amplifies unwanted signals caused by noise. Use the lowest ISO setting possible for your target to reduce the amplification of noise.

Do-it-yourselfers can try actively cooling the camera's sensor to reduce noise caused by random electron motion in the sensor. Variations of the technique include using frozen cold-packs, cold-water circulation tubes, small air-cooling fans mounted on the camera and electric Peltier coolers as described in Sections 12.4 and 12.5.

You can also reduce most noise using various image processing techniques. These include image averaging and stacking to smooth out random noise by combining many images of the same object. Non-random noise, such as stuck or hot pixels, can be subtracted with dark-frame subtraction. The more repeatable the non-random noise is, the more easily it can be eliminated with image processing. Factors that will render images easier to work with in noise-reduction processing are: the use of high-image quality settings with no compression; and disabling any camera options that cause image processing to be done in the camera before the image is saved to the memory card.

3.4 Image Sensors

The Charge Coupled Device (CCD) image detector was developed at Bell Laboratories by George Smith and Willard Boyle in 1969. The original device was composed of Metal Oxide Semiconductor (MOS) capacitors, and to this day most high-performance CCDs are MOS architecture. Complementary Metal Oxide Semiconductor (CMOS) arrays have been around almost as long as CCD arrays; but until recently, they produced low-quality images.

The major operational difference between the CCD and CMOS sensor is that the CCD transfers the charge from one row of pixels to the next until it reaches the edge of the CCD where it is converted into a digital value. A CMOS sensor has transistors at each photodiode site to amplify and read the charge instead of having to move the charge across the sensor before amplifying it.

The CCD manufacturing process creates a near-lossless means of transporting electrical charges across the sensor without introducing distortion that leads to the creation of high-light sensitivity and a high-quality image. CMOS sensors, on the other hand, rely on standard manufacturing techniques used to make computer microprocessor chips and can be fabricated on any modern large-scale silicon production line. They use up to 100 times less power than CCD sensors and usually have more and higher-quality pixels than a CCD. However, historically CMOS sensors have had higher noise levels leading to lower-quality images.

On the other hand, CCDs do not lend themselves to large-scale chip integration. Rather than having all the electronics on one chip, other chips are required to handle support functions. While there has been some progress in integrating other electronics functions into the CCD they still require a considerable amount of supporting electronics. Depending upon the camera design, anywhere from three to eight chips are incorporated in the camera's image capture and conversion process. On the other hand, CMOS sensors can be made of the same silicon material as other computer chips. That means all the electronics can be incorporated onto one chip, thus reducing production costs, space requirements, and power usage. With CMOS, it's possible to produce entire digital cameras on a single chip.

Fig. 3.4 *Shown here is the Kodak full-frame 35-mm KAI-11000CM CCD image sensor. It is a high-performance 11-million pixel sensor designed for professional digital still-camera applications. Image courtesy of Kodak.*

3.5 Sensors Are an Evolving Technology

It has only been in the last few years that the limitations of the technology have sufficiently been overcome to make CMOS a viable alternative to CCD. The quality gap between images that are captured with CCD sensors and images being taken with CMOS sensors is rapidly narrowing. That is especially true as digital-camera resolutions climb because CMOS sensors don't suffer from a decrease in their signal-to-noise ratio as resolutions increase. One advantage of this is that higher-resolution CMOS cameras can be produced without having to significantly increase the supporting electronics.

A major advantage for CMOS sensors is that they have one amplifier for each photodiode and therefore can perform signal amplification simultaneously across the array. CCD sensors have to read all the photodiodes and then perform the amplification and that consumes more power and produces more heat.

Because of their basic architecture, CMOS sensors transfer image data faster than CCDs. Canon achieves even higher data transfer with its CMOS sensors by incorporating a system that simultaneously reads each image element line through two color channels for a 2x speed advantage. This sensor also gives greater bit depth to each pixel, allowing more dynamic range in the gradations from image highlights to shadow detail giving it near equality with color reversal film.

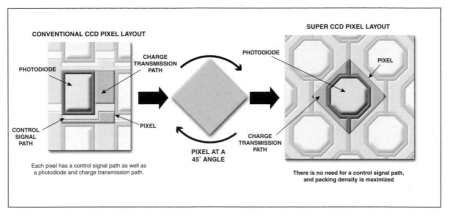

Fig. 3.5 *To increase both the resolution and dynamic range of their CCD imaging sensors, Fujifilm has created octagonal photosites that can be placed closer together than standard square photosites and also placed a second less sensitive photodiode at each photosite to preserve detail in bright image highlights. Drawing courtesy of Fujifilm.*

Fujifilm has concentrated on the CCD sensor by addressing a basic problem—the fact that higher pixel counts on the same size substrate produce higher noise levels and reduced sensitivity to light. Fuji's solution is to turn each photodiode at a 45-degree angle and then clip the edges of the square photodiodes creating octagonal photodiodes. The octagonal spacing allows 1.6 times more photodiodes in a given area of the sensor so the larger the sensor area the greater the benefit from the octagonal spacing. This has two benefits. First, the round micro-lens over each photodiode more closely matches the actual shape of the photodiode and is thus more efficient at gathering light. Second, the center-to-center distances between pixels are now less, allowing greater image resolution, while gaining a 50 percent larger photon-collection area on the photodiode. The larger light collection area increases ISO speed and has the benefit of greater signal-to-noise ratio within the sensor. Fuji also claims that the image acquisition speed is increased because the octagonal shape of the photodiodes allows the use of a honeycomb structure in the CCD chip that speeds the readout of the image.

Fuji's fourth generation Super CCD SR (Super Dynamic Range) addresses another shortcoming of the CCD imaging sensor—limited dynamic range—by adding a second smaller photodiode to each octagonal sensor area (photosite). The second photodiode is less sensitive and thus preserves detail in bright objects, preventing highlights from becoming overexposed. Fuji claims this adds two *f*-stops of extra dynamic range compared to conventional CCDs and more closely simulates the performance of color slide film. In astrophotography, however, we are rarely presented with the problem of overexposure; so while this new development may benefit terrestrial applications, it will be of little use in celestial photography with the exception of the Moon.

Fuji also incorporates software into their cameras that interpolates the data from a six-megapixel Super CCD into a 12-megapixel image. While there is some additional detail gained by the clever use of closer-spaced octagonal photosites the resulting images have only increased detail by a third but with the penalty of image file sizes that are twice as large.

Digital sensor design will be in a state of flux for the foreseeable future, and you should expect to see a constant stream of improvements that may or may not be significant to the astro imager. The potential consumer market is huge, and vast sums are being spent to develop new technologies to do the job quicker, better, and cheaper. Very little of this effort will be expended specifically for astronomical imaging where noise reduction and resolution go hand in hand. The good news is that while specific efforts have not been made on our behalf, improvements have happened that now allow us to capture images with a digital camera and process them with powerful software that a few years ago would be impossible to do with film.

3.6 Color

We live in a colorful world. The cone-shaped cells in the retina of our eye are sensitive to three colors—red, green and blue. A traditional digital image sensor, however, does not have three separate color sensitive layers; it is a monochrome device. It only registers intensity of light, not colors, and thus it is a grayscale sensor. Today, there are three primary ways to create a color image with a digital image sensor:

1. Tri-color filters
2. Bayer Pattern filter
3. Foveon sensors

3.6.1 Tri-color Filters

The problem with early digital cameras was how to get the three primary colors from the single monochromatic sensor. One early approach was to use a beam-splitter to divide the image into three identical ones, each falling on a separate image sensor employing either a red, green or blue filter. The advantage of this system was that the camera recorded each color at the same pixel location, and then recombined the three images into one color composite. The disadvantage was that the camera was bulky and ex-

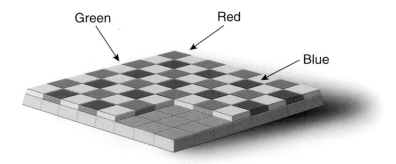

Fig. 3.6 *To create a color image using a black and white sensor, a mosaic of red, green, and blue color filters, called a Bayer Pattern array, is placed over the sensor so each photosite is covered by one individual color filter. The camera's software creates a color image by sampling the signal from other colors surrounding each single color photosite.*

pensive. Another approach was to use a rotating filter disk to take three images in succession, each through a different red, green and blue filter much like the way today's astronomical CCD cameras record color. Again this process provided all three colors at the same pixel location, but the disadvantage was that the camera and subject had to remain stationary during all three exposures, making it impractical for handheld photography or photography of moving objects.

3.6.2 RGB Bayer Pattern Filters

Most digital cameras create a color image with a Bayer Pattern filter called the color filter array (CFA) that is laid over the sensor. This filter is a grid of colored filters that looks much like the pattern of a checkerboard. Two types of color filter arrays can be used—a GRGB that passes red, blue and green, or a CYGM that passes cyan, green, magenta and yellow. The Bayer Pattern filter allows only a single particular color to reach each individual photodiode, and the camera's software determines the "true color" of that pixel by interpolating the color values of the surrounding pixels. The pattern of pixels forming the image is said to be "Bayer-interpolated" using the data from overlapping clusters of four photodiodes to create each pixel. This interpolation software creates a pattern of different colored pixels that constitutes the visual image we see in the photograph.

The GRGB pattern Bayer filter is the most commonly used and attempts to mimic the human eye by passing more of the green portion of the spectrum. To accomplish this, the filter contains twice as many green filters as red and blue. For example, on a 2048x1536 pixel GRBG sensor, the pattern would be:

1024x768 RED photodiodes (786,432)
1024x768 BLUE photodiodes (786,432)
1024x1536 GREEN photodiodes (1,572,864)

Sony improved the visual "daylight" performance of their Bayer filter by substituting an emerald filter for one of the green filters, thus creating an RGBE variation. The idea is similar to the addition of two lighter colors of yellow and magenta to CYM color cartridges in ink jet printers to achieve improved color accuracy.

The CYGN pattern filter uses the cyan, magenta and yellow colors but also adds green to add weight to the middle portion of the spectrum where the human eye is most sensitive. The pattern for a 2048x1536 pixel CYGM sensor would be:

1024x768 CYAN photodiodes (786,432)
1024x768 GREEN photodiodes (786,432)
1024x768 MAGENTA photodiodes (786,432)
1024x768 YELLOW photodiodes (786,432)

The downside of the Bayer Pattern filter is that it limits the amount of light reaching the sensor by filtering each photodiode to capture just one color; thus mosaic filter sensors capture only 25% of the red and blue light, and just 50% of the green.

The camera's internal software combines the three (or four) sets of color intensities with the color values of each photodiode's surrounding colors and produces the combined color pixel. The image is then processed in a demosaicing algorithm that combines the color value of each pixel with its eight surrounding pixels to produce a 24-bit color value for that pixel. This system has inherent drawbacks because no matter how many pixels an image sensor contains each photodiode can only capture one-third of the color. Interpolating color data from surrounding photodiodes leads to color artifacts and a loss of image detail. Some digital cameras intentionally blur pictures to reduce these artifacts.

3.6.3 Foveon Sensors

The Foveon sensor stacks three separate color-sensitive layers of photodiodes instead of spreading them on a single plane like the Bayer-type sensors. As each layer absorbs a particular color, the lower layers receive the remaining colors. This has the effect of allowing each of three distinct photodiode arrays to be selectively exposed to one of the three primary colors instead of interpolating colors from selectively sensitive photodiodes on the same chip layer.

Because each color-sensitive layer receives all of the light in each color, Foveon sensors are capable of producing sharper, more life-like im-

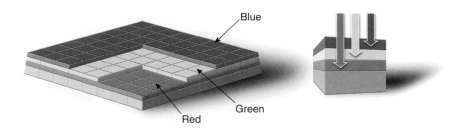

Fig. 3.7 *The Foveon sensor is unique in that each pixel receives all colors, not just a single color like a Bayer Pattern array. The blue, green and red colors are separated as the light beam penetrates progressively deeper into the sensor.*

ages for a given chip size. This is due to the fact that each pixel in the final image is the combination of three stacked color photodiodes instead of interpolated color information gathered from a larger four-photodiode array. This results in three times the color information per pixel than with standard sensors without the need for software interpolation of colors. Thus a three-megapixel Foveon sensor performs like a much larger standard sensor that must combine color information from adjacent pixels in order to interpolate colors. Although the Foveon chip produces better color, it is currently produced only in one size—the X-3—that has photodiodes that are 10.2 micron square. Each diode is thus larger than the six-micron photodiode used on the Canon 10D or Nikon D100. However, Foveon sensors suffer from poor low-light performance because when the three color layers are stacked, their effective output is like a 3.4-micron photodiode (10.2 microns divided by 3 layers) and therefore these sensors are not yet considered a viable candidate for serious astrophotography.

3.7 Sensitivity

Digital sensors react differently to light than film emulsion. If film is exposed to light that is too weak, it has no effect until the light reaches a certain intensity called the threshold. Whether film has been exposed or not, it has a certain density when developed because of the natural density of the film material and a small amount of chemical-induced emulsion fog. Together these are known as film base plus fog. Above the threshold level, exposure starts to rise very gradually, then more steeply. A graphical plot of a film's response to light intensity produces a characteristic curve, see **Figure 3.8**. The initial slow rise in this curve is called the toe of the curve. The middle part where it rises steadily is called the straight-line section. At the top of the curve, where the film has recorded its maximum exposure,

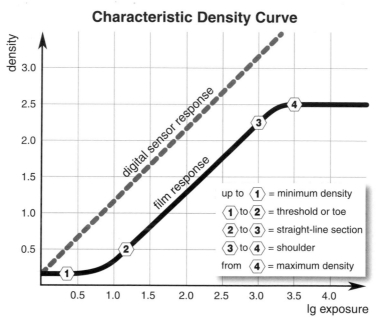

Fig. 3.8 *A typical characteristic density curve for film.*

the curve starts to level off again in an area called the shoulder. The contrast range of the film image depends on the slope of the curve; the steeper the slope, the higher the contrast. Film photographs tend to have low contrast in the shadows because they normally fall in the shallow toe of the curve. Similarly, highlights usually have little detail if they are in the shoulder of the curve.

Digital sensors also have a certain threshold below which they do not react to light. Above the threshold, sensors react to light increases in a directly proportional way. This means there is no toe to the response curve; it begins a straight climb up the slope until the sensor becomes saturated with photons and abruptly reaches a point where further photons have no effect at all. Thus a plot of the response for a digital sensor is essentially a straight line without either a toe or shoulder.

With film, underexposure or overexposure of the negative respectively result in thin shadow details or dense highlights, that are hard to print. The consequence of digital's linear response to light is that with underexposure there is no image at all and with overexposure the detail in the saturated pixels is also gone. There is nothing with which to work. Above the threshold point however, we can achieve better shadow detail with digital

Table 3.1 Digital Resolution vs. Print Size

Megapixels	Array Size in Pixels	Print Size in Inches
<1	800x600	3x5
1	1154x842	4x6
1.5	1280x1024	5x7
2	1600x1200	Sharper 5x7
3	2048x1536	8x10
6	3072x2048	Cropped 8x10
8	3264x2448	11x14

because of its more linear response.

Professional-level digital cameras can be set to warn of overexposed, fully-saturated highlights by flashing that area of the image when it shows on the LCD preview screen. This feature is also useful when imaging the Moon and bright planets. If a portion of the image is saturated, no detail will be visible and no additional detail can be gleaned through image processing steps.

3.8 Resolution

Image sensors are generally identified by the total number of pixels contained on their surface, such as a three-megapixel sensor or a five-megapixel sensor. A sensor is further distinguished by its resolution, such as a three-megapixel sensor has a 2048x1536 pixel resolution. Theoretically, the greater the number of pixels, the greater the resolution of the sensor. **Table 3.1** shows the relationship between pixel resolution and print size.

3.9 Fill Factor and Microlenses

The area of the light-sensitive photodiode compared to the area of the pixel on the image-sensor surface is small. This is because of the electronics on the sensor required around each photodiode. With Active Pixel Sensors, like the Canon CMOS and Nikon JFET sensors, this problem is worse because of the added circuitry needed to perform processing functions within the imaging chip. As a result, the light collecting area, or "fill factor" of the photodiode only intercepts a portion of the light that falls on the entire pixel area. To channel more incoming photons onto the actual photosensitive areas of the imaging sensor, a microlens array is overlaid the pixel array to funnel light more directly onto the photodiodes.

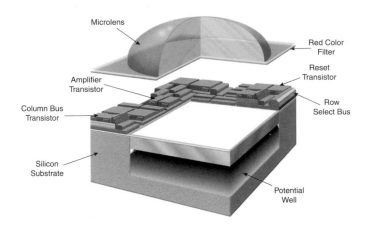

Fig. 3.9 *To increase the sensitivity of an imaging sensor, a microlens is placed over each photo-site to concentrate the incoming light onto the smaller photodiode within the photosite.*

3.10 Sensor Size, Aspect Ratio and Perspective

Unlike the single size format used with 35-mm film cameras, digital cameras can have any of about 10 different sensor sizes. As long as the correct lens is matched to the sensor, the camera will produce images of proper aspect and perspective. But the differing sensors can be confusing to a photographic enthusiast who has grown up with only the single 35-mm format. As we have discussed earlier, imaging sensors can be either a CCD or a CMOS sensor; and both use size designations described by a fraction such as $\frac{1}{1.8}$ or $\frac{2}{3}$. This designation dates back to the 1950's when the size of television vidicon tubes were measured with this system. The designation does not actually refer to the size of the imaging surface, but to the outer diameter of the round glass envelope of the tube, expressed as a fraction of an inch. However, it was soon found that the usable area of the vidicon was approximately $\frac{2}{3}$ of the diameter of the tube, and this thinking has carried over into the digital imaging field. There is no specific mathematical model between the diameter of the imaging circle and today's digital imaging sensor size, but it always roughly works out to $\frac{2}{3}$. Most photographers feel this size description system is outdated for digital-imaging sensors, but it is so deeply embedded into the electronics industry mentality that it is unlikely to change.

Manufacturers are continuously increasing the pixel counts in small consumer cameras. Fixed-lens point-and-shoot cameras now have counts as high as eight-megapixels. With this increased resolution comes reduced individual pixel size in order to place more pixels on the same imaging sur-

Table 3.2 Common Image Sensors (All dimensions in millimeters)

Type	Aspect	Circle	Diagonal	Width	Height
1/3.6"	4:3	7.056	5.0	4.0	3.0
1/3.2"	4:3	7.938	5.680	4.536	3.416
1/3"	4:3	8.467	6.0	4.8	3.6
1/2.7"	4:3	9.407	6.592	5.270	3.960
1/2"	4:3	12.7	8.0	6.4	4.8
1/1.8"	4.3	14.111	8.933	7.176	5.319
3/3"	4:3	16.933	11.0	8.8	6.6
1"	4:3	25.4	16.0	12.8	9.6
4/3"	4:3	33.867	22.5	18.0	13.5
APS-C	3:2	N/A	30.1	25.1	16.7
35 mm	3:2	N/A	43.3	36.0	24.0
645	4:3	N/A	69.7	56.0	41.5

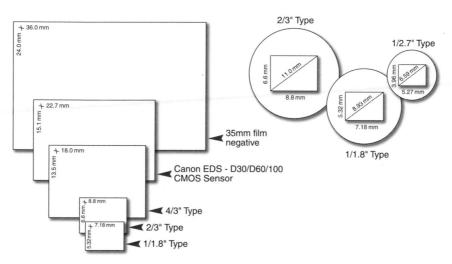

Fig. 3.10 *The dimensions of the imaging sensor used in a majority of digital cameras are much smaller than the area of a standard 35-mm film frame. This illustration also demonstrates how small the sensor area of compact cameras is compared to sensors used in DSLR models, whose sensors are also smaller than the 35-mm film frame.*

face area. Smaller pixels are less sensitive to incoming light and more prone to noise with longer exposure. Some argue that the increased pixel count with such small sensors results in little increased actual resolution in the finished picture because the individual pixels are now smaller than the lens can effectively use.

Table 3.2 describes some of the common image sensor sizes, and Table 3.3 shows how pixel size shrinks with increased chip resolution.

Table 3.3 Digital Camera Sensor Pixel Size and Area

Type	Sensor Pixels (Mp)	Dimens (mm)	Pixels Size (Microns)	Area (Square Microns)
$\frac{1}{1.8}$" CCD	3.3	7.2x5.3	3.5	12.3
$\frac{1}{1.8}$" CCD	4.1	7.2x5.3	3.1	9.6
$\frac{1}{1.8}$" CCD	5.2	7.2x5.3	2.8	7.8
$\frac{1}{1.8}$" CCD	6.3	7.2x5.3	2.5	6.3
$\frac{2}{3}$" CCD	5.2	8.8x6.6	3.4	11.6
$\frac{2}{3}$" CCD	8.0	8.8x6.6	2.7	7.3
$\frac{5}{3}$" CCD	5.6	18.0x13.5	6.8	46.2
CMOS	6.5	22.7x15.1	7.4	54.7
CMOS	11.4	36.0x24.0	8.8	77.4
CMOS	13.8	36.0x24.0	7.9	62.4

3.11 Four-Thirds System to Standardize DSLR Cameras

While DSLR cameras have standards for lens mount size (either Canon EOS or Nikon) there has been, until recently, no standard image sensor size and that has led to variations in the field-of-view between different cameras with the same focal length lens. In film photography, an entire camera format was based upon an established film format: 35-mm motion picture film. But electronic imaging became a reality before a standard imaging sensor size was universally adopted, leading to image scale chaos. To bring order to this situation Olympus, Kodak, Fuji, Panasonic, Sanyo and Sigma have adopted the four-thirds standard. The benefit to users is that the lens mounting, sensor size and lens backfocus are universal among all participating manufacturers.

As CCD and CMOS sensors began to replace the vidicon it was natural for the ratio of their physical dimensions to be similar; such as the $\frac{3}{1.8}$, $\frac{3}{3}$, or $\frac{4}{3}$. Using this standard, a one-inch (24.4 mm) vidicon tube will have a diagonal imaging area of about 16 mm. Using the same formula, a sensor with an actual imaging area of 13.5x18 mm has a diagonal measurement of 22.3 mm, or a $\frac{4}{3}$-type sensor. Coincidentally, $\frac{4}{3}$ is also the aspect ratio of this new sensor format where the first number is image width and the second height. Image sensors with this ratio fully fill a standard 8x10-inch print without cropping. These dimensions also closely match those used by professional photographers (who use medium format cameras) and the dimensions of most computer monitors. The $\frac{3}{2}$ aspect ratio of 35-mm film means that it had to be cropped almost three inches to fit an 8x10-inch print. But are these factors sufficient to move away from nearly a century of 35-

Table 3.4 Light Incidence Angle at the Edge of the Field

Crop Factor	Maximum Light Angle from 52-mm exit angle	Maximum Light Angle from 80-mm exit angle
1.0	19.1	12.7
1.3	14.9	9.8
1.5	13.0	8.5
1.6	12.2	8.0
1.7	11.5	7.5
2.0	9.8	6.4

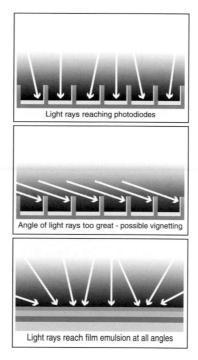

Light rays reaching photodiodes

Angle of light rays too great - possible vignetting

Light rays reach film emulsion at all angles

Fig. 3.11 *Film is sensitive to light no matter at what angle it falls upon the emulsion. The photodiodes in digital imaging sensors, however, are recessed and will be partially shadowed if the incoming light rays are at too great an angle to the sensor. To prevent vignetting because of shadowing, incoming light rays must be as perpendicular as possible to the sensor.*

mm imaging and establish a new standard? Before we delve into that discussion we need to briefly consider vignetting.

Vignetting is the phenomenon where an image gradually fades out to the edge. Vignetting occurs because obliquely-incident light is clipped by some obstruction within the lens after entering the front element, see Figure 3.11. But what does that have to do with digital sensors? Most of these devices use a microlens, or a tiny lens in front of each sensor element, in order to increase sensitivity. Because these lenses are directionally sensitive, sensors such as the Kodak KAI-11000CM "full frame" and KAF-5101CE "four-thirds" sensors lose 50 percent of their sensitivity if light strikes them at greater than 15 degrees from perpendicular. To avoid noticeable vignetting, light must arrive at these sensors within 12 degrees of perpendicular.

3.12 Film and Digital Sensors Record Light Differently

The four-thirds standard came about for DSLRs not because it was more convenient, but because as we have just seen, there is a fundamental difference in how electronic sensors and photographic film record light. With film, light can strike the photochemical surface at a reasonably acute angle and still properly record the image. The digital sensor does not have this latitude. The sensor is an array of photodiode sites and the "vision" of each of these sites is, for all practical purposes, limited to light arriving perpendicular to the sensor. Oblique rays striking a sensor at the periphery of an image will be vignetted and register colors inaccurately. This is one reason why DSLRs that can use lenses designed for 35-mm film cameras all have imaging sensors smaller than 35-mm film and are not useful with extremely wide-angle lenses. (Cost is the other—large arrays are expensive.)

However, if there is anything constant in the digital world, it is the expectation of ever increasing capability. It is just a matter of time before consumer-grade imaging arrays will be made that exceed the size of 35-mm film format. But if the digital sensor size were standardized to equal the 35-mm film format, the lens mount opening would have to be about as large as that used in medium-format film cameras, and by necessity would become larger and more costly. This would no longer be what people have come to expect with digital cameras.

It is axiomatic in the film-based world that for a given emulsion, larger formats produce sharper pictures. However, over time the resolution of film was steadily increased. Will that occur with digital image sensors? The four-thirds concept standardizes certain dimensional factors between cameras, but it does not standardize the pixel count. The microelectronics industry has continuously demonstrated an ability to increase circuit density, and there is no reason to think that will be different with image sensors. There will be pixel wars between manufacturers as they compete for higher image resolution.

What could limit the steady march of increased pixel density? The answer is the lens. Today, the average pixel size in a four-thirds sensor is 6.8

Fig. 3.12 *The natural high blue-sensitivity of digital cameras makes them a good choice for imaging a bright reflection nebula like IC-2118, the Witch Head Nebula in Orion. 19 four-minute exposures taken with a Canon 10D at ISO 800 through a Takahashi Epsilon 160 at f/3.3 were stacked to create this image. Photo by Rick Krejci.*

microns; that approximates the limit of consumer-grade photographic lens resolution. Until manufacturers achieve significantly better lens resolution, there is little point in raising the pixel count.

So what is the current "state-of-play"? After all, you intend to image now not later.

1. The DSLR $^4/_3$ image sensor today has four to five times the imaging area of the 2:3 imaging sensor used in fixed-lens cameras, giving it the ability to surpass film imaging in overall photo quality.

2. Since the $^4/_3$ sensor is smaller than a 35-mm film frame, shorter focal length lenses achieve the same magnification. A 300-mm lens used with a four-thirds sensor will have the same field as a 600-mm lens used with 35-mm film.

3. Most long-focus lenses in your present inventory should work if the lens mount is compatible or there is an adapter.

4. Lenses will improve in the years to come. Since it is desirable that the image contain at least the same detail as that of a 35-mm film frame using a lens with twice the focal length, manufacturers will be under pressure to produce top-notch optical components for this new system.

5. Camera bodies and lenses made by any of the manufacturers that adhere to the four-thirds standard will be fully interchangeable. The market for an individual lens will be much larger and should spur competition.

6. Optics produced under the $^4/_3$ standard will be smaller and lighter.

7. On the downside, smaller optics, while producing sharp images, will also have smaller overall apertures. An $f/2.8$ lens will still be an $f/2.8$ lens but because the focal length is shorter, yet covers the same field-of-view as the longer focal length optics used with 35-mm cameras, the diameter of the lens' front element will be smaller. Since the limiting magnitude of stars in an image is governed by aperture, a large aperture is required for wide-field astrophotography. The four-thirds standard will therefore reduce the size of wide-open apertures.

3.13 Effective Number of Pixels versus Actual Pixels

The number of pixels on the final image and the number of pixels on the image sensor are different. This is because there are several different methods of measuring the number of pixels on a sensor. The actual pixel count is higher than that on a finished image, but manufacturers prefer to advertise the total number of pixels. However, for the photographer, the number that really counts is the number of actual non-interpolated pixels in the raw image. The difference between the actual and effective number of pixels is not really marketing subterfuge because the extra sensor pixels that do not end up on the final image do play an important part in creating the image.

Table 3.5 Sony ICX252AQ CCD Array Pixel Usage

	Pixel Width and Depth	Pixel Count (Mp)
Total number of pixels in the sensor	2140x1560	3.34
Number of pixels read	2088x1550	3.24
Number of active exposed	2080x1542	3.21
Number of effective pixels in the image	2048x1536	3.15

Let's look at a typical 3.34 megapixel sensor such as Sony's ICX252AQ that is used on a large number of three-megapixel cameras. The total number of pixels is 2140x1560, or 3.34 million. The number of pixels that is read by the camera's operating system is lower—2088 x 1550. The outer four rows of pixels of this 2088 x 1550 subset is covered with a black dye to create what is called "Video Signal Shading" to give the camera's operating system a true black so it can take a dark current reading needed to calculate what is completely black. Within this shaded frame of four rows of pixels is the 2080x1542 pixel sensor area actually exposed by the camera's lens. Within this area, manufacturers usually select 2048x1536 pixels to create the final image using a 4:3 aspect ratio; however, some pixels just outside this area can be used in the Bayer mosaic color interpolation. Table 3.5 shows how the differing pixel counts reduce a 3.34 Mp sensor to an effective 3.15 Mp image area.

3.14 Interpolated Pixels

Some cameras do not present a true pixel-per-pixel reproduction of what the sensor sees. Assuming a three-megapixel sensor as described above, let's look at two ways of increasing its pixel count to seemingly increase the resolution. One way is for the camera to use software to interpolate six million pixels from the three million captured by the sensor. This is essentially an in-camera image processing enlargement of the original image before it is converted to the JPG format. By doing the upsizing interpolation before converting to JPG, the process eliminates JPG artifacts that are visible when an image is resampled at a large scale. The process is, however, more of a marketing ploy than actual improvement in resolution. The increase in image quality is minor, and the resulting file sizes are twice as large without any real improvement.

3.15 AD Converter

The analog-to-digital (AD) converter is the brain of a digital camera's imaging system. It converts the signal from the image sensor into the digital data that creates the final image. The CCD or CMOS sensor in a digital camera outputs a voltage, or analog signal, for each of the pixels in the sensor's array. This voltage is proportional to the amount of light the sensor receives. The voltage is amplified and processed by the camera's analog-to-digital converter that changes the voltage in to a binary number. This binary number is stored in a file containing an array of other binary numbers representing the other pixels on the image sensor. Most consumer digital cameras use an 8-bit AD converter. This means that the differing voltage produced by the pixel can have up to 256 distinct values of brightness ranging from completely black to completely white. More advanced "prosumer"[1] (between consumer and professional) cameras can have either 10-bit or 12-bit AD converters that divide variations in the pixel's voltage into tinier increments resulting in either 1024 or 4096 separate brightness steps between black and white.

The differing AD converters do not affect a pixel's dynamic range—that is determined by the number of electrons the pixel can hold. The 10- and 12-bit AD converters merely allow the pixel's voltage signal to be resolved into finer graduations of brightness. Prosumer cameras using 10- and 12-bit converters gain advantage over less-advanced cameras because they do not compress the dynamic range to fit eight bits of data. The higher bit count allows the camera to save images in the RAW format that will record the greater range of data inherent in the 10- and 12-bit data from each pixel.

3.16 Image Buffer

Early digital cameras transferred the image from the sensor directly to the storage card and thus had to wait for the image to be completely written to the card before another image could be taken. This process introduced a specific delay in the time before another image could be shot with the camera. To shorten the time between shots and allow the camera to be used on fast-paced action involving rapid-fire successive images, an image buffer was added. This is basically an amount of RAM in the camera that temporarily holds an image, freeing the image sensor for the next shot. The buffer transfers the image to the storage card more slowly than the image sensor can shoot another exposure; thus, it has a limited capacity for rapid-fire, or burst mode, images before the camera has to wait for the images to finish writing to the card. If a specific camera processes the image to reduce the image resolution and size, or to compress it to JPG format before the image is moved to the buffer, the image buffer can hold more images. However,

---

[1] So-called because their capabilities straddle those required by professionals and consumers.

Table 3.6 Transfer Speed Comparisons

Type of Device	Approximate Transfer Speed in Kb/sec
Camera serial cable	115
Camera USB	350
Camera firewire	500
USB card reader	500
SCSI card reader	1000
PCMCIA card adapter	1300
Firewire card reader	2200

if the processing is done as the images are transferred from the buffer to the card, the buffer will hold a reduced number of images because it will be holding the raw unprocessed and uncompressed image. A three to six image burst mode capacity is normal for consumer cameras while advanced DSLRs like the Nikon D1 have enough RAM to burst up to 21 images at 4.5 frames per second.

Burst mode is often used in lunar and planetary photography to allow the camera to take a series of exposures without being disturbed by triggering the shutter.

3.17 Image Transfer Speed

A digital camera is useless unless the images can be transferred from the camera to a computer or printer. This usually involves connecting the camera to a computer with a cable attached either to the camera or a docking station into which the camera is placed. Table 3.6 shows the various transfer speeds for typical interfaces. Early digital cameras through the late 1990s used an RS 232 serial connection cable and were notoriously slow. As digital camera image size grew, the need for faster transfer rates was satisfied with the adoption of the universal serial bus (USB) connection. Today, professional cameras with very large image files use the Firewire (IEEE 1394) connection for even faster image transfer, but not all computers are compatible with Firewire.

3.18 White Balance

Every light source has an inherent colorcast associated with them. The human eye has an amazing capacity to accommodate wide variations in lighting color. If it didn't, the world would seem very bluish at noontime and very reddish at sunset. In film photography we are used to considering the "color temperature" of the light source when choosing which film to use.

We had the choice of "daylight" or "tungsten" films that are balanced for bluish daylight sources or reddish artificial light sources. If the color of the illuminating light did not match the sensitivity of the film we often had to use color-correcting filters to prevent color shifts in the resulting images. This color shift was most prominent in objects that should appear as white.

Digital cameras have what is called the "white balance," a parameter that gives them a large advantage over film cameras. Since digital cameras cannot change sensors to accommodate a different color of illumination, we have to modify the sensor's perception of the off-color illumination in order to remove any colorcasts present. The white balance is an electronic compensation for variations that exist between different light sources. It adjusts the white balance by varying the gain for the red, green and blue channels so white objects are truly white and there is no colorcast in the image.

Most digital cameras perform the white balance selection automatically, but many newer cameras now allow the user to override the control and set the white balance manually if desired. Cameras do a reasonably good job of automatically setting the white balance; but in some cases, they can be fooled into producing off-color images. For most users, the true white balance is a personal judgment call.

For astronomical photography, we get the best results when using films balanced for normal daylight illumination. The same is true with digital cameras. For the most part, we can leave the white balance setting on "auto." If your camera allows specific selection of color temperature for the white balance, use the 5700K setting that approximates normal daylight illumination.

3.19 Shutter Lag

Digital cameras offer many advantages for creativity and ease of use, but most consumer cameras possess one trait that receives many complaints: the time lag between when the shutter button is pushed and when the shutter actually triggers. This delay is usually caused by the mechanics of the autofocus mechanism working to achieve optimum focus and can take up to a second. This may not seem like a long time, but for photographers used to the instant reaction of film cameras, this lag can be annoying. Care must be taken to not disturb the camera or telescope until the image is actually finished.

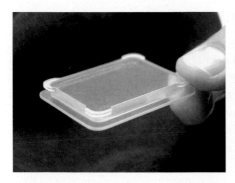

Fig. 3.13 *LCD preview screens on digital cameras can be easily scratched. A screen hood (left), attaches over the LCD with self-adhesive (right) and can be quickly replaced if it is damaged. Photos courtesy of Hoodman.*

3.20 LCD Preview Screen

The LCD screen on a digital camera is used to preview, review, and even act as a finder on fixed-lens cameras by displaying a live-video feed from the imaging chip. The LCD cannot act as a finder on DSLR cameras because the reflex-flip mirror is blocking the imaging sensor until the shutter is tripped and the mirror rises out of the optical path. Most LCD screens are attached to the back of the camera body, but on some models there are flip-out or flip-up screens as well. The screens are usually 1.5- to 2-inch rectangles. The better ones have a reflective layer behind them to make them visible in daylight.

The LCD screen on most digital cameras will provide confirmation that the image has been taken and is reasonably exposed, but it is far too small to critically examine focus. Some cameras allow zooming in on the LCD image by varying amounts depending on the model. Advanced cameras like the Canon EOS offer up to 12x zoom on the LCD screen, but this is no substitute for examining the image at full resolution on a computer screen.

The LCD screen on most cameras is relatively hardy; but if it is damaged, it will make any preview difficult. The most popular way to protect it is with a screen hood. Hoodman makes screen covers, screen hoods and screen magnifiers that fit all professional grade cameras (see Figure 3.13).

Some cameras also have a tiny LCD display in the viewfinder eyepiece that reproduces what is displayed on the external LCD preview screen. These tiny LCDs measure a little more than half an inch diagonally and have the advantage of showing exactly what the lens sees. They may have a brightness advantage with dim objects; however, the resolution of the tiny LCD makes focusing astronomical objects difficult.

3.21 Is It Time to Go Digital?

The time is ripe to jump in and buy a digital camera now. The technology is mature and relatively inexpensive, the accessory availability is excellent, and the user experience and support base is well established.

It is easy to understand why enthusiasts who are taking up photography for the first time will likely choose a digital model for their first camera. The corporate momentum of film manufacturers like Kodak and Fuji is clearly swinging toward digital imaging products instead of furthering research on new film emulsions. But what about the dedicated film photographer who has a considerable investment in film equipment? Digital cameras are following the replacement cycle similar to computers—more advanced and less costly models are being introduced every time you turn around. So is now the time to buy a digital camera?

Early digital models were hopelessly inferior to film cameras in image detail and the ability to take long exposures. But the gap between overall film and digital image quality has steadily narrowed. The arbitrary "crossover point" where 35-mm film and digital quality become equal has been unofficially defined as the point where digital imaging achieves a 50 lines-per-millimeter resolution on a 2400 x 3200 pixel imaging sensor with a $\frac{4}{3}$ aspect ratio. In fact, that crossover point was reached in 2003 with the prosumer Nikon 5400, Sony V1, and Canon G-5 cameras, all of which used the five-megapixel Sony ICX456AQF sensor. Indeed, I can't tell the difference between scans of medium-speed 35-mm film and the images produced by my Canon 10D. Should the newcomer buy digital now, or wait knowing that future updated, more capable camera models will soon appear, often selling for less than today's?

Cameras today are so capable and prices so low that the choice of pixel counts and lens zoom range is not really an issue. What should be considered is how the camera will be used in your non-astronomical imaging. Only the most dedicated astroimagers use a separate high-end camera exclusively for celestial imaging. The majority of users simply adapt existing cameras to astrophotography.

Digital cameras come in an amazing array of designs from full-sized SLR models to compact "shirt pocket" cameras that are no larger than a small cell phone. Depending on the application, all will perform well at some form of celestial photography. However, be aware that the extreme convenience of shirt pocket cameras can cause a problem. If it will slip into your pocket for snapshot use, it can just as easily slip out of your pocket. Digital cameras do not bounce very well and often fail to function after doing so.

Fig. 3.14 *Entry-level cameras such as the 3.9-megapixel Nikon Coolpix 4500 (left) and the 5-megapixel Olympus Camedia C-5060 (right) are both capable and affordable, retailing for about $350 and $500, respectively. Photos courtesy of Nikon and Olympus.*

Fig. 3.15 *Five-megapixel cameras such as the Nikon Coolpix 5400 are regarded as "crossover point" models in that their output, without extreme enlargement, is indistinguishable from film cameras. The Coolpix 5400 is capable of 10-minute time exposures. Photo courtesy of Nikon.*

Entry-level cameras can be amazingly cheap, yet quite capable. A three-megapixel model can take very satisfying images of solar system objects. But what features should you look for? Really inexpensive digicams have no removable storage media. These cameras should be avoided. They are cheap, but do not have the image capacity for serious picture taking. If you buy an inexpensive point-and-shoot camera, make sure the lens contains no plastic elements. Non-zoom cameras cost about $150; but as will be discussed in detail later, zoom helps to eliminate vignetting which is a common problem in afocal astronomical imaging. A zoom lens can add about $100 to the price of a camera, but they are worth it. Do not be fooled by claims about "digital zoom." This is nothing more than electronic cropping and does not add image detail.

When buying a first digital camera, think ahead to what you might want to be doing in a year or two once you have the basics down. A particular model may suit your current needs; but as your expertise grows, will you be able to grow into the camera? I have found that it is always best to

Table 3.7 Digital SLR Camera Price Trends

Make and Model	Year Introduced	Pixels (mb)	Cost in US $
Kodak DSC460	1995	6	12,000
Contax N Digital	2001	6.13	7,000
Canon EOS-1Ds	2002	11.4	7,999
Kodak DCS 14n	2002	13.89	4,995
Nikon D1X	2001	6	3,650
Nikon D100	2002	6.1	1,999
Sigma S-9	2002	3.4	1,500
Canon EOS 10D	2003	6.3	1,499
Nikon D70	2004	6.1	1,299
Canon EOS 300D	2003	6.3	939

purchase one with more features than you might initially need because as you learn and experiment more, you will need the added capacity. You don't want your camera to become obsolete for your needs.

The key is to match the camera's capabilities to the type of astrophotography you expect to be doing. If short-exposure solar, lunar and planetary imaging is your primary interest, then a fixed-lens camera will do. However, if you have been doing long-exposure deep-sky work with film cameras using either piggyback or prime focus techniques, then a DSLR model will be needed to continue this type of work in the digital realm.

The cost of jumping to digital will, of course, be the price tag of the new camera. But over several years, that price will slowly be paid back in both time and money. In my household we have nine 35-mm cameras. Within a year of purchasing our first digital model, a three-megapixel Olympus 3020, the film cameras were retired except for use in specialized meteor-patrol work requiring a large number of simultaneous exposures. The lone digital camera quickly paid for itself with zero film costs, and hours saved by not having to scan negatives for image processing. Indeed, my lunar photography alone "paid" for the camera within a year with the savings in film. Previously, a lunar photo session would cost a 35-exposure roll of film and processing, followed by time-consuming scanning of images that looked like good candidates for printing. With the digital camera, the best images can be selected quickly and reject images become recycled electrons with a push of the delete button.

The price of digital cameras, at both the consumer and professional level, has fallen sharply over the past few years. Table 3.7 lists the introductory price of many well-known high-end digital cameras. The early models cost as much as a new car; today's are within amateur affordability.

Chapter 4
Digital Astrophotography's Special Considerations

Digital photography presents a set of unique characteristics that must be dealt with in order to image the sky. In this chapter, we will explore these and how to work with and around them. This is nothing new. Film astrophotography has its share of special considerations that had to be mastered with understanding and practice, so in this regard, digital astrophotography is no different.

4.1 Care of Digital Cameras

The basic, no-frills film camera used in astrophotography, tends to be more rugged than digital cameras. While some compact fixed-lens digicams appear to be well protected when their lens is retracted and covered, the reality is that inside they are basically delicate electronic instruments. Digital cameras do not respond well to being abused. Heat and humidity will affect all electronic devices and digital cameras are no exception. If exposed to heat a camera will display increased amounts of noise in the image. This noise is the primary limitation when imaging deep-sky objects, and can even degrade daytime snapshots if the camera gets too hot, as when left on the dashboard of a locked car in the sunlight.

Digicams don't tolerate dampness well, so they must be protected from dewfall that is common late at night. Always use an anti-dew device on the lens, but try to avoid getting heat near the camera body where it will affect the imaging sensor. Cover the camera when it is not in use and leave it powered up at all times to maintain internal heat. This will, of course, increase the noise in each exposure, but taking twice as many shots, each with half the exposure, may beat both the dew and the additional image noise. If there is heavy dew, it may be time to change the evening's photography plan.

Leaving the camera on all evening will deplete its power supply, so several sets of spare rechargeable batteries are a must, or an AC adapter

Fig. 4.1 *If dew is a problem, camera electronics can be protected by enclosing the camera body in a plastic bag with the lens protruding through the bag opening and sealed with a rubber band. Photo by Robert Reeves.*

powered by a 12-volt DC to 115-volt AC inverter can be used to power the camera. It is also best to seal the camera inside a plastic Ziplock freezer bag to protect it from condensation when it is brought back indoors after a cold night. Once the camera is warmed back up to room temperature, it can be removed from the bag.

4.2 Dust and Dirt on the Imaging Sensor

Fixed-lens cameras have little problem with dust and dirt getting on the sensor and creating spots on the image, but DSLR cameras are another story. Dust is the bane of removable-lens digital cameras. Since astrophotography is done outdoors where there is often a breeze, dust will inevitably settle onto a camera's sensor. With 35-mm film cameras, the closed shutter affords some protection from dust and any that gets onto the film surface during an exposure will be swept away when the film is advanced. With DSLR cameras, we do not have these two natural dust protection mechanisms. Any dust that accumulates on the sensor will stay there, and may create a spot on all future images.

Technically, dust particles do not actually lie on the sensor itself, but on the infrared filter that lies just in front of the imaging sensor. The fact that there is a slight separation between the plane where dust actually lies and the imaging sensor itself is actually fortunate for astrophotographers. Astronomical imaging with camera lenses is usually done at wide apertures. This has the effect of hiding small dust particles because their shadows will be fuzzy. Dust is mainly visible on images shot at high f/ratios

Fig. 4.2 *Normally, the shutter is closed on a DSLR camera. To clean the sensor, the camera must be placed in "sensor cleaning mode" in order to hold the shutter open without actually taking a picture. Photo by Robert Reeves.*

where the image sensor is illuminated by a small aperture which forms a tight, constricted, incoming beam of light that will create sharp shadows of anything in its path. At $f/22$ the tiniest dust particles will create a visible spot on the image. At wide "astronomical" apertures, the incoming light cone from the lens is wider and creates more diffuse shadows. Only the largest "chunks" would normally show on the image. But astrophotography adds another factor—celestial images are usually extensively processed, and as a result, the dust spots that would normally be inconspicuous may be enhanced to the point where they show up on the image.

The first line of defense is to always keep the sensor clean by avoiding as much dust as possible. Minimize the time the camera is open with no lens or body cap. Since electrical charges may attract dust, some people power down their cameras when changing lenses. This option, of course, will be difficult for those who use their cameras at prime focus on open-tube Newtonian telescopes where the camera is open for extended periods. A fix would be to create a seal over the open-camera lens mount using a light-pollution filter.

DSLR models that use a CMOS imaging sensor are less prone to dust contamination than those that use a CCD. The reduced power consumption

Fig. 4.3 *The Olympus E1 is one of the first cameras to actively combat dust on the imaging sensor. An ultrasonic "shaker" dislodges dust on the surface of the sensor every time the shutter is triggered. Photo courtesy of Olympus.*

of the CMOS sensor may be the contributing factor to the reduced electrostatic attraction to dust, but whatever the reason, the consensus of experienced digital camera owners is that the CMOS sensor suffers less from dust.

Most DSLR cameras have no built-in dust control, but two manufacturers are taking action against dust on the sensor. Sigma cameras approach the dust issue directly on the SD-9 and SD-10 models by placing a glass sensor protector behind the lens mount. This effectively seals the camera interior and can be cleaned or even removed if the sensor itself needs cleaning as well. Olympus is the only manufacturer that has an active dust-control system in their camera. The E-1 uses an ultrasonic vibrator to shake dust off the sensor before each exposure.

Dust on the sensor does not create any physical damage, like scratching a negative does with film images; but it does call for spot removal from each image with the clone or healing brush tools of an image processing program. Eventually this will become drudgery, and more direct action will be needed to eliminate the offending dust blob.

Do not remove the lens on a DSLR and peer inside to look for dust. This act is guaranteed to simply put more dust inside the camera. Most is too small to be seen by the unaided eye anyway. The best way to find the dust is to record it on an image taken with the camera. Using any light, monochrome, background, such as an evenly illuminated white wall, take

Fig. 4.4 *Our closest large neighbor galaxy, M31, is always a spectacular target in a wide-field telescope. This image is a combination of 13 five-minute exposures with a Canon 10D at ISO 400 taken through a Takahashi Epsilon 160 at f/3.3. Photo by Rick Krejci.*

an out of focus but properly exposed image. Load the image into a processing program and look at it under 100 percent magnification. Use the sharpening tool to accentuate the edge of any spot on the blank image. If there are an objectionable number of dust spots, it is time to remove them.

The first thing people are tempted to try is a blast from a canned air blower, but this may be too powerful for some of the camera's internal mechanisms. Liquid-powered blowers that are tipped too much may spray cold fluid and cause damage to the sensor's cover. Besides, this will usually not remove enough dust to satisfy a critical camera user. Some manufacturers are specific in that the sensor must be cleaned by a factory authorized repair facility, while others give advice on how to safely clean the sensor. Canon, for instance, recommends using a squeeze bulb blower with the tip held no closer than the lens mount. The problem with this is that if there is more dust around the sensor than on it, then the puffs of air may result in even more dust on the sensor than there was in the beginning. A solution that works in some cases is to place a vacuum cleaner nozzle near the open-lens mount while using the blower. The theory here is that dust dislodged by the blower will be sucked away by the vacuum instead of resettling on or near the sensor.

Canon cautions against using a blower brush. Thus, use of any sort of

Fig. 4.5 *M33, the Pinwheel Galaxy, is another fine galactic target for short focal length telescopes. This image is a combination of 16 four-minute exposures with a Canon 10D at ISO 800 through a Takahashi Epsilon 160 at f/3.3. Photo by Rick Krejci.*

brush that touches the sensor surface, including the small vacuum brush such as those sold as computer keyboard vacuum cleaners, would go against Canon's cleaning policy. The experience of those who have tried it anyway was that there was either more dust after cleaning or at best, the dust was concentrated into a smaller area.

Beyond these relatively gentle cleaning methods, the camera owner is advised by the manufacturer to seek professional cleaning at a repair facility. This leaves users on their own to address heavy dust problems when a trip to a camera repair shop is not an option. The alternative cleaning methods discussed here are presented with a full disclaimer that a mistake can scratch the sensor and lead to a repair bill nearly as high as the cost of the camera itself. Any cleaning steps that involve actually touching the sensor are performed entirely at the user's own risk.

"Large" lumps of debris—specks big enough to be seen with the eye— can be picked off the sensor with a device called the SpeckGRAB-BER that is marketed by Kinetronics. This is a thin pencil-like device with a small soft washable pad on the tip that provides a high-adhesion surface that sticks to contaminant particles more strongly than the particles stick to the surface they are contaminating. The offending particles can then be lifted off and the SpeckGRABBER washed with soap and water for reuse.

Cleaning kits that purportedly will restore sensors to a dust-free condition are available from various sources. The "Sensor Swab" and "Eclipse Optic Cleaner" are two products recommended by Kodak and Fuji. They even include an instructional video that plays on your computer. Each cleaning swab is used only once and then discarded.

Be aware that not all sensor-cleaning kits are recommended by camera manufacturers. For instance, Canon and Nikon do not approve of the swab sold by Photo Solutions for use with their cameras; and even approved kits must be used carefully so as to avoid sensor damage.

Some inventive photographers devise their own sensor cleaning swipes to ease the cost of repeatedly cleaning sensors in dusty conditions. Among the most popular of these is the SensorSwipe. Additional details and user techniques for homemade sensor cleaners can be found in articles written by Peter Marshall and Bob Atkins (see Appendix B).

4.3 The Recording Medium Never Changes

The light-recording medium used by film cameras comes in wide varieties: black and white or color, with ISO speeds ranging from 25 to 3200. With digital cameras, however, we are confined forever to the sensor that the camera was built around. We cannot change it for finer resolution sensor like a film camera can change to a finer-grain emulsion. With this limitation in mind, a question often asked is if a five-megapixel camera can be purchased for $400, why spend three or four times that much to purchase a professional-level camera with the same pixel count? After all, we would have the same five megapixels of resolution with either camera. The answer is increased flexibility, lens selection, more rugged construction, and better image quality. In digital imaging, size does count. The larger sensors in professional cameras have the same pixel count as the smaller ones in cheaper models, but the larger pixel size in the pro camera allows each pixel to receive more photons per exposure, and thus can achieve greater dynamic range.

4.4 Overcoming Heat in Digital Cameras

As good as today's digital cameras are becoming, they all suffer from the same major limitation in long-exposure applications—electronic noise in the sensor, which produces a snowy pattern of multicolored spots on the image if the exposure exceeds a certain limit. The closer pixel spacing of a compact consumer camera's small sensor both introduces electrical interference between the pixels, and reduces the total number of photons that reach each pixel. The resulting electrical signals from the pixels thus have to be amplified more than with larger professional camera sensors. These factors introduce a higher level of noise into images taken with consumer models than is present with the larger sensors in professional-level DSLR cameras. The noise difference between same-length exposures in a consumer and a professional camera, both having similar pixel counts, is very noticeable. This noise is sometimes called "dark current" because it will be present and show as specks on the image even if the sensor was exposed in total darkness.

The electronic noise in a camera's imaging sensor forces astrophotographers to be creative in devising ways around its limitations. In film photography, various chemical methods of hypersensitizing the film helped control reciprocity law failure. Also, cooling the film to very low temperatures helped improve its low-light sensitivity. Similarly, cooling the imaging sensor in a digital camera will mitigate the noise that builds during a long-exposure image. Some specialized astronomical CCD cameras use the same CCD sensors as found in digital cameras, but they are actively cooled to temperatures below zero Centigrade to control electronic noise. Cooling an astronomical CCD chip is accomplished with equipment that is impractical to add to a consumer camera that is originally designed to take snapshots. So, astrophotographers are forced to devise procedures or create homemade devices to cool digital cameras for their specialized use. Additional post-exposure processing can also eliminate much of the image degradation caused by noise. These concepts are explored in detail in Chapter 14.

Some models of the Nikon Coolpix cameras have an imaging head that is isolated from the main body of the camera. The heat-producing batteries and LCD preview screen are mounted in a module that swivels independently of the imaging head. This has the advantage of keeping the heat generators away from the imaging chip and thus reducing noise. It also allows the viewfinder to rotate to a convenient viewing position no matter how the telescope is pointed.

As the market for digital imaging has grown, image sensors have been introduced that have lower noise levels. For now, these "cleaner" imaging sensors are to be found in the more costly cameras, but this technology will eventually trickle down to consumer models. The CMOS sensors used in the latest generation of Canon and Nikon cameras produce noticeably less noise than their predecessors.

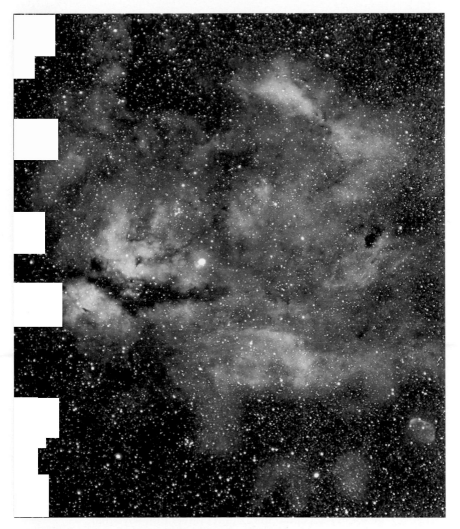

Fig. 4.6 *The nebula complex around Gamma Cygni, including the Butterfly Nebula, imaged by combining 8 ten-minute exposures with a Canon 10D at ISO 800 through a 180-mm lens and a UHC filter at f/2.8. Photo by Johannes Schedler.*

It is generally agreed that three-megapixel image sensors are capable of producing "film-like" results. Most entry-level cameras achieve this and indeed, these models can outperform film cameras for lunar and planetary imaging. Moderately priced "point-and-shoot" models routinely now have five megapixel sensors. Prosumer DSLR cameras, the advanced models based on professional grade 35-mm cameras, now operate in the six- to eight-megapixel range and produce images indistinguishable from those of film cameras.

A higher pixel count satisfies the consumer's desire for greater resolution to eliminate the grainy, pixilated look of early digital images, but packing more pixels in a given area of sensor creates a problem for the astrophotographer. The smaller pixels are more prone to electronic noise.

4.5 Hot, Stuck and Dead Pixels

There is a lot of discussion on Internet discussion sites about "hot pixels" in digital cameras. Hot pixels are single pixels that have higher than normal rates of charge leakage compared to the average of all pixels on the sensor. These appear as individual pixel-sized points of light on a long exposure. Some users regard a hot pixel as a camera defect, but the basic point is that all digital cameras have them. There is not much you can do about them because the problem is essentially built into the sensor. Digital cameras became popular very quickly; and as a result, a misconception evolved among new users that long exposures could damage a camera by heating the sensor and thereby resulting in hot pixels. This is not true. The fact is every pixel on a sensor has some charge leakage and if exposed long enough, any pixel will eventually brighten even if it is not struck by any photons. If a particular camera does not have one or more hot pixels now, it will eventually with repeated use. Faint, hot pixels contribute to the non-random noise present in long-exposure images, and charge leakage is greater at higher temperatures. Hot pixels are only an issue when there are so many that they become objectionable and image processing is needed to reduce or remove them. It should be noted that the topic of hot pixels refers to those on the camera's imaging sensor and not those that may be part of the camera's LCD display.

There is no formal terminology for the hot-pixel phenomenon and sometimes the term "stuck pixel" is used interchangeably with "hot pixel." However, generally accepted unofficial convention says that:

1. A hot pixel is one that reads high on longer exposures.
2. A stuck pixel is one that always reads maximum on all exposures.
3. A dead pixel is one that always reads zero and registers as black.

Hot pixels can appear in different colors because each individual photodiode is covered by one of the red, green, or blue colors of the image sensor's color-filter array (see Section 3.6.2). In early Nikon Coolpix cameras, hot pixels are cyan, magenta, yellow, or green. A hot pixel can be any color if two of them occur next to each other, each with its own color. Green hot pixels are the most common because there are twice as many green sites in the color-filter array. Some cameras have internal noise-reduction soft-

Fig. 4.7 *A combination of bright stars and blue reflection nebula make M45, the Pleiades, a favorite astrophotography target. This image is 22 four-minute exposures taken with a Canon 10D at ISO 800 through a Takahashi Epsilon 160 at f/3.3. Photo by Rick Krejci.*

ware that attempts to suppress hot pixels. Sony and Minolta cameras turn the hot pixels white.

There are freeware packages available on the Internet that will test various cameras for hot pixels. It is best to ignore these and simply rely on your eyes to tell you if there are objectionable hot pixels in your camera. These software packages will show you hot pixels you didn't even know existed but provide no guidelines on how to interpret the results of their scans. In astronomical photography exposures are routinely so long that hot pixels will always be present, so dealing with them is an integral part of post-exposure image processing.

While it is useless to check for hot pixels because they will always be present in long exposures, you can check your camera for stuck pixels. To do so, take short exposures of solid white or black backgrounds. It is a popular misconception that you test for stuck pixels by exposing for several seconds with the lens cap in place. These will show red and green hot pixels scattered across every image in what is known as "Christmas tree artifacting." Some consumer digital cameras such as the Nikon Coolpix and Sony models begin to use automatic dark frame subtraction after several seconds, even as short as 1.3 seconds with the Canon G1. The Minolta D7 series uses the feature automatically on all "bulb" exposures. This in-cam-

Fig. 4.8 *The Cocoon Nebula in Cygnus is an often-overlooked deep-sky gem. This image is 16 stacked five-minute exposures with a Canon 10D at ISO 400 through a Takahashi Epsilon 160 at f/3.3. Photo by Rick Krejci.*

era processing will confuse the interpretation of such test attempts.

Stuck pixels or dead pixels will repeat at the same place on all images, regardless of long or short exposures. To test for these, simply shoot a white or black background in normal lighting. If the pixel is dead, it will show as a black spot against the white background. If the pixel is stuck, it will show as a white spot against the black background. View the images at 100 percent size in an image processing program so you can see all the individual pixels. The only real criterion for passing or failing the test is whether the resulting stuck pixels (if any) are personally objectionable enough to ruin images. Bear in mind that there is no such thing as a "perfect" imaging sensor. If you exchange the camera or have its sensor replaced, there is a good chance of getting another sensor with the same problem.

A good retailer will certainly let you take some test exposures with a camera before you buy it, but examining the images for stuck pixels on the camera's own LCD viewfinder is a waste of time. It simply does not have the resolution needed for this, even at full zoom. A buyer could take a laptop to the store and download the test image with a USB card reader. However, at best, this exercise will only allow you to select a camera that has the least amount of hot and stuck pixels at that time.

Manufacturers have set up proprietary criteria for use when they must decide whether or not to repair or replace a camera with hot or stuck pixels. If it is still in warranty, this will be done at no cost to the customer and you may or may not get back your original camera. If it is out of warranty, a pixel-mapping routine is usually run through the camera, the camera's internal software is updated with the new pixel mapping data, and a modest service fee is charged. Before you decide to send a camera in for service due to hot pixels, remember their leading causes, and the fact that astrophotography naturally aggravates these causes:

1. Long exposures (we are doing astrophotos, so of course the exposure is long)

2. High ISO (celestial targets are dim, we need high ISO settings)

3. Dark background (this is astrophotography, it *is* nighttime)

4. High temperature (the camera is on for long periods of time for long exposures)

Under these specialized circumstances, every camera will show some hot pixels; it is unavoidable. The "cure" is in post-exposure image processing discussed in Chapter 14.

As camera electronics and internal software become more advanced, many manufacturers have included software features that automatically map the sensor's pixels to locate stuck or dead pixels when the cameras batteries are installed. This was introduced in 2001 with the Olympus E-10 and E-20 cameras and was quickly adopted by the Sony 707 and 717 cameras. Other makers are now incorporating pixel-mapping routines into their cameras, which provide fresh information about the location of stuck or dead pixels every several hours of camera operation. This software interpolates the image data surrounding the anomalous pixels and either smooths out a bright pixel or fills in the dark spot from a dead pixel.

4.6 Dealing with Afocal Vignetting

The most common problem with the afocal coupling technique is image vignetting. Vignetting is the effect where the sensor is not fully illuminated by the light cone exiting from the lens or eyepiece. The result is an image that has its corners lopped off, or worse, an image that looks like it was taken through a circular mask.

Some vignetting should be expected when using fixed-lens digital cameras afocally on a telescope. Fortunately, an entire industry has evolved dedicated to perfecting products and techniques that minimize vignetting in this situation. By choosing the proper eyepiece, camera mount and to some extent the proper camera itself, vignetting can be controlled to where it is not objectionable. Planets and many bright deep-sky objects are small enough that they can be centered in a field that has some vignetting, but solar and lunar photography usually require the full image field. Less vignetting also makes it easier to frame objects with a telescope.

With afocal photography, we are basically substituting the camera for our eye at the eyepiece, so the same factors that govern whether an observer wearing eyeglasses experiences vignetting come into play when using a camera at the eyepiece. There are ways to minimize vignetting. The first is to use a camera with a physically small lens. This sounds counter to the standard advice that states we should use large wide aperture lenses for celestial photography. But for afocal photography smaller lenses are better. For instance, the tiny lens on a Nikon 995 can shoot vignette-free images through an eyepiece with a small exit pupil that a camera with a large lens like the Sony 717 cannot. Large lenses can be used, but the imager must be careful about the camera attachment and eyepiece selection.

For the best afocal photography results, use the following parameters for eyepiece selection:

1. Use an eyepiece with a flush-mounted eye lens. The top glass element should be at the very top of the eyepiece. Designs with a recessed lens may be good for blocking stray light during visual observation, but they should be avoided for afocal photography.

2. Ideally, the eyepiece should have a large eye lens. If the choice is between a low power eyepiece with a large lens or a higher-power one with a small lens, use the one with the larger lens and a Barlow or the camera's zoom to increase the magnification.

3. If possible, the eyepiece should be compatible with adapters such as the Scopetronix Digi-T system that keeps the camera square with it, and centered on the field-of-view. Cameras with very large lenses can use specially designed eyepieces such as the Scopetronix MaxiView 40 that are designed to minimize vignetting with large camera lenses. See Section 8.5.

4. Keep the camera to eyepiece spacing as close as possible. Near glass-to-glass contact is ideal.

5. Zoom in on the eyepiece field-of-view with the camera's own zoom feature. For models with a small lens, this will likely eliminate vignetting. For models with large lenses there will likely still be some vignetting even at maximum zoom.

6. If your camera is so equipped, try the macro mode. It may perform better this way, but there is no guarantee that it will do so because there are so

Fig. 4.9 *The 24-mm Brandon (left) is an excellent eyepiece for visual observing. However, its recessed eyelens and relatively narrow field-of-view will cause vignetting in a digital image, while the 26.6-mm eyepiece (right) with a flush-mounted eyelens will not. Photo by Robert Reeves.*

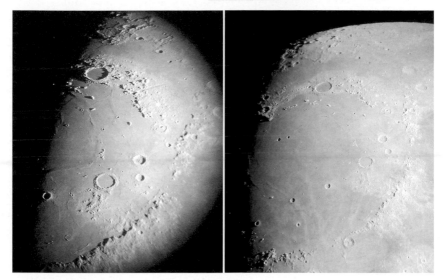

Fig. 4.10 *These two images of Mare Imbrium demonstrate the value of using an eyepiece design that is compatible with afocal photography. The image on the left was exposed through the recessed-eyelens ocular shown in Figure 4.9 and experienced significant vignetting. The full-field image on the right was exposed through the eyepiece with the flush-mounted eyelens. Photos by Robert Reeves.*

many different features on different camera models. The only sure advice is to experiment with all your camera's settings to see which ones work best with your telescope.

4.7 Predicting Vignetting

Understanding the principles of the vignetting phenomenon allows us to predict how camera and telescope combinations will react. The causes of vignetting are simple and if just two requirements are satisfied, vignetting will be eliminated. Four parameters govern these requirements. They are the camera's field-of-view in degrees, the eyepiece apparent field-of-view in degrees, the eyepiece eye-relief distance and the camera lens entrance pupil location. These parameters interact to give us the following two requirements:

1. The apparent field-of-view of the eyepiece must be the same or larger than the camera's field-of-view. Thus a camera with a zoom lens can zoom in and narrow the camera's field-of-view and eliminate vignetting with all but the most narrow field-of-view eyepieces.

2. The eye relief of the ocular must extend far enough from it to meet the camera lens entrance pupil. If the eye relief is too short for the camera lens to get close enough, vignetting may occur.

When applying the above parameters, the unknown factor is usually the location of the camera lens entrance pupil. Generally, this is some distance within the camera lens and getting the eyepiece close enough to the entrance pupil can be a problem if there is little eye relief. Think of the camera in terms of someone who views through a telescope with eyeglasses on. The collision between the eyepiece and the glasses prevents the wearer from getting close enough to see the entire eyepiece field-of-view, thus he or she sees a vignetted field just like a camera does if short eye relief prevents the camera from getting close enough.

Data on the other three factors are readily available or can be calculated. Most good makers publish the apparent field-of-view and the eye relief distance for their eyepieces in catalogs or web sites, so getting these data is not difficult. What the data may not show, however, is whether the eyepiece lens is recessed within the barrel, thus reducing the usable eye relief. An ocular with an advertised 30 mm of eye relief, but having its eyelens recessed 10 mm inside the barrel will only leave 20 mm of usable eye relief. A camera lens field-of-view in degrees is also readily available from its manufacturer's advertising media, and is simple enough to calculate if that source is not available. (See Section 10.18 for the formula to calculate the camera field-of-view.)

Fortunately, the fixed lenses on compact digicams are relatively small, and thus their entrance pupil is not very deep within the lens. This means that an eyepiece with sufficient eye relief to be used by eyeglass wearers (many manufacturers advertise this fact for certain eyepieces) will most likely work. The main governing factor will thus be the camera's field-of-view. If it exceeds the lenses' apparent field-of-view, then the porthole effect of vignetting will be present. However, most digicams have a zoom lens so zooming can narrow the camera's field-of-view until vignetting is eliminated.

As a zoom lens varies its focal length, the position of its entrance pupil will shift. Moving the camera lens entrance pupil can cause the eyepiece's apparent field-of-view to become narrower but at the same time the camera's field-of-view also narrows with greater zooming. As long as the camera's field-of-view narrows faster than does the eyepiece's, there will be no vignetting. Ideally, we would change the camera position to maintain optimal position of its lens' entrance pupil relative to the eyepiece eye relief. But in practice, the camera is mounted in a fixed position relative to the eyepiece, and we simply accept the results as long as vignetting is not objectionable.

There are exceptions to every rule; and if the eyepiece being used happens to have much more eye relief than the camera needs, then vignetting can still occur. Being too close with extremely long eye relief can be just as bad as being too far with very short eye relief.

Cameras with very large front lens elements do not necessarily need afocal-projection eyepieces with larger eye lenses. The vignetting occurring on cameras with very large front lens openings most likely arises because the large front element is part of a complex lens with a high zoom factor, and the entrance pupil is mismatched with the eyepiece eye relief. Very good images have been taken with a camera lens three times the diameter of the eyepiece eye lens. As long as the eye relief and entrance pupil factors are in agreement, the image will have little vignetting.

4.8 Fixed-Lens Camera Limitations

The majority of cameras on the market are the fixed-lens variety. Most are equipped with moderate-range zoom lenses that make them capable of a wide range of snapshooting needs. Although these cameras are quite capable of doing a variety of astrophotography tasks they are inherently less versatile than SLR cameras with removable lenses. This is not a condemnation of fixed-lens cameras for astrophotography, but it is an acknowledgement of the inherent limitations that astro imagers must learn to circumvent. With the proper techniques, many fixed-lens cameras have produced stunning celestial images.

Zoom lenses have smaller apertures and are by nature slower than fixed-focal length varieties. They produce dimmer images when doing wide-field work. The lack of removability prevents their use at prime focus on a telescope and forces shooting through an eyepiece using afocal-projection, even for low-power views. Also, most of the less-capable fixed-lens cameras do not offer ISO speeds higher than 400. Fixed-lens cameras often have limited time-exposure capability, usually a maximum of 30 seconds to one minute. Some of the latest 8-megapixel pseudo-SLR cameras from Olympus and Nikon allow 8- and 10-minute exposures respectively, but they are the exception. The small imaging sensors of the fixed-lens cameras are more prone to noise and will limit the long-exposure performance of these cameras even if they do allow longer exposures.

4.9 Digital Camera Viewfinder Limitations

Experienced astrophotographers who have spent a lot of time using SLR film cameras know how important it is to be able to see the sky through a camera's viewfinder. With film SLR models a lot of effort has been placed into designing viewfinders that are as bright as possible to aid in manually focusing the image. Most film SLR cameras present a bright, life-sized field-of-view with a "normal" lens and astrophotographers go a step further by using special brighter-than-normal focusing screens and magnifying viewfinders to help see dim stars through the camera. Often these accessories, such as the clear C focusing screen and DW-2 magnifier used with the Nikon F2 camera, make the view through a camera comparable to viewing through a telescope eyepiece.

With the experience of using such film-camera viewfinders fresh in one's memory, it can be a shock when switching to a digital camera for astrophotography. Small point-and-shoot digital models have tiny optical viewfinders that are hard enough to use in daylight, much less under the stars. The digital photographer's shooting pose, camera outstretched at half-arm's length while peering at the LCD preview screen through the lower part of bifocal glasses, didn't evolve because it looks cool. Viewfinders on small digital cameras are hard to see through. Fixed lens quasi-SLR digital cameras merely place a small LCD screen in the camera's eye-level viewfinder and do not actually use a reflex mirror and pentaprism to show what the lens is seeing. These cameras are often even harder to use for wide-field imaging because the stars virtually disappear on the small LCD screen. Even high-end true digital SLR cameras like the Canon 10D present a difficult image in their viewfinder. They were designed to use motorized autofocus lenses and theoretically do not need bright viewfinders just for the sake of focusing. An auxiliary magnifying viewfinder, such as the Canon Angle Finder C, will improve the view; but only cameras with expensive 35-mm sized full-frame sensors can produce eyepiece views like the old film SLR cameras.

The dimmer and smaller viewfinder on digital cameras has led astrophotographers to resort to auxiliary video monitors, large external LCD remote displays, or laptop computers to aid in focusing and composing their shots. These focusing aids are discussed in detail in Section 11.2.

Fig. 4.11 *To focus and frame dim targets through the smaller viewfinder in DSLR cameras with less than full-frame image sensors, a magnifying finder, such as the Canon Angle Finder C used on the EOS 10D, is a helpful accessory (left). The finder has variable magnification and can be swiveled for convenient viewing (middle and right). Photos by Robert Reeves.*

Interestingly, the Canon Angle Finder C shown in Figure 4.11 will also fit the Nikon D70 as well as Nikon's own DR-6 finder. To fit the Nikon, the Angle Finder C requires the ed-c adapter (which it comes with). The Canon finder is physically larger and heavier than Nikon's but it has higher magnification. A minor inconvenience is that at 2.5x, not all of the Nikon exposure data is visible. More, but not all, of the data is visible through the Nikon finder. Both finders vignette about 20 percent of the field-of-view at highest magnification.

4.10 DSLR "Crop Factor" and Lens Focal Length

Another DSLR problem is the loss of wide-angle capability when using wide-angle lenses that were designed for film cameras. As we discussed in the previous chapter, the majority of digital cameras have imaging sensors that are smaller than a standard 35-mm film frame. In most cases this benefits astrophotographers because the smaller sensor views a smaller portion of the image and produces a larger image scale. So, for example, when a 200-mm film lens is installed on a DSLR with a "crop factor" of 1.3, the effective focal length is increased to 260 mm. You are in effect getting extra focal length and higher magnification for free. But there is a downside to this. While telephoto lenses act like longer-focus lenses, so do wide-angle film lenses, and a longer focal length wide-angle lens is no longer a wide-angle lens. It is a "normal" lens. Astrophotographers who enjoy sweeping wide-angle views of the sky must now use even shorter focal length lenses on a DSLR to retain wide-angle views, and because of mechanical limitations, they almost always have to be specially designed for digital cameras. Table 4.1 shows how common focal length lenses used with 35-mm film cameras respond with the different DSLR crop factors.

Fig. 4.12 *The crop factor present in DSLR camera sensors effectively increases the focal length of existing lenses. In cameras with a 1.6 image-sensor crop factor, a 50-mm lens performs like an 85-mm (left), an 85-mm performs like a 135-mm (center), and a 105-mm performs like a 170-mm (right). A major advantage is that as the shorter focal length lenses act like longer focal-length optics, they retain the wider aperture of the shorter focal length optics. Photos by Robert Reeves.*

Table 4.1 Equivalent "Digital" Focal Length of Existing SLR Lenses (mm)

Film SLR	Crop Factor			
	1.3x	1.5x	1.6x	2.0x
28	36	42	45	56
35	46	52	56	70
50	65	75	80	100
85	110	128	136	170
135	175	203	216	270
200	260	300	320	400

4.11 Electrical Power Requirements

A digital camera with a dead battery is nothing more than a nonfunctional conversation piece. In controlled studio situations, digitals can be powered with AC adapters, but in the field where astrophotography is done, this is sometimes not practical. If your observing site does have electrical power, an AC adapter will solve the problem of reliably powering a camera for lengthy exposures.

If electrical power is not available, extra batteries are called for. Cameras that use ordinary AA batteries can be powered with Nickel Metal Hydride (NiMH) rechargeable batteries. These batteries typically last twice as long as similar-sized alkaline models and are relatively inexpensive. If your camera requires a proprietary-size lithium ion battery, then spare factory batteries are a necessity.

See Section 12.2 for details on how I built a battery-powered 115-volt inverter power supply to run cameras with an AC adapter while at remote observing sites.

4.12 Limited Red Spectral Response

Some of the most colorful celestial objects are those that shine in the crimson light of the 656-nanometer hydrogen-alpha emission line. The CCD and CMOS sensors used in digital cameras are naturally very sensitive to this portion of the spectrum. Unfortunately, when these sensors are installed in a digital camera, two factors significantly reduce their actual sensitivity to the far-red portion of the spectrum where these celestial objects are the brightest. First, these sensors have twice as many green-sensitive pixels as red and blue. This automatically reduces their red sensitivity, but makes them more closely matched to the spectral response of the human eye—an important factor for normal daylight snapshot photography. Second, camera manufacturers insert an infrared blocking filter in front of the sensor. This filter is designed to eliminate infrared light that will ruin the color balance of the image. Also, the infrared portion of the spectrum does not focus as sharply as the visible portion so, if it is present, the autofocus mechanism will respond poorly during normal terrestrial photography. The net effect is that blue and green wavelengths reach the sensor, but only a small portion of the red wavelengths gets through to it. For instance, measurements show that the Canon 10D and 300D cameras record only 15 percent of the incoming hydrogen-alpha wavelengths.

Because of the infrared blocking filter, sensors in digital cameras have a spectral response similar to the current crop of re-engineered color negative films that feature reduced-red sensitivity so as to not render skin-tones as too pinkish. These newer film chemistries do not record the hydrogen-alpha wavelengths as well as previous generations of color films did. Digital cameras still have excellent blue, green and yellow response; so they can image constellations, star clusters and galaxies quite well. But the views they provide of red emission objects are actually closer to the way the human eye perceives these nebulae. However, one of the reasons we enjoy astrophotography is to see the unseen, and the astrophoto community has become accustomed to the enhanced red capability of some color films. In order to achieve similar results with digital imaging it is necessary to use image processing to enhance hydrogen-alpha emission objects.

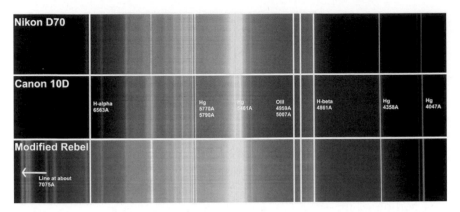

Fig. 4.13 *The spectral response of the sensors in the Nikon D70, Canon 10D, and a modified Canon 300D that have the IR blocking filter removed are compared in this spectrogram. The modified camera's extended red sensitivity is evident. Photo courtesy Dennis di Cicco.*

If deep-sky imaging featuring long exposures of emission nebulae is one of your goals, there are a number of digital options to achieve a better red balance. The first is to simply expose longer to accumulate more signals in the image's red channel. This will require image processing to separate the additional red image information while reducing the even greater information from the blue and green channels. Alternatively, a portion of the cumulative stack of images that make up the total exposure can be taken through a hydrogen-alpha filter to build up the red portion of the image. hydrogen-alpha filters shift the focus of a camera lens; so once the filter is removed to expose colors in the rest of the spectrum, the lens must be refocused.

Astrophotographers have never let technical challenges get in the way of recording a desired celestial image, and the digital infrared filter is no exception. When the sub-$1000 Canon 300D appeared on the market, it didn't take long for creative sky shooters to begin experimenting with the removal of the astronomically unfriendly infrared filter. The same thing has happened with the Nikon D70. Removing the IR filter is not trivial. It must be understood that performing this modification voids the manufacturer's warranty, and there is a chance that the camera may not function at all after the modification.

For those who do not have the expertise to dismantle and modify the camera themselves, it is possible to purchase one that has already been specially modified. Hutech Corporation sells a modified Canon Digital Rebel camera body that has the red filter removed. Two versions are available—one with no red filter at all, but the camera can not be used for terrestrial photography, or one with an astronomically-friendly red filter installed

Table 4.2 Sensitivity Increase for Modified Canon 300D

	Red	Blue	Green
No IR cutoff filter	2.2x	1.2x	1.04x
Baader filter installed	2.0x	1.2x	1.0x

that passes hydrogen-alpha wavelengths, but blocks the infrared wavelengths that prevent daylight snapshot imaging.

Terry Lovejoy has performed the infrared modification to his 300D and documented the results. His measurements of how much the stock Canon infrared filter attenuates light in all three color channels can be seen in Figure 4.16. He found that with the infrared filter removed, the Canon 300D displays a significant improvement in the red channel as well as some increased sensitivity in the other channels. Installing a Baader High Transmission UVIR blocking filter to eliminate the unneeded infrared wavelengths beyond the hydrogen-alpha wavelengths retained the significant improvement in the red sensitivity and allowed more light to reach the sensor, boosting the sensitivity in the other color channels as well.

The good news is that for astrophotography, the improvement in red sensitivity is better than the numbers in Table 4.2 indicate. The red sensitivity trails off toward shorter wavelengths beyond 656 nanometers; thus, at the hydrogen-alpha wavelength, there is a massive 4x increase in sensitivity.

The bad news is that modified cameras using the astronomically-friendly Baader filter, while capable of producing wonderful celestial images, will produce poor results on terrestrial subjects. Even with a custom white-balance setting, snapshot images will be too red, and the autofocus function may not set the lens properly. Those with mechanical expertise can reset the autofocus through internal adjustments, and using an appropriate thread-on color filter over the lens can restore the color balance. A B+W 081 blue filter closely matches the original Canon filter.

Another "side effect" of removing the stock filter from a digital-camera sensor is the loss of the anti-aliasing filter. This actually allows images to be sharper, but at the expense of "jaggies" becoming apparent along lines of sharp contrast, and moiré patterns appearing under certain circumstances. However, most modern-graphics programs can take care of these problems.

So the ultimate question is, do the results justify the time and expense (and possible damage to a camera) needed to modify a modern digital camera specifically for astrophotography? The performance gains shown by Lovejoy's results, displayed in Figures 4.17 through 4.19, would indicate

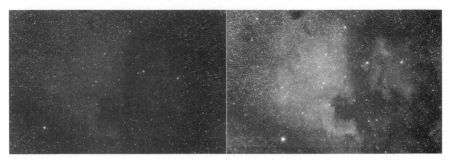

Fig. 4.14 *These before and after images of the North America Nebula are a combination of 4 four-minute exposures through a Borg 76EDGF4 astrograph. The image on the left was taken with a standard Canon 300D. The one on the right was taken with an AstroHutech modified 300D and shows the camera's enhanced red sensitivity due to IR-blocking filter removal. Photos by AstroHutech.*

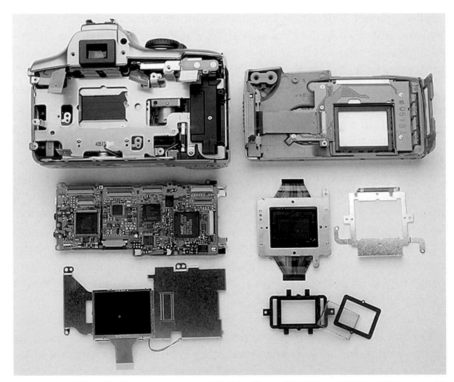

Fig. 4.15 *AstroHutech offers an astronomically-friendly modified Canon 300D which has the IR blocking filter removed to increase sensitivity in the red portion of the spectrum. This picture shows the disassembly of the 300D required to access the filter (lower right). Photo by Astro-Hutech.*

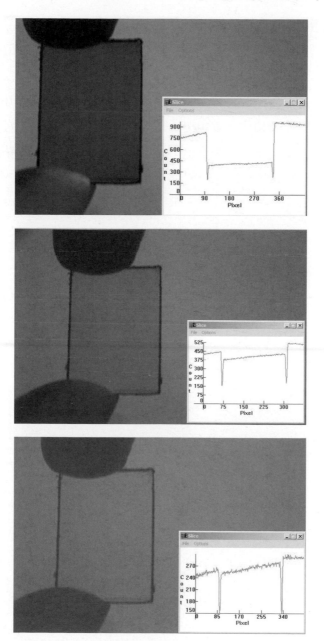

Fig. 4.16 *The reduced red-wavelength transmission through the factory-installed infrared blocking filter in the Canon EOS 300D is apparent in these three images of the red, blue, and green channels (top to bottom). Photos by Terry Lovejoy.*

Fig. 4.17 *This view of the Tarantula Nebula is a combination of 6 three-minute exposures taken through a Takahashi Epsilon 160 at f/3.3 at ISO 400 with a modified Canon 300D. The red and blue nebulosity is clearly shown. Photo by Terry Lovejoy.*

Fig. 4.18 *The Lagoon and Trifid Nebula have both red and blue nebulosity that is captured in this image taken with a modified Canon 300D set at ISO 400 through a Takahashi Epsilon 160 at f/3.3. This is a combination of 4 ninety-second exposures. Photo by Terry Lovejoy.*

Fig. 4.19 *This image of Zeta Scorpii taken with a modified Canon 300D records the rich red hydrogen-alpha emission and the bright blue stars. The camera was set at ISO 400 and 2 five-minute exposures were taken through a Takahashi Epsilon 160 at f/3.3. Photo by Terry Lovejoy.*

yes. Bear in mind that the new product introduction cycle with electronic items is measured in months, not years. Manufacturers will soon release newer and more advanced models, and today's excellent cameras capable of outstanding astrophotography will become very affordable eBay and Astromart items. This will lead to wider availability of inexpensive cameras for experimentation with modification procedures like the filter removal discussed here.

Chapter 5
Types of Astrophotography

5.1 Introduction

In this chapter, we will look at the various types of astrophotography that we can do with digital cameras. Some models have limitations, such as restricted ability to take long exposures or the inability to remove the camera's lens, and these constraints will factor into the type of astrophotography that can be accomplished with various cameras. For instance, if your camera is a fixed-lens unit like the Olympus Camedia or Canon Powershot series, excellent images of the Sun, Moon, and planets may be captured. However, minutes-long exposures of wide-field views of constellations and the Milky Way or deep-sky objects through a telescope will require a more advanced DSLR camera like the Nikon D70 or the Canon 10D.

The types of astrophotography can be categorized as follows:

1. Stationary camera mounted directly on a fixed tripod for stability.

2. Piggyback camera mounted atop a guided telescope to allow longer exposures without the stars trailing.

3. Prime focus with a camera body attached, without eyepiece, to the focuser of a telescope.

4. Eyepiece projection with a highly magnified image projected on to the camera's image plane.

5. Afocal-projection using the camera's own lens to image through a telescope eyepiece.

6. Wide-field photography with a homemade "barn door star tracker."

With the night sky truly filled with everything in the universe, it may seem odd to ask what is there to photograph? Matching a specific celestial target to your camera may be a bit of a mystery to the novice skyshooter. But the upside is that with digital photography, there is no cost in experiments that do not work. We don't have to wait for the photo lab to show us the results—they are available immediately. After determining what is needed to fix a shot that didn't turn out as expected, the delete button becomes our friend. Don't be afraid to try different exposure times, and if possible, different focal lengths on your targets.

Above all, keep a log of the objects you shoot. Keeping such records may seem like a drudgery at the time, but take it from me that it is worth the effort. I wish I had been more diligent in my record keeping in my earlier efforts, because such information is vital to later astrophoto success. More than once I have taken what turned out to be an excellent astrophoto; but, strange as it seems, I later had no idea of the technical details of the exposure. Repeating a similar shot was once again a trial-and-error effort.

Just as a star chart is an important astronomy accessory, an exposure logbook should also accompany your astrophoto gear. In it record the following: the target, exposure length, lens settings, the camera's ISO setting and white balance (if it was changed from normal daylight settings), sky and weather conditions, and the results achieved by the exposure.

Let's examine the needs and possibilities of each of the different astrophoto techniques.

5.2 Fixed-tripod Imaging Introduces the Camera to the Sky

As the name implies, this type of astrophotography involves simply placing the camera on a tripod to hold it steady during the exposure. This technique can be used for short exposures showing little or no star trailing, or it can use lengthy exposures that allow the stars to trail and are limited only by the brightness of the sky and the exposure-time limitations of your camera model. This type of photography can even be done without a tripod as long as there is some method for aiming and holding the camera rigidly steady, such as a clamp and ball-head camera adapter (see Figure 5.1).

The Moon and planets are easy targets for short exposures several seconds in length with the camera mounted on a fixed tripod. The crescent Moon or bright Venus in deep twilight is always a pretty picture—especially if you place some foreground objects such as recognizable buildings or trees in the field-of-view. This transforms an ordinary image into a nocturnal landscape.

Similarly, the bright constellations can be recorded with a fixed camera using short exposures of up to 15 seconds. By using wide-angle or normal lens settings, the stars will remain pinpoints, and the exposure is brief enough that the apparent motion of the night sky will not cause the stars to appear as streaks. Again, placing familiar landscape or architectural objects in the foreground will add interest to an otherwise starkly technical image. Try to photograph the bright constellations such as Orion near the

Fig. 5.1 *A simple tripod-mounted camera (left) is capable of taking astrophotos such as star trails and short-exposure subjects like auroras, satellites and Iridium flares, and planetary conjunctions. If weight or space are a concern while traveling, a variety of clamps can be used with a swivel ball-head adapter to enable a camera to be aimed almost anywhere. In the middle image, the adapter has been attached to a C-clamp while on the right a truck CB antenna clamp has been pressed into use. Photos by Robert Reeves.*

celestial equator, or the asterism of the Big Dipper in the northern sky. Try longer exposures if your camera will allow more than 15 seconds. Compare the movement of the stars in Orion to the movement of the stars in the Big Dipper using equal exposure times. This comparison will show that the closer to the pole the target is, the less its apparent motion in a given time. Objects near the celestial pole can be exposed half again as long as those near the celestial equator without increased star trailing because of the geometry of their circular path around the celestial pole.

Also examine these longer exposures to see how the sky background is brightening with increased exposure time. In large cities the night sky can be 50 or more times brighter than in a rural setting. In urban areas sky glow caused by city lights and waste lighting reflected by atmospheric dust particles will cause the sky to be too bright for exposures more than 30 to 60 seconds in length. This sky glow will fog the image to the point where the stars are no longer visible. In darker areas away from the city exposure times can increase, and may be limited only by the capability of your camera. Take some long exposures of the sky in your area and become familiar with just how long you can leave the shutter open without overexposing the background sky glow. The maximum you can expose without sky glow degrading the image is called the sky fog limit. As we shall see in later chapters, there are methods of suppressing sky glow by using filters to block out portions of the spectrum where streetlights dominate, and by use of image processing tricks.

Fig. 5.2 *A brief exposure using a Nikon Coolpix mounted on a tripod captured the serenity of a lunar eclipse over Albuquerque, New Mexico. Photo by Becky Ramotowski.*

Fig. 5.3 *Auroras are good targets for cameras mounted on fixed tripods. These two images were taken in Scotland using a Canon 10D. The camera faithfully recorded reds, blues, and greens in various portions of this auroral storm of October 29, 2003. Photos by Phil Hart.*

Of course, there are times when you may want star trails, or don't mind if they are present. If your camera allows exposures of 60 seconds, or has a "bulb" setting where the shutter can be locked open for many minutes, then other possibilities exist for fixed-camera photography. Circular star trails arcing around the celestial pole are always a pretty sight. Digital cameras, because of the dark current that is always present in imaging sensors, cannot make exposures of up to hours in length like film cameras can. However, we shall see later that successive digital images of short star-trails can be combined using an image processing program to create star-trail images that are superior to those produced by film cameras.

Photography of orbiting Earth satellites is also an interesting offshoot of astrophotography. A photographic record of your observation of the passage of the International Space Station, the Hubble Space Telescope or perhaps a bright flare of an Iridium communications satellite will be an interesting keepsake. The technique for photographing various bright satellites is as simple as photographing constellations with a fixed camera. The trick is predicting when a certain satellite will appear and what part of the sky will it pass through so the camera can be aimed at that area to await the passage of the target. Once the satellite is visible, or the Iridium flare begins to happen, the shutter is opened for the duration of the event.

Fortunately, the Heavens-Above web site (see Appendix B) simplifies the satellite prediction process. Once you log onto the site, you can register your exact location on the Earth. This places a harmless cookie in your web browser so that later visits to the same web site will allow the generation of satellite pass predictions for your location without having to reenter your map coordinates. The predictions are so accurate that I have stopped calculating with other software. With a few mouse clicks the web site produces local pass predictions for any visible satellite, including the time, azimuth and elevation, and expected brightness of the satellite's pass or Iridium flare.

Bright comets and meteors present other possibilities for a fixed camera. If a comet is visible to the naked eye, it can be photographed with an exposure short enough that the background stars do not trail. Meteor photography on the other hand needs lengthy exposures to maximize the time your camera is acting as a "trap" to bag the few bright meteors that happen—by chance—to pass through its field-of-view. In this case you want to keep the shutter open as long as the sky conditions, and the camera's capabilities, allow.

Auroras are often not thought of as an astrophoto target, but they occur in the sky and often stars are simultaneously visible. Those living in

Fig. 5.4 *Circumpolar star trails were captured by combining 20 ten-minute exposures with a Canon 10D at ISO 100 and a 20-mm lens set at f/4. Photo by Johannes Schedler.*

middle latitudes rarely see aurora, but astrophotographers in high northern or southern latitudes will eventually encounter one when they are under a dark sky. Capturing an aurora used to be a matter of luck and patience, but today with advanced monitoring of the Sun by specialized satellites, the prediction of auroras—and even their expected intensity—is becoming quite accurate. Visit Spaceweather on the web to see aurora predictions for your area and be ready with your camera. Aurora vary in form and intensity—but shorter exposures generally will reveal the ripple effect of curtain aurora while longer ones will show the color intensity of the display.

5.3 Piggyback Adds Tracking Capability

Digital cameras that allow exposure times longer than about 15 seconds offer the ability to take long, timed exposures of the stars and to record objects that are too dim for the eye to see. However, these longer exposures present the problem of having the camera track the diurnal movement of stars in order to prevent star trailing. This is most easily accomplished by the piggyback method of astrophotography which involves attaching a camera to an equatorially-mounted, polar-aligned telescope. The camera is fastened to the telescope tube by its tripod-mounting socket with a bracket that either bolts to the tube or is secured with clamps around it. The camera uses its own lens to image the sky and uses the telescope and its clock drive solely as a tracking platform to follow the moving stars.

Most fixed-lens digital cameras are limited in exposure length to 15, 30, or sometimes as long as 60 seconds by the camera's electronic controls. Guiding is not critical with these short exposures because most telescope drives are accurate enough, if properly polar aligned, to produce trail-free exposures in that short time. Fortunately, longer exposures than those allowed by the camera's limited shutter timer can be achieved by digitally combining repeated short exposures with an image processing program. This technique will be explored in Chapter 14. Some more-advanced DSLR models have a "bulb" setting that allows the shutter to remain open as long as you want. Longer exposures with these cameras will require guiding.

For exposures longer than a minute, a camera will need to be guided to compensate for small errors in tracking caused by imprecision in polar alignment, imperfections in the clock-drive worm and gear, or by atmospheric turbulence. Guiding can be accomplished either manually—by monitoring a guide star in a special crosshair eyepiece and correcting the telescope drive by hand; or automatically—with an autoguider that follows a guide star and automatically issues commands to correct telescope track-

Fig. 5.5 *An Iridium flare is captured in this 16-second ISO 400 exposure with an Olympus 3020Z. The bright trail extending east from Cassiopeia is a jet vapor trail illuminated by moonlight. Photo by Robert Reeves.*

Fig. 5.6 *Because the imaging sensors of most DSLR cameras are smaller than a 35-mm film frame, a moderate telephoto used with piggyback astrophotography will begin to show detail even in small objects like M13. This image is a combination of 9 one-minute exposures with a Canon 10D at ISO 400 and a 200-mm f/4 lens. Photo by Robert Reeves.*

Fig. 5.7 *The author's piggyback astrophotography setup is typical of those using a DSLR mounted on a small telescope. A laptop at the telescope is a normal part of digital imaging. Photo by Robert Reeves.*

ing. Discussion about guiding techniques and proper polar alignment is found in Chapter 9.

A word of warning: piggyback photography requires an accurately polar-aligned equatorial mounting that pivots the entire telescope tube assembly around an imaginary line that intersects the celestial pole. Not all telescopes that track the stars across the sky use this method. Some recent models of computerized telescopes operate on an alt-azimuth mounting that uses electronics and control motors to automatically follow the motion of the stars across the sky. This technique utilizes a small computer built into the telescope's mounting to continuously calculate the location of the stars relative to your location on Earth and the local time. The computer then commands motors on the telescope's altitude and azimuth axes to slew the instrument in order to follow the sky. This will accurately track a celestial object and keep it in the field of an eyepiece, but in reality, the telescope is simply following the object through a series of very small up or down and side-to-side zigzags. The telescope may tip up and down while tracking, but the orientation of the telescope tube remains constant relative to the Earth. This causes a phenomenon called field rotation, in which the field-of-view of the eyepiece slowly rotates as the sky seemingly

Fig. 5.8 *Piggyback astrophotography involves attaching a camera to an equatorially mounted telescope with a suitable adapter. As the telescope follows the stars, it allows the camera to also properly track the stars. (Left) A camera can be mounted on top of the telescope for horizontal "landscape" views or mounted sidesaddle to record vertical "portrait" oriented views. If the telescope is too small to allow convenient sidesaddle mounting, a 90-degree angle bracket can be made from ¼-inch steel to orient the camera for vertical views.(Right) Photos by Robert Reeves.*

Fig. 5.9 *A swivel ball-head adapter will keep the front end of the telescope from intruding into the image by enabling the camera to aim slightly above the normal field-of-view. The short exposures inherent to digital astrophotography prevent flexure in the adapter from affecting individual images. Photo by Robert Reeves.*

Fig. 5.10 *The International Space Station is typically as bright as Jupiter and is visible from any location on Earth at least once every several weeks. The station makes an easy target for a tripod-mounted camera. Photo by Robert Reeves.*

Fig. 5.11 *The Sagittarius region of the Milky Way was captured with a Canon 10D at ISO 800 using a Voigtlander 19 to 35-mm zoom lens set at 19 mm. Five separate four-minute exposures were combined to produce this view that also captured a bright meteor. Photo by Antonio Fernandez.*

pivots around the celestial pole while the telescope remains level relative to Earth. Similarly, the image in a camera attached to the telescope will also show field rotation, even though the telescope's computerized alt-az mount is accurately tracking the target. The resulting image would resemble a traditional polar star-trail image taken with a tripod-mounted camera, except the star pattern would not match those around the true celestial pole.

Piggyback photography turns a camera into a powerful astronomical instrument because it can now record objects in beautiful astronomical scenes that are invisible to the unaided eye. Sweeping views of constellations or the Milky Way are possible with wide-angle lenses while telephoto lenses allow us to get close-ups of star clusters and nebulae that approximate the same field-of-view as a pair of binoculars.

5.4 Prime Focus Enables Deep-Sky Imaging

Prime focus imaging is an astrophotography technique where the camera body is attached without its lens, to the telescope, without its eyepiece. The telescope's objective mirror or lens is used to focus the image directly onto the camera's focal plane. Because the camera body is attached to the telescope with a device called a T-mount, this photo method is limited to users of DSLR cameras that have removable lenses. The T-mount screws onto a coupler that in turn threads onto the rear cell of a Schmidt-Cassegrain telescope in place of the eyepiece or star diagonal or screws onto a collar that fits into the focuser of a Newtonian or refractor. In effect the telescope itself becomes a giant telephoto lens for the camera. A standard 8-inch Schmidt-Cassegrain telescope (SCT) becomes a 2000-mm focal length *f*/ 10 telephoto lens.

A T-mount is a universal camera lens adapter that was standardized by the photo industry after its introduction by the lens maker Tamron in 1957. The concept was that aftermarket lenses would have a standard rear threaded flange and would be attached to a camera brand-specific adapter, the T-mount, at the time of purchase. In its original concept the T-mount consisted of two parts: a T-adapter that threads onto the aftermarket lens, and T-ring that mates to the specific brand of camera body the lens is being adapted to. The T-adapter and T-ring are designed to mate together with several setscrews. With this system a particular aftermarket lens could be adapted to many different makes of camera body by simply changing the T-ring to match the proprietary lens mount on the camera body. The internal thread size of the T-adapter is a metric thread 42 mm in diameter with a 0.75-mm thread pitch. This is referred to as M42-.75 threads. In non-metric equivalent measure, this is 1.654 inches internal diameter with 33.866

Fig. 5.12 *A T-ring, or T-adapter, is used to mount different camera bodies to accessories that use a common mounting thread. The camera side of the T-adapter (left) is specific to the type of camera body used while the threaded accessory side of the adapter (right) uses the same threads for all T-adapters. Photo by Robert Reeves.*

threads per inch. The lens barrel or camera-to-telescope adapter must thread into the T-adapter and have sufficient wall thickness to be structurally sound. This restricts the internal diameter of the optical opening to about 1½ inches. This same T-mount system has been adapted by major telescope manufacturers like Celestron and Meade to allow a camera body to attach directly to the focal point of the telescope. By adopting this system using existing standardized mounts, the telescope manufacturers allowed the art of amateur astrophotography to flourish. Fortunately, today manufacturers of DSLR cameras agreed to adopt only two types of lens mounting systems, regardless of camera make. DSLRs thus either accept the Canon EOS lens mount, or Nikon lens mount, making the selection of T-mounts much easier.

The magnification achieved with prime focus as compared to the normal view with a conventional film camera used to be simple to calculate; simply divide the telescope's focal length in millimeters by 50, the focal length of a "normal" 35-mm camera lens. Thus, a telescope with a 2000-mm focal length provided a prime focus magnification of 40. Digital cameras throw a curve into this simple calculation because a majority of them have imaging surfaces smaller than standard 35-mm film. For instance, the Canon 10D has an imaging sensor that is about 15 by 22 mm, and thus all image magnification calculations have to be multiplied by 1.6 when comparing the field-of-view to normal 35-mm cameras. The 10D would thus achieve a magnification of 64 when used at prime focus on a 2000-mm focal length telescope. Consult the individual camera manufacturers image sensor specifications in order to calculate prime focus magnification with your particular camera and telescope.

Fig. 5.13 *Coupling a camera directly to a telescope with a T-adapter essentially turns the instrument into a giant telephoto lens allowing close-ups of small celestial targets. Photo by Robert Reeves.*

Deep-sky imaging of galaxies, star clusters and nebulae is the principal aim of prime focus work. However, the high magnification achieved with this technique dictates that due care be given to the accurate guiding of the telescope mount. Any imperfection in the tracking accuracy will immediately trail star images into egg-shaped blobs. Fortunately, two factors help reduce the guiding burden for the digital astrophotographer. First, the maximum useful exposure for even the best of today's digital cameras is just a few minutes before electronic noise becomes objectionable. This forces the photographer to take a series of shorter exposures and electronically combine them in an image processing program. Since the response of digital cameras is linear—they keep recording at the same rate throughout the exposure—multiple images can be combined to create the effect of a much longer exposure. Thus, instead of taking a single half-hour exposure as a film astrophotographer would, the digital astrophotographer would take 15 separate two-minute exposures of the same object. Breaking the exposure into segments has two advantages: first, any single exposure with a serious guiding problem can be eliminated from the final image stack. Second, many of today's good electronically-controlled telescope

Fig. 5.14 *The Scopetronix MaxView DSLR Variable Projection Adapter can be used as a prime focus coupler by removing the eyepiece and collapsing the projection body. Photo by Robert Reeves.*

mounts are capable of accurately tracking an object for two minutes without the need for constant guiding corrections if the mount is well polar aligned.

If the telescope mount you use is less accurate and needs frequent guiding corrections to maintain the pointing accuracy needed for long exposures, it is recommended that polar alignment and guiding experience be gained first with piggyback photography before jumping into prime focus work. Piggyback is far more forgiving of tracking and guiding errors because the camera's lens has far less magnification and resolution than a telescope does, and minor jiggles in tracking will not affect the image. Such small errors will have devastating effects on a prime focus shot, as it will record every deviation of the star image.

Because the camera is attached in place of the eyepiece on a telescope being used for prime focus photography, we cannot directly view through the telescope in order to guide it during the exposure. There are two ways to get around this limitation: a separate guidescope attached piggyback to the main telescope, or a device called an off-axis guider.

The separate guidescope guiding method allows viewing a guide star

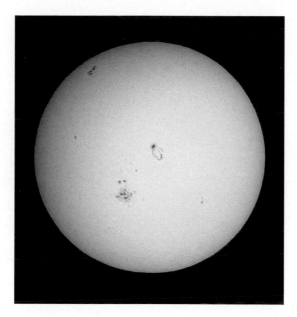

Fig. 5.15 *Full-aperture solar filters make imaging the Sun no different than shooting the Moon. The ever-changing patterns of sunspots presents a different solar view every day. Photo by Becky Ramotowski.*

directly through a smaller telescope and allows some flexibility in choosing a bright guide star as the guidescope can be shifted slightly within its mount, thus not disturbing the framing of the camera's view. However, this system has the disadvantage of being prone to differential flexure, or slight movement between the main telescope and the guidescope caused by gravity as the telescope combination changes orientation while tracking an object across the sky. This flexure causes guiding errors as the guidescope and main telescope fields-of-view diverge slightly. As the aim of the guidescope changes slightly relative to the main one, corrections made on the moving guide star are actually moving the main scope off its target. Resulting images have unexplained trailings that do not fit the known accuracy of the guiding. Stars can be trailed in unusual directions, not just along the expected right ascension axis, or display bowed or hooked trails. Using a very rigid well-engineered mounting between the main and guidescope can cure differential flexure. However, this adds weight to the load carried by the telescope mount, and a very massive and costly mount is needed for proper stability.

With visual guiding, a guidescope should ideally be of equal or longer focal length than the primary scope in order to accurately track the slightest deviation of the guide star before it can register on the photographic image

Fig. 5.16 *The Leo Triplet, consisting of M65, M66, and NGC 3628, was imaged with a Canon 10D at ISO 800. Fifty-nine separate sixty-second exposures taken through a Vixen RS200 at f/4 were combined to create this image. Photo by Antonio Fernandez.*

as a trailed star. However, portability, mounting, and cost restrictions often make this impractical for the amateur. However, the optical quality of the guide scope, and thus its cost, can be less than the primary instrument. Most practical guidescopes have a focal length about half that of the primary instrument and achieve the needed focal length for visual guiding by using a Barlow lens. Additionally, the primary telescope can use a focal reducer to decrease its focal length, with an apparent gain in photographic speed, and further increase the ratio of guidescope to primary instrument focal length. Electronic autoguiders, discussed in Section 9.16, can by design use shorter focal length—and thus cheaper—guidescopes, somewhat offsetting the added cost of the autoguider.

The off-axis guiding method is cheaper than using a separate telescope and eliminates the differential flexure problem by allowing the guiding device to view through the same telescope as the camera. This is accomplished by attaching the camera to the telescope with a T-adapter mounted on an oversize extension tube attached to the rear cell of a Schmidt-Cassegrain telescope, or the focuser of a refractor or reflector. Inside the adapter tube, a small mirror or prism is inserted into the edge of the light cone exiting the telescope before it enters the camera. Often called

Fig. 5.17 *The Pleiades was the "first light" target for Jay Ballauer's Canon EOS 300D. This image is a combination of 8 five-minute exposures through a Takahashi FSQ-106. Photo by Jay Ballauer.*

the pick-off prism, it is either just outside the camera's field-of-view or is so far ahead of the focal plane that it is unrecognizably out of focus and essentially invisible on the image. The light beam from the pick-off prism is then directed 90 degrees off-axis and into an eyepiece mounted on the side of the off-axis guider. Either an illuminated-reticle eyepiece or an electronic autoguider is mounted at this location, and it monitors the guide star reflected by the pick-off prism at the edge of the telescope's field-of-view while the camera views the central portion of the telescope's field-of-view. Off-axis guiders eliminate differential flexure, but present their own unique challenges. Since the pick-off prism is at the very edge of the telescope's view, it is gathering star images that suffer from comatic aberrations common to the extreme edge of a telescopic image. Further aggravating the problem is the fact that off-axis guiders are used most often with Schmidt-Cassegrain telescopes. SCTs use a spherical primary mirror, compounding the coma problem. This, coupled with the small area of the pick-off prism, presents us with extremely dim, fan-shaped guide stars. Autoguiders, like human eyes, prefer to view sharp round stars. The adjustable parameters in autoguider software need to be configured to take into account the smeared, comet-shaped stars at the rim of the telescope's

Fig. 5.18 *The problem of how to guide a telescope while a camera is mounted at its focus can be solved by using an off-axis guider. An off-axis guider allows the camera to view the target while a small "pick-off" prism reflects the edge of the telescope field-of-view to the guiding eyepiece or autoguider. The prism is out of the camera's field-of-view and does not affect the image. Photo by Robert Reeves.*

field-of-view.

The mechanics of using an off-axis guider can lead to some late-night searching for a suitably bright guide star. First, the camera is focused, and then the desired field-of-view is centered and framed on the object being photographed. A guiding eyepiece or autoguider is then inserted into the tube on the side of the off-axis guider. It is focused by sliding it in or out of its tube to focus the guider separately after the camera is focused. If a suitable guide star is not visible, the search begins. Either the entire off-axis guider assembly rotates or that part of the assembly that makes up the pick-off prism and the guiding eyepiece rotates until a suitable star is found at the edge of the telescope's field-of-view.

Often, the guide star lands at the edge of the eyepiece field and the entire telescope must be moved slightly to align the star with the eyepiece cross hairs, with the resulting loss of desired camera framing. Use of a projection reticle, a device that fits between the eyepiece tube and the eyepiece and projects a virtual crosshair onto the field-of-view, can eliminate this

Fig. 5.19 *Traditionally, with 35-mm cameras, the full Moon would fit into the field-of-view of popular 8-inch f/10 Schmidt-Cassegrain telescopes when exposed at prime focus. Most DSLR cameras, however, have imaging sensors that are smaller than a 35-mm film frame, and thus the whole lunar disk does not fit into a single exposure. The image above is a composite of two prime-focus images taken with a Canon 10D through a Celestron-8 f/10 Schmidt-Cassegrain telescope. Photo by Robert Reeves.*

dilemma. Two advantages of the projection reticle are that any eyepiece can now be used for guiding, and that the virtual crosshair is movable and can be shifted to a star instead of moving the entire telescope to move the star to the reticle. More details on guiding methods are in Chapter 9.

Bright objects like the Sun, Moon, and brighter planets can often be focused directly through the DSLR's view screen. For these targets, the exposure is brief and focus can be easily confirmed by viewing the image on the camera's LCD preview screen. Most cameras allow zooming in on the preview image to check small detail. However, dim objects requiring sev-

Fig. 5.20 *Twelve separate eight-minute exposures taken with a Canon 10D at ISO 400 through an 8-inch f/5.9 Maksutov-Newtonian were combined to create this image of M27, the Dumbbell Nebula. Photo by Paul Hyndman.*

eral minutes to build up an image are entirely another matter. Often you cannot even see the object you intend to image, and you will have to focus on a nearby bright star, and then slew the telescope to the target's location. Then a sequence of short exposures is usually taken to verify the focus. The exposures can be viewed with a stand-alone monitor or laptop computer. To speed things up, the camera's ISO setting can be temporally increased to brighten the image. Increasing the ISO will, of course, increase image noise, but this is not a problem when focusing; once focus is achieved the ISO can be reduced before you start imaging.

Most DSLR viewfinders are too dim to adequately focus a star on the focus screen. Conventional film cameras featuring a removable viewfinder made it possible to exchange the view screen for one that presents a brighter, easier to focus image, and to use a substitute viewfinder with a built-in magnifier. This is not possible with today's DSLRs as they do not have removable viewfinders or view screens. The easiest way to accurately focus on stars for prime focus photography is to sight on a bright star and use a software program called DSLRFocus and take a succession of quick images at different focus settings that are analyzed by the software to determine best focus. If a computer at the telescope is unavailable, then a device called a knife-edge focuser can be temporarily substituted for the camera

Fig. 5.21 *The Stiletto knife-edge focuser marketed by Stellar Technologies takes the guesswork out of focusing on stars by using a Ronchi grating to show a telescope's exact focus point for a given brand of DSLR camera. Photo by Robert Reeves.*

body to focus the telescope; then the camera replaced after focus is achieved.

A knife-edge focuser can be constructed from an inoperative camera body of the same make as the camera being used for photography, or can be purchased from manufacturers such as Stellar Technologies International. The device is so called because it was originally perfected using the same knife-edge principle used by the Foucault test for astronomical mirrors. However, most focusers these days use a Ronchi grating that has 180 to 300 lines per inch. The grating performs the same function as a true knife-edge, but is easier to use.

The theory and construction of a knife-edge focuser is simple; a thin flat straight edge, in most cases something like the edge of a razor blade, is placed at the same distance from the focuser's mounting flange as is the imaging plane of the camera being used. The idea is that a focused star is a very small pinpoint, but an out-of-focus star is a large blob. A target star is then viewed through the focuser and maneuvered close to the knife-edge. When the star is moved in a perpendicular direction across the knife-edge, it reacts in a predictable fashion. If the image is out of focus, it vanishes slowly as the knife-edge occults it. If the star is in focus and is a small pinpoint, it will seem to vanish instantly.

The Stiletto knife-edge focuser attaches to the telescope using the same style of T-mount that the camera uses, see Figure 5.21. Looking similar to a camera with a magnifying focuser attached, the unit places a Ronchi grating the same distance from the camera mount on the telescope as the camera places the focal plane from the camera mount. A bright star is viewed through the focuser, and the telescope focus is adjusted until the bars in the Ronchi grating start to spread out on the star's image. When the bars have disappeared completely, the telescope is focused. Without disturbing the focus mechanism, the knife-edge focuser is removed, and the camera is reinstalled on the telescope. The scope is then re-aimed at the desired target for the exposure.

Prime focus astrophotography with a DSLR attached directly to the telescope expands our capability by allowing close-ups of the Moon's phases. Similarly, the Sun can be imaged through a special filter that fits over the front of the telescope to remove more than 99 percent of the solar radiation which will harm human vision and camera optics. Tracking the day-to-day movement of sunspots as they move across the face of the Sun is an interesting daytime astrophoto project. Planetary images are generally too small to be useful with prime focus. Even Jupiter is still relatively small at 2000-mm focal length. However, the daily, and even hour-by-hour, movement of the bright Galilean moons of Jupiter is a natural target for prime focus photography.

Deep-sky photography is where prime focus is most useful. Guided long exposures of galaxies, star clusters, and nebulae are best achieved with this system. DSLR cameras are quite efficient at recording galaxies and star clusters; however, their red sensitivity is low compared to some conventional films, and digital processing to emphasize the red emission nebulae will be required.

5.5 Eyepiece Projection for Lunar and Planetary Close-ups

Prime focus will capture views of the entire lunar globe, but even higher magnifications of the Moon and planets require a different photographic technique called eyepiece projection. With this method the camera is again mounted to the telescope with a T-adapter, but this time the tube attaching the T-adapter to the telescope contains an eyepiece that projects the image onto the camera's focal plane. The tube, often called a tele-extender, can be as long as practical, but its most common application is 6-inches long when used with Schmidt-Cassegrain telescopes. The principle is that the image projected from the eyepiece is enlarged on the focal plane in much the same manner as a slide is enlarged when projected onto a screen. The further away the camera is from the eyepiece, the bigger the projected image is on the focal plane. However, with increased projection distance comes reduced photographic speed as the image dims with greater enlargement. The effective focal length of eyepiece projection systems with average-sized amateur telescopes can approach 15,000 to 20,000 mm, depending upon the eyepiece used. However, photographic f/ratios increase to between f/50 and f/100, resulting in exposures lasting up to several seconds on the Moon and bright planets.

Fig. 5.22 *Eyepiece projection is a tool useful for achieving high-magnification images. The same adapters used for film-camera eyepiece projection are used with a DSLR camera and the proper T-adapter. Photo by Robert Reeves.*

Fig. 5.23 *The Moon is a natural target for high-magnification, eyepiece projection photography. The highlands area south of the Straight Wall was imaged with a Canon EOS 60D at ISO 100 through an 8-inch Maksutov-Newtonian and a pair of stacked TeleVue Powermates (2x and 4x). Photo by Paul Hyndman.*

The following three formulae will give us basic photographic parameters when using the eyepiece projection technique:

$$M = \frac{D - F_2}{F_2}$$

$$T = F_1 \times M$$

$$F = \frac{T}{A}$$

Where: A = telescope aperture

D = distance of projection from eyepiece to focal plane

F = focal ratio of projection system

F_1 = focal length of the telescope

F_2 = focal length of the eyepiece

M = projection magnification factor

T = total projection system focal length.

If we plug in the numbers for a common 8-inch f/10 Schmidt-Cassegrain telescope using a 6-inch projection tube and a 25-mm eyepiece, we come up with the following:

$$M = \frac{D - F_2}{F_2} = \frac{150 - 25}{25} = 5$$

$$T = F_1 \times M = 2000 \times 5 = 10,000 \text{ mm}$$

$$\dot{F} = \frac{T}{A} = \frac{10,000}{200} = 50.$$

We find that our 8-inch Schmidt-Cassegrain telescope using a 25-mm eyepiece with 6-inches of eyepiece projection now operates as a 10,000-mm focal length f/50 photographic system.

The enormous magnification with eyepiece projection and resulting high f/ratios require accurate telescope tracking even though the exposures are far shorter than those used on deep-sky targets. The object of such photography is to record minute lunar and planetary detail. Even the slightest tracking drift can soften the focus of very high-magnification views of the Moon and planets. There is no practical way to guide a telescope during eyepiece projection photography, so accurate polar alignment and a good tracking rate are a must.

Steady atmosphere is also essential because heat waves and turbulence in the air will blur and distort high magnification images. Even if the camera focus and telescope tracking are flawless, normal atmospheric turbulence will reduce the success ratio of lunar and planetary images to a low

Fig. 5.24 *The Scopetronix MaxView DSLR Variable Projection Adapter allows eyepiece projection with DSLR cameras using standard T-adapters. The MaxView is big enough to hold virtually any eyepiece, even the huge 31-mm Nagler or 41-mm Panoptic. The adapter is available in 1.25-inch and 2-inch versions and can be used without an eyepiece as a prime focus adapter. If lack of backfocus limits the usefulness of MaxView, an included telenegative lens can be used to increase the focal length of the system to where focus can be achieved. Photo by Robert Reeves.*

number. Experienced planetary photographers know to beware of high-resolution photography limitations when a weather front or the jet stream passes through. Also, local topographic conditions have to be considered. It is impractical to attempt high-resolution planetary photography from a parking lot where a large expanse of asphalt or concrete is radiating the heat it absorbed during the day.

Eyepiece projection photography uses the same focus techniques as prime focus. Since DSLRs do not have removable viewfinders or interchangeable focus screens, focusing software like DSLRFocus or a knife-edge focuser will work the best with this system.

5.6 Afocal Photography Allows Use of Fixed-Lens Cameras

As advanced as modern digital cameras are, they are also in some respects much like the simple fixed-lens box cameras of the 1930s and 1940s. The vast majority of digital cameras are fixed-lens point-and-shoot models; thus, we don't have the luxury of removing the lens and attaching the camera body directly to the telescope to shoot at prime focus or use eyepiece projection methods like we can with SLR film cameras. The only way around this limitation is to take images through the telescope using the afocal method. This involves aiming the camera's own lens directly into the telescope eyepiece. In the photographic era before the SLR camera became popular in the early 1960s, afocal-projection was the primary way to use consumer cameras with a telescope. Since the old cameras were rangefinder or twin-lens reflex models, the user was essentially blindly aiming the camera into the eyepiece. When the removable lens 35-mm SLR camera arrived on the photographic market, astrophotographers happily abandoned the afocal-projection method in favor of techniques that allowed them to not only view and frame the image but also mount the camera directly to the telescope instead of using an adapter bracket. Now fixed-lens digital cameras are reviving interest in this once dormant astrophoto technique, but digital cameras have the advantage of being able to frame the image with either the camera's LCD display or an external video monitor.

Although it is seemingly a primitive throwback to the old days, the afocal method produces surprisingly good results and is simple to do. One advantage the system has over lensless SLR camera bodies that couple directly to the telescope is that the fixed lens digicams have no reflex mirror slap to shake the telescope at the beginning of the exposure.

We can use the following formula to calculate what the resulting equivalent focal length of the afocal telescope/camera lens combination will yield:

$$\text{Effective Focal Length} = \left(\frac{F_1}{F_2}\right) \times F_3$$

where: F_1 = telescope focal length

F_2 = eyepiece focal length

F_3 = camera lens focal length.

Note that $F_1 \times F_2$ will yield the telescope's magnifying power, a number that is useful in determining the camera's approximate field-of-view through the telescope. To calculate this, simply divide the camera's field-of-view in degrees by the telescope's magnifying power.

For instance, if we were using an Olympus C-5050 camera featuring a 7.1- to 23-mm zoom lens with an 8-inch Celestron *f*/10 SCT with a 25-mm eyepiece and the camera's lens zoomed to full telephoto, the numbers would work like this:

Fig. 5.25 *A Texas thunderstorm illuminates the horizon while Ursa Major rises in the springtime sky in this two-minute exposure taken with a Canon 300D at ISO 400. Photo by Michael Howell.*

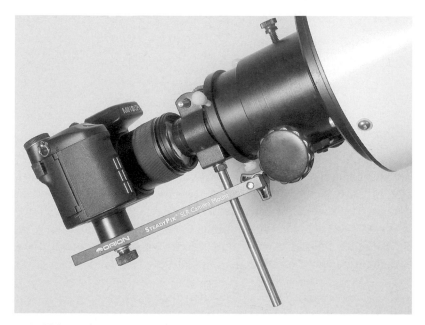

Fig. 5.26 *If the camera design does not allow an adapter to screw directly into the lens filter thread or camera body, an adapter bracket can be used to position the camera at the eyepiece. This Orion SteadyPix model clamps onto the eyepiece drawtube. Photo by Robert Reeves.*

Fig. 5.27 *This excellent image of a lunar eclipse was taken by manually holding a compact digital camera to the eyepiece of a Takahashi refractor. Photo by Becky Ramotowski.*

$F_1 = 2000$ mm
$F_2 = 25$ mm
$F_3 = 23$ mm

$$\frac{2000}{25} \times 23 \ = \ 1840\text{mm Effective Focal Length.}$$

The frame field-of-view as seen through the same afocal system is equal to the camera lens field-of-view divided by the telescope magnification. In this case, the 25-mm eyepiece gives an 8-inch $f/10$ system a magnification of 80. Using the camera lens field-of-view calculations shown in Section 10.18, we see that the Olympus C-5050 has a field-of-view measuring 13 by 17 degrees at maximum zoom. We have to convert the degrees to arc minutes to simplify the calculation:

$$\frac{1020}{80} \ = \ 12.75 \text{ arc-minutes horizontal field-of-view}$$

$$\frac{780}{80} \ = \ 9.75 \text{ arc-minutes vertical field-of-view.}$$

Fig. 5.28 *The Nectaris impact basin was imaged with a Canon 10D at ISO 100 through an 8-inch Maksutov-Newtonian and a pair of stacked TeleVue Powermates (2x and 4x). Photo by Paul Hyndman.*

Remember that digital cameras with their smaller imaging detectors, often less than half the size of a standard 35-mm film frame, "see" a smaller area of the image produced by a given focal length. The resulting image will be of higher magnification than one produced by a conventional 35-mm camera.

The effective focal ratio of an afocally-coupled camera and telescope system can be determined by dividing the system's focal length as determined above by the telescope's aperture. In the case of the Olympus C-5050 on the Celeston-8, the system's 1840-mm focal length is divided by the telescope's 200-mm aperture to yield an $f/9.2$ photographic system.

With brilliant targets such as the Sun, Moon and brighter planets, the camera can be simply handheld at the eyepiece as illustrated in our first astrophoto experience in Figure 1.14. However, thin lunar crescents or higher magnification views of the planets will require additional camera support for steady images. In some cases, for exposures of a second or less, the camera can simply be placed on a tripod and aimed into the eyepiece. This low-tech method of mounting has some inherent advantages. It lowers the mass your telescope has to carry, and operating the camera does not shake the telescope. If this is not practical, various manufacturers make adapters or brackets that attach the camera to the telescope and hold it to

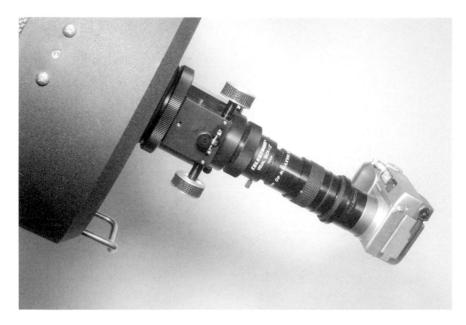

Fig. 5.29 *Threaded adapters such as the Digi-T from Scopetronix thread directly into the camera and couple to the eyepiece. This option is only good for relatively lightweight cameras. In the above illustration, the JMI auxiliary rack-and-pinion focuser makes it easier to focus high-magnification images by eliminating the image shift common when focusing Schmidt-Cassegrain telescopes. Photo by Robert Reeves.*

the eyepiece. Today, the favored method is to couple lightweight digicams to the telescope with adapters that attach the camera directly to the eyepiece.

When digital astrophotography was in its infancy, those who wished to mount a camera to their telescope had to sort through a jumble of threaded filter and camera adapters to find the right combination of pieces to assemble into a rigid coupling. Ideally, the combination would attach to the camera via the lens's filter thread, or if the lens was motorized to extend out of the camera body, the adapter threaded into the camera body. Sometimes, to gain the proper eyepiece-to-camera lens distance to minimize vignetting, existing components had to be cut, sawed or modified, taxing the astrophotographer's handyman abilities. Fortunately, there is no more need for the frustrating search for compatible adapter parts. Ready-made adapters are now available from companies like Scopetronix that thread into virtually any camera and attach it directly onto the telescope eyepiece. Another option is the Orion SteadyPix Universal Camera Mount that clamps to the telescope and holds the camera to the eyepiece.

When using adapters that attach the camera directly to the telescope's

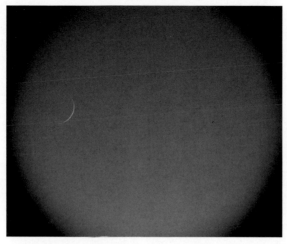

Fig. 5.30 *Afocal photography is not limited to telescopes. Here, the crescent Moon and Venus were imaged through one side of a pair of binoculars! Photo by Becky Ramotowski.*

eyepiece, do not use the telescope's star diagonal, if possible. This may make it inconvenient to view the camera's preview screen, but the weight of the camera being offset from the central axis of the telescope will make it harder to balance. Additionally, the camera may not be secure in a slip-fit diagonal where the eyepiece is not held in place with a setscrew. Be sure to check the clearances between the attached camera and any telescope mount parts, especially with motorized goto mounts, or an unintentional collision may occur. Above all, loop the camera's neck strap around the finder scope for added safety. Accidents happen in the dark, and it would be a shame if a camera fell off its eyepiece adapter and smashed on the ground.

The afocal-projection method is similar to the eyepiece-projection method in that it dramatically increases image size; but like eyepiece projection, it does this at the expense of increasing the telescope's f/ratio. This translates into the loss of photographic speed. High-magnification views of the Moon and planets can require exposures of up to a second and steady mounting and telescope tracking become important issues.

Because the camera is separate from the telescope, stray light can enter the lens from the side while it is aimed into the eyepiece. This can cause flare or loss of image contrast. If needed, a dark cloth or cardboard mask can be used to shield the lens from stray light.

Image distortion or only part of the image being in focus may show up in full-frame images of the Moon when using the afocal-projection method. If present, it can be caused by two things. If the central portion of the image is in focus while the outer portions are softer, the problem is like-

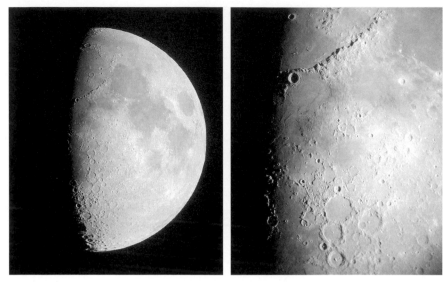

Fig. 5.31 *Most compact digital cameras have a built-in zoom lens that will allow varying magnification to be achieved with the same eyepiece. Both images above were taken with an Olympus 3020Z though a 40-mm Plössl on an 8-inch Schmidt-Cassegrain. That on the left was taken at full wide-angle zoom while the image on the right was taken with full telephoto zoom. Photos by Robert Reeves.*

ly a curved focal plane produced by the telescope optics. This is common with Schmidt-Cassegrain systems. If one side of the image is in focus and the other is not, the problem is that the camera is not square with the eyepiece.

With the traditional eyepiece-projection method, vignetting is not a problem because the camera's imaging surface is entirely inside the light cone exiting the eyepiece. With the afocal method, however, the selection of eyepiece and its apparent field-of-view has a large effect on whether the image will fill the frame or appear to be cut off around the edges. Vignetting is of little concern when imaging small targets like planets since the outer edges of the image are normally black anyway. However, solar and lunar imaging often requires the entire field-of-view.

Eyepieces with generous eye relief—the distance from eyelens to where the image is formed—will work best with afocal-projection photography. Traditionally, shorter focal length eyepieces will have less eye relief. Some eyepieces have good field-of-view, but the eye relief is so shallow that the camera lens cannot be placed close enough to the eyepiece, and thus it only images the central portion of the view.

Eyepiece selection for afocal astrophotography is driven not so much by the desired magnification, but by the desire to minimize vignetting. Im-

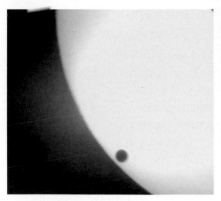

Fig. 5.32 *Imagination often leads to innovations in astrophotography. Here, the 2004 transit of Venus (left) was imaged using the familiar afocal technique—but this time the camera was the lens of a cell phone (right)! Photos by Thomas Schumm and Aaron Thul.*

aging a full-frame view through an eyepiece not matched to the needs of afocal photography will result in the image looking like it was taken through a round porthole. Compare the apparent field-of-view for all your eyepieces. The widest-angle one will perform better with the afocal-projection method. Think in terms of taking pictures through the small end of a cone-shaped funnel. If that funnel is too narrow, it will block the edges of the field-of-view. If the funnel is wider than the camera's field-of-view, it will not block the edges of the image. It is best to set up your telescopic photography system in daylight and image a brightly illuminated distant scene to test for vignetting.

Try all your eyepieces and go through the camera's entire zoom range to see what combination produces the least amount of vignetting. Above all, avoid those with poor eye relief and small or recessed eyelenses. If your camera has a zoom lens, don't worry about the power of the eyepiece. The camera's zoom feature will allow higher magnifications with any eyepiece than it yields when used visually. For example, if using a typical compact camera such as a Nikon Coolpix 990 with an 8-inch *f*/10 SCT and a 30-mm Plössl eyepiece, you can achieve about 200 power by zooming the camera to 3x zoom. With a Barlow lens, this jumps to 400 power and approaches the limits of practicality with an 8-inch telescope. Using a higher-power eyepiece with a small eyelens will introduce a lot of vignetting. The eyepiece selection boils down to one that will work well with your camera lens and mounting setup. Because magnification can be controlled with the camera zoom, it is truly a case of one-size-fits-all in that the same, relatively low power, eyepiece can cover a wide range of magnification.

5.7 Barn Door Star Tracker Photography

If an equatorially mounted telescope is not available, Section 12.9 shows how to build an inexpensive simple tracking platform that will enable a camera to accurately follow the stars for several minutes. These platforms, called "barn door" mounts because they consist primarily of two hinged pieces of wood, are easy to build and make ingenious use of common hardware store items to solve a celestial tracking problem.

The bottom board is the base of the tracker and is attached to an adjustable mounting, such as a photo tripod. The hinge attaching the two pieces of wood is the polar axis that is aimed toward Polaris by using the tripod tilt-head. The hinged upper board is the camera platform. Ideally, the camera is attached to this board with a swiveling ball-head adapter so that it can be aimed anywhere in the sky while the platform is tracking the stars.

The genius behind this device is that an ordinary ¼x20-threaded rod can be used as a clock-drive mechanism. If we place the pushbolt in the proper location relative to the hinge, turning the pushbolt at one revolution per minute will drive the mechanism in time with the rotation of the sky. For a pushbolt with 20 threads per inch, this separation from the polar hinge is 11.43 inches. In the northern hemisphere, if we place the hinge on the left side of the apparatus and aim it at Polaris, the threaded rod on the right side will act as a pushbolt and raise the upper board in synchronization with the rotation of the sky. In the southern hemisphere, these positions have to be reversed; or the positions can remain the same but the device has to start tracking with the upper board already raised. Then use the pushbolt to lower the upper board in time with celestial rotation. If polar aligned, this mechanical setup allows accurate tracking of the stars for about five minutes with nothing but slow hand turning of the pushbolt in time with the sweep of the second hand on a wristwatch.

Barn door trackers can be as simple, or as complex, as the builder desires. They can be manually operated or electric, and advanced designs are available that allow tracking for up to an hour with incredible accuracy. Additional details of these advanced designs are available in the book *Handbook for Star Trackers* by Jim Ballard. This book has been out of print for a number of years, but an early 2004 Internet search revealed over 700 sources for used copies of this book.

Chapter 6
What Camera to Choose?

There is no correct, long lasting answer to this question. Digital cameras are still in their infancy; they have been on the consumer market for only a decade and the technology is still heavily influenced by electronic innovation. You can expect that today's breakthrough is tomorrow's old-technology cast-off. In spite of all the hype from various manufacturers about new features and greater megapixel counts for their newest models, for the deep-sky imager it still boils down to one single point: how little noise is there on long exposures? Noiseless long exposures are the holy grail in digital astrophotography—all other camera features are negotiable. So, in a field where even the experts have a hard time keeping up with the latest releases, and where new, more capable cameras cost as much as a down payment on a new car, how does the astrophotographer make the choice?

The main problem in choosing a digital camera for astrophotography is that all the consumer publications and web sites are geared toward "normal" terrestrial photography. Reviews of new models and comparisons of one camera to another are based upon features relevant to "snapshot" photography. Nothing is ever mentioned about the astronomical suitability of a particular camera. You will see pertinent data such as image size and resolution, daytime color balance of the imaging sensor, various exposure modes, memory cards used, performance of lenses offered, and so forth. But missing are the astronomically important items like the performance of the camera's imaging sensor in low light, how long it can expose without objectionable noise, and what the sensitivity of the camera is to the far red portion of the spectrum. In the world of film, these factors were never addressed when discussing the camera; it was the film that determined these parameters—in the digital world, it is the camera.

As astrophotographers, we are left to our own devices to explore the suitability of a particular digital camera for our pursuit. Various web sites exist that compare the long exposure suitability of various types of photographic films, but so far no one has come forward and created a web site that does side-by-side comparisons of digital cameras for astronomy use. This would be a daunting task and very expensive. Purchasing a roll of

Fig. 6.1 *Although small enough to cradle in the palm of your hand, modern digital cameras are an amazing combination of precision optics, high-speed computers, and precision metal and plastic machining. This exploded view of a Sony Cybershot DSC-F828 demonstrates the complexities of a compact digital camera. Photos courtesy of Sony.*

each film type is one thing, but it is doubtful that any person could afford to do this with digital cameras. This leaves us to rely on the experience of others to steer us toward a good choice.

In order to narrow the choice of cameras to a proven astrophoto performer, I recommend the reader join the *digital_astro* Internet list (see Appendix B). Regardless of one's budget, the worldwide input from this forum will prove valuable in helping to select a good digital astrocamera. Here, people who have experience with a certain camera share their knowledge with everyone. If you monitor these discussions for a month or two and search the archives you should be able to conclude what works and what does not.

6.1 Desirable Features in a Digital Astrocamera

6.1.1 Fixed-Lens Point-and-Shoot Cameras
Here are some thoughts on what to look for in a compact fixed-lens point-and-shoot digital camera destined for use in astrophotography:

1. An LCD preview screen is essential with non-SLR cameras; otherwise, you cannot see the target for framing when the camera is mounted on the telescope. Fortunately, nearly all-digital cameras today have a preview screen.

2. It is desirable to have filter threads on the camera lens, or on a collar on the camera body if the lens is retractable. This makes it easier to attach the camera to the telescope by allowing the use of commercially made adapters.

3. If the camera is destined mainly for use on a telescope with an afocal coupling, a small lens is desirable to minimize vignetting and allow use of a wider variety of eyepieces.

4. A zoom lens is desirable for controlling vignetting and increasing magnification. The lens should preferably zoom by internally moving the lens elements rather than extending the lens out from the camera body. The latter option can be used, but it will require an additional lens tube adapter to mount the camera body to the telescope independently of the lens itself.

5. A swivel body, or at least a flip-out LCD preview screen or tilting viewfinder, allows you to frame the image when the telescope is aimed in a position that makes viewing the preview screen inconvenient.

6. Remote shutter operation is a must for shake-free exposures on a telescope. Not all digital cameras have this option, and it is nearly impossible to manually trigger the shutter on a scope-mounted camera without inducing vibration. A workaround on cameras without remote capability is to use the self-timer to delay the exposure for about 10 seconds after the shutter is tripped to allow vibration to damp out.

7. Without electrical power, a digital camera will not function. Extended imaging sessions can drain a camera's battery; therefore, it is desirable to be able to operate the camera with a remote transformer-powered cable that can plug into a 120-volt outlet. This has the added benefit of eliminating heat-producing batteries that will warm the sensor and contribute to image noise.

8. Of less importance is the ability of the camera to store exposure information making it easier to determine what exposure combinations worked and those that did not.

9. Also of less importance is internal camera software for noise reduction on long exposures. Ideally, noise reduction is done in post-exposure image processing, but novices will find this feature useful until they learn advanced image processing techniques.

You may notice one parameter that is not mentioned in the above list—the camera's pixel count. Many factors beside the pixel count affect the ability to resolve fine details. The aperture of the telescope, the quality of the lens and telescope optics, the telescope mount and the seeing conditions all have a greater impact on the final image quality than the pixel count of the sensor.

Marketing specialists make a big deal about the pixel count but for the most part, as astrophotographers, we are more worried about the quality of the pixels used, not the quantity. A higher pixel count usually means improved resolution on normal snapshots, but if this increase is achieved

Fig. 6.2 *Some useful features of fixed-lens digital cameras are displayed on this Minolta DiMage A1. Both the LCD display and eye-level viewer eyepiece tilt and swivel. This allows more convenient preview of lunar and planetary images while the telescope is aimed toward zenith. Photos by Robert Reeves.*

without increasing the size of the imaging sensor, the result may be detrimental for astrophotography because all the pixels are smaller pixels, and smaller pixels are less sensitive to light. Small pixels are also more prone to electronic noise because smaller the pixel, the more it will be "brightened" by heat, stray electrons in the sensor, or even cosmic rays. Smaller imaging sensors are also susceptible to blooming, the phenomenon where the electrical charge from a pixel that has been saturated with light, such as from a bright star, spills over and brightens an adjacent pixel that should normally be less exposed. For these reasons, a camera with a large three-megapixel imaging sensor will likely outperform a camera with a small six- or eight-megapixel sensor when used for astro imaging. The "megapixel wars" played out in camera advertising should not be used as a determining purchase parameter. Depending on the lens quality, a camera in the five-megapixel range is at the crossover point where image quality is essentially "film-like" anyway.

Eventually, the marketing emphasis placed on pixel counts will fade and consumer emphasis will be on how good the images look. After all, does a film user go to great lengths to measure what size the grain is on a piece of film? No, they are more concerned about how the resulting image looks. Megapixels alone will not tell the story because lighting conditions, lens exposure settings, image contrast, the display medium, and a host of other issues influence an image's appearance. Today, this issue seems to be resolving itself with the general acceptance of the "film-like quality" of six- to eight-megapixel sensors.

If the camera has a fixed lens make sure that it will handle the range of pictures you wish to take. Avoid those that have "digital zoom" or interpolated resolution; neither of these features gathers more image data.

Another feature to look for in a camera is memory card capacity— make sure the camera will accept a large capacity memory card. Compact-Flash cards have steadily increased in capacity, and their cost dramatically declined; but older cameras using SmartMedia have a limitation of 128 Mb.

The vast majority of cameras you will find today use a USB or Firewire connection to download pictures into a computer, both of which are much faster than older serial or parallel interfaces. A camera will also usually have its own proprietary software to install on your computer to recognize the camera once it is connected with an interface cable. Normally, just connecting the camera to the computer automatically triggers the file transfer process, and the images are deposited into a default file folder set up in advance when you initially install the software. With a USB or Firewire connection download speed is usually not a problem. However, if you are using huge gigabyte-capacity memory cards, a separate card reader can further speed things up (see Section 8.21).

6.1.2 Digital SLR Cameras

Most compact point-and-shoot digital cameras have a time exposure limit—usually in the 15-to-30 second range. DSLRs, like the Canon 10D or 20D and the Nikon N70, allow time exposures that are limited only by image noise which for unmodified models is about 10 minutes. DSLRs are more expensive and feature-laden, but their added performance makes them the best choice for astronomical photography. They allow both wide-field imaging with a camera lens, and telescopic work using the same kinds of adapters and imaging techniques familiar to 35-mm film users.

All the major manufacturers of film cameras now also make DSLRs. These cameras have features that give you more control than a point-and-shoot fixed-lens model:

1. **Manual Focus.** All but the cheapest fixed-focus digital cameras have an autofocus feature that will automatically focus the lens just as the image is taken. Dim astronomical objects may not be bright enough to trigger the built-in autofocus. This is especially true if the focus mechanism uses a proximity sensor on the outside of the camera body instead of viewing through the lens. The ability to manually focus the camera on a bright star, a planet or the Moon will enhance its ability to produce pleasing astrophotos.

2. **Control of Exposure.** The ability to adjust the shutter exposure times is

important in astronomical imaging. Cameras with automatic exposure control will not be able to properly expose a wide-field constellation shot, a small bright object like a planet against a black background, or an image of the lunar terminator where much of the image is black.

Manual exposure means being able to fully select the shutter speed independent of any camera parameters such as shutter priority (it will try to vary the aperture to maintain proper exposure at a given shutter speed), or aperture priority (which will also vary the shutter speed to maintain proper exposure at a given f/stop). In principle, aperture priority can be useful for lunar imaging where the camera's through-the-lens light meter calculates the proper shutter speed, but it can lead to varying image density for different areas of the Moon as more or less dark area is included in the picture. I have found that it is always best to use full manual shutter control and find the proper exposure by experimenting and viewing the results on the camera's preview screen.

3. **Time Exposure Capability.** The ability to take extended time exposures is perhaps the single most important aspect of the increased versatility allowed by a DSLR camera.

4. **Lens.** Fixed-lens digital cameras usually have zoom lenses that allow the camera to image from moderately wide views through short telephoto scenes. These may be fine for snapshot photography, but their apertures are small and do not admit much starlight. Wide aperture lenses like those used on standard 35-mm film cameras gather far more light, allowing dimmer stars and nebulae to be imaged. Also the ability to completely remove the lens greatly increases the astronomical usefulness of a camera because it can then be coupled directly to a telescope for prime focus photography or high-magnification eyepiece projection photography.

5. **Image File Format.** The default image file setting on most cameras is a full resolution medium quality JPG. A menu selection in the camera's operating system allows raising this to a high quality JPG or choosing lower resolution images to fit more pictures in the memory card. But for serious celestial imaging requiring extensive image processing, the JPG file format is not good enough. A format is needed that does not compress the image and preserves all of the original image data. Some cameras allow saving non-compressed images in TIF format; but most DSLR cameras now offer the RAW format, which may be (but not in all cases) a true bit-for-bit copy of the original data as read off the imaging sensor, with no compression or processing applied to a RAW format image.

6.2 Vast Selection of Digital Models Available

There are hundreds of different models of digital cameras, and newer more capable ones are being released continuously. In mid-2004 I made an informal survey of available digital models from the major manufacturers and found an astonishing 316 different ones in production. In the decade since digital cameras became accepted consumer items, these manufacturers have discontinued 213 camera models. These numbers do not include numerous off-brand cameras that are also on the market. The vast majority of these are fixed-lens, point-and-shoot models equipped with a moderate zoom lens. Many are SLR-like fixed-lens cameras. They have an eye-level viewfinder that shows the image as viewed by the camera lens, but what is displayed is actually a small LCD screen inside the viewfinder. Only a limited number of high-end, true interchangeable-lens SLR digital cameras are available, but in the future they will become mainstream cameras just as film SLR cameras have been for the past four decades.

Only the most affluent astroimager will have the ability to buy a new high-end camera dedicated solely to astrophotography. The rest of us will have to do a little research and rely on the experiences of other users before committing substantial money to a camera purchase. This may leave us one step behind the leaders in using the latest technology cameras, but our choices will more likely match our astrophoto needs.

If solar system photography from bright urban locations is the primary goal of your astroimaging, the choice of cameras is less restricted because noise-free long exposures are not a priority when imaging bright solar, lunar and planetary targets. Ease of camera use, the ability to quickly and accurately focus, and a proper interface for downloading many images to a computer are of more priority than the amount of noise on a multi-minute exposure. Here, the choice of cameras is vast. Virtually any consumer camera of three-megapixel resolution or higher will perform well with solar system objects. This leaves the buyer free to choose an inexpensive, used older camera or a newer, feature-laden model that will also serve well in a daytime "snapshooting" capacity. The constantly changing list of very capable cameras in both of these categories is too vast to give here. The camera that "feels comfortable" to the user and properly interfaces with their telescope and computer is the right one for solar system work.

6.3 Fixed-Lens Cameras

Today's digital astrophotographer needs a present-day camera, not one that is still on the drawing boards. The question is which one to choose when newer models are becoming available all the time? If you already own a digital camera, the choice of cameras to use for astrophotography is simple—use the one you already have. However, a person entering the

Fig. 6.3 *Two 8-megapixel cameras. The Olympus C8080 Wide Zoom (left) has a 5x zoom lens that at its widest is equivalent to a 28-mm lens on a 35-mm film camera. The camera is capable of time exposures up to eight minutes. The Coolpix 8700 (right) has an 8x zoom lens and is capable of 10-minute time exposures. Photos courtesy of Olympus and Nikon.*

Table 6.1 Popular 8-megapixel Fixed-Lens Cameras

Model	Equivalent Focal length	*f*-ratio	LCD Viewfinder	Maximum Exposure (sec)
Canon PowerShot Pro 1	28–200	2.4–3.5	235,000	15
Minolta DiMAGE A2	28–200	2.8–3.5	922,000	30
Sony DSC-F828	28–200	2.0–2.8	235,000	30
Olympus C-8080	28–140	2.4–3.5	235,000	480
Nikon Coolpix 8700	35–280	2.8–4.2	235,000	600

field for the first time is faced with a bewildering array from which to choose. Which digital camera to buy should be based on two factors: what you would like to do in astrophotography, and how will the camera be applied to daytime snapshot photography.

Assuming your choice will also be a "family" camera, the selection today includes some powerful 8-megapixel fixed-lens cameras in the sub-$1000 range that can perform double duty. Table 6.1 is a representative list of what is available as of late 2004, but newer models are always being introduced. This list is therefore presented as a representative example of the types of features to look for in a fixed-lens digital camera destined for both family and astronomical photography. It is not meant as a recommendation for a specific model. All the cameras listed use the Sony ⅔-inch 8-megapixel CCD sensor and produce 3265 x 2448 pixel images. They all possess a swiveling external LCD preview panel that makes composing a telescopic image easier. All the listed cameras also feature an eye-level viewfinder using a miniature LCD display instead of a true SLR-style through-the-lens view.

Those preferring nature and sports photography may be drawn to the longer focal length zoom lens available with the Nikon Coolpix 8700, but that camera also possesses the slowest lens for astrophotography. The Minolta DiMAGE A2 is capable of only a 30-second maximum exposure; but for lunar and planetary imaging, the higher resolution 922,000-pixel LCD display of its eyelevel viewfinder will prove a blessing. The Olympus and Nikon models are capable of multi-minute exposures, but the noise level of small ⅔ sensors will limit the quality of longer deep-sky exposures. Afocal-projection lunar and planetary imaging can be performed nicely with the 28- to140-mm equivalent zoom lens of the Olympus C-8080, while the higher zoom ratio of the Coolpix 8700 will reduce vignetting with narrower field-of-view eyepieces.

All five cameras have pros and cons, depending on the desired level of astrophotography. However, since the ⅔ sensors are small, resulting in higher noise levels than the larger sensors used in DSLR cameras, they have limited application to deep-sky photography. All will work well for afocal-projection photography, and in this application the selection of brand should be driven by which camera "feels best" for you. Handle as many different models as possible and make your choice based on which provides the best level of comfort and intuitive operation.

6.4 Interchangeable Lens DSLR Cameras

The DSLR market is so new that even the original DSLR models, now discontinued, are not very old. The earlier cameras are now on the used market and can be good performers. Because newer models are now available at attractive prices, the older cameras are often reasonably priced. Lets look at a brief history of the DSLR camera and the models that have proved popular with astrophotographers.

Minolta, now merged with Konica, was one of the first manufacturers to offer a true interchangeable lens DSLR. The $4000 Minolta RD3000 was introduced in 1999 and used all Minolta V-mount lenses. This camera demonstrates how a piece of cutting-edge technology can be rendered obsolete within a year. The RD3000 is now out of production, but it featured a 2.7-megapixel image that was considered large when it was introduced. As with most early digital cameras, the RD3000 suffered from a lack of long exposure capability. Maximum exposure was limited to 30 seconds with a warning from the manufacturer that exposures longer than several seconds would be noisy. To get near a three-megapixel image when three megapixel sensors were not yet in production, a beam-splitting prism was used to divide the image into a left and right component. Each half of the image was directed toward a separate 1.5-megapixel CCD. The camera's

Table 6.2 Evolution of the Canon DSLR Camera

Model	Year	Price ($)	Array Type	Crop Factor	Array Size (mm)	Array Pixels millions
EOS 30D	2000	2,800	CMOS	1.6	22.7x15.1	3
EOS-1D	2001	7,000	CMOS	1.3	27x17.6	4
EOS-60D	2002	2,000	CMOS	1.6	22.7x15.1	6
EOS-1Ds	2002	8,000	CMOS	None	35-mm sized	11
EOS 10D	2003	1,499	CMOS	1.6	22.7x15.1	6.3
EOS 300D	2003	900	CMOS	1.6	22.7x15.1	6.3
EOS 20D	2004	1,499	CMOS	1.6	22.7x15.1	8.2

internal software then seamlessly recombined the two images into a single 1984x1360, 2.7-megabyte image.

6.4.1 Canon

Canon DSLR cameras have garnered a huge following among celestial imagers by producing a series of astronomy-friendly cameras at a price many amateurs have been willing to pay.

The EOS 30D was Canon's first DSLR. It had a 3-megapixel sensor measuring 22.7x15.1 mm in size (1.6x crop factor), and used Canon EOS lenses. Priced at $2800 it was beyond the means of most amateurs.

Next was the EOS-1D. This was Canon's first attempt to produce a "professional grade" DSLR. It also used the popular EOS lenses and featured a 4-megapixel sensor that measured 27x17.6 mm, but had only a 1.3x lens crop factor. Priced at $7000, it remained out of reach for most astrophotographers.

The EOS-1Ds was more than a simple upgrade of the 1D. It had nearly double the number of pixels and used a full 35-mm film sized CMOS sensor producing an 11-megapixel 4064x2704 pixel image. It is reputed to produce extremely high quality results in commercial photography, rivaling the output of 120-format film cameras. However, at $8000, it is too expensive for the consumer market.

The need for a capable, high-end consumer camera led to the $2000 6-megapixel Canon D60 with a CMOS sensor producing 3072x2048 pixel images. The lens crop factor was 1.6x. The D-60 was based on the popular EOS Rebel G (EOS 500N in Europe). This camera is very useful for both short-exposure solar system photography and long-exposure deep-sky applications.

The 6-megapixel CMOS sensor Canon 10D and Digital Rebel 300D were priced at $1499 and $900, respectively, and produced images with significantly less long-exposure noise. These cameras retained the same

Fig. 6.4 *The Canon EOS 10D (left) and the Canon EOS 300D (right) were "breakthrough" 6-megapixel DSLR cameras. They had a sufficiently high level of image quality and affordability that convinced many people to move from film-based SLRs to digital imaging. Photos courtesy of Canon.*

3072x2048 pixel image size with a 1.6X crop factor as the 60D, but sensor noise reduction and their lower price immediately made them popular among a significant number of trend-setting astrophotographers. The 300D broke the sub-$1000 DSLR barrier, making the DSLR a viable film-replacement camera for astrophotography. Indeed, the 300D is less than $900 if purchased without a lens, making this camera the most important model to enter the market since the Nikon D1 began the DSLR era by bringing full-featured functionality to an affordable DSLR.

The 10D is a hefty camera, tipping the scales at 1.9 pounds, including battery. As a 40-year user of Nikon F and F2 film cameras, the author found the bulk of the 10D to have a familiar and comfortable feel. But for owners of small telescopes, it may be too massive for easy balancing, so they may find the lighter 300D, at only 1.4 pounds a better match. Both cameras use the same imaging sensor, accept the same lenses and are capable of similar exposures. The 10D allows mirror lock-up to reduce shutter release vibration—the 300D lacks this feature. Both allow unlimited time exposures on "bulb."

The 20D is externally similar to the 10D. The difference is a higher pixel count. Although the individual pixels are smaller in order to squeeze 8.2 megapixels on the same size CMOS sensor, the camera has similar noise levels as the 10D and 300D. A special version of this camera, the 20Da, is designed to be used for both standard photography and astrophotography. Three major modifications make it a better deep sky camera than the 20D.

1. In addition to the standard shutter settings, the 20Da has two "live focus" modes—FC1 and FC2 where the mirror is locked up and the LCD preview screen displays a real time image. FC1 displays the central 20 per-

cent of the imaging area and FC2 the central 5 percent. This magnified real-time image makes it possible to focus much like looking through an eyepiece but only the brightest objects are visible on the LCD display, so focusing on the Moon or bright planet like Jupiter is much easier than focusing on a star. The high-magnification live focus mode produces results similar to traditional methods using software measurement of a focused star's full-width-half-maximum. When using either of these modes the focus must be achieved within 30 seconds or noise will build up and complicate the focus procedure. The live focus preview can be monitored on an external television monitor which can be useful when the aim of the telescope places the camera's LCD preview screen in an awkward position. After focusing with live preview, Canon recomments waiting an additional 30 seconds before exposing to prevent excess noise in the image. To further prevent excess noise, Canon also recommends a maximum ISO 400 exposure length of 16 minutes at 0C and 8 minutes at 20C. The live preview focus method does not work with some Canon EF autofocus lens models.

2. The standard infrared (IR) filter has been replaced with one that allows 2.5 times more of the astronomically important Hydrogen-alpha wavelength to reach the imaging sensor yet normal terrestrial pictures still produce acceptable color.

3. The 20Da's noise reduction function has been modified to work better on astronomical images but the total exposure time is doubled. During the second half of the expose the shutter is closed while it performs noise reduction. This feature can be disabled and noise reduction by dark frame subtraction can be done with image processing software later.

6.4.2 Nikon

Nikon's first DSLR offering, the $5000 Nikon D1, is another illustration of the rapidly changing market. The 2.6-megapixel image was considered huge when it was released in 1999. It was one of the few digital cameras in its time that allowed long exposures with the "bulb" setting.

The D1-X followed two years later and featured a 6-megapixel CCD sensor and a lower $3600 price. The D1-X sensor was 23.7x15.5 mm for a 1.5x crop factor. "Bulb" exposure allowed long exposure astrophotography, although above ISO 800, the images are very noisy.

The 6-megapixel Nikon D100 (released in 2002) was the first Nikon consumer DSLR priced at less than $2000. Like its professional predecessors, it was based on the Nikon N80 film camera body and featured a 1.5x lens crop factor. Sony supplies the low noise CCD sensor (ICX413AQ) that is capable of excellent long-exposure astrophotography.

The D70 (2004) was Nikon's first sub-$1000 full-featured DSLR. Competitively aimed at the Canon 300D/10D, the 6-megapixel D70 al-

Table 6.3 Evolution of the Nikon DSLR Camera

Model	Year	Price ($)	Array Type	Crop Factor	Array Size (mm)	Array Pixels millions
D1	1999	5,000	CCD	1.5	23.7x16.7	2.62
D1-X	2001	3,600	CCD	1.5	23.7x15.5	6
D1H	2001	4,500	CCD	1.5	23.7x15.5	2.7
D100	2002	1,999	CCD	1.5	23.4x15.6	6.1
D70	2004	999	CCD	1.5	23.7x15.6	6.3

lowed a choice of brands in a field dominated by Canon. Maximum image resolution is 3008x2000 in either JPG or RAW formats. The RAW format can be converted to TIF using the optional Nikon Capture software. The D70 has similar performance characteristics when compared to the Canon 300D/10D. Almost all of the D- or G-mount Nikkor lenses fit this camera and have a 1.5x lens crop factor which is the current Nikon standard. Older Type F Nikkors will also work but they lack features such as autofocus capability and electronic communication with the camera. The D70 also uses the new DX lenses that are specifically designed for use with digital cameras. These lenses are smaller and lighter. One of these, the 18- to 70-mm AFS-DX zoom lens provides reasonable performance; however, one would expect less vignetting for a lens specifically designed for the D70's sensor. At 18 mm and full aperture, this lens has a 38 percent light drop-off at the edge of the field. At 70 mm and full aperture, this improves to 28 percent. Unlike the Canon 300D, the D70 does allow mirror lockup to reduce camera vibration when the shutter is tripped.

The D70 can be configured for remote time-lapse photography by con-

Fig. 6.5 *The sub-$1000 6-megapixel Nikon D70 is capable of outstanding astrophotography and allows the use of existing high-quality Nikkor lenses. Photo courtesy of Nikon.*

Fig. 6.6 *The Olympus E1 is the first camera to adapt the new ⁴⁄₃ sensor standard. The E1 produces 5-megapixel images and is capable of time exposures up to eight minutes. The camera accepts a new line of Zuiko Digital lenses designed specifically to maximize the effectiveness of the camera's digital imaging sensor. Photo courtesy of Olympus.*

necting the camera to a computer with a Nikon-supplied cable and the optional Nikon Capture software. Unlike Canon's Remote Capture software bundled with the 10D, Nikon's software must be purchased separately.

In 2005 Nikon introduced the 6-megapixel D50, priced below $600 and the 10-megapixel D200 at about $1600, both produce images with a 1.5x lens crop factor. The D50 is slightly smaller and lighter than the D70. It uses of Secure Digital memory card, a larger 2-inch LCD preview screen, and USB 2.0 port. The D200 has a 2.5-inch LCD preview screen, USB 2.0 port and standard Compact Flash memory. The D200 is built with a magnesium alloy frame instead of the D50's plastic. Overall, the D200 is closer to a professional-level camera than the enthusiast-level D50.

6.4.3 Olympus

The ⁴⁄₃ format represents the next step in the evolution of the DSLR. The first such model available was the Olympus E-1. This camera is noticeably smaller and lighter than the Canon 20D and Nikon D70. This might be a factor to consider if you are using a small telescope.

The E-1 uses a ⁴⁄₃ Kodak CCD sensor (KAF-5101CE) producing images up to 2560x1920. It is capable of programmed exposures up to one minute and time exposures up to eight minutes on "bulb." It saves images in TIF, RAW and JPG file formats. Remote operation is possible with either wired or wireless operation. It also has a video out port so the camera can be hooked up to a video monitor for rapid check of camera focus and image composition.

While the E-1 has rubber seals making the camera weatherproof, it is *not* waterproof—but the weatherproofing feature makes it more resistant to heavy dewing at night. It has Supersonic Wave Filter to remove dust from the sensor.

Chapter 7
Camera Lenses for Astrophotography

7.1 The Ideal Lens for Astrophotography

The ideal astrophotography lens would be aberration-free, fully color corrected, produce a flat, fully illuminated field, with sharp images, even at its widest aperture. Mass production camera lenses take many factors into account; the ease and cost of design and manufacture, the compromise between image quality and lens price, and a host of market-force issues. For the most part, lens manufacturers do a remarkably good job of producing moderately priced cameras and lenses. However, a lens that performs well with terrestrial photography may not image the stars well. The reason is that terrestrial photography involves imaging extended objects while astronomical photography is concerned with imaging pinpoint light sources. A star is thus one of the most critical tests of optical quality. If a lens is not sharp at the edges in daytime photography, you may never know it. But for an astrophoto to be considered good, the stars must be sharp all the way across the field.

Excellent lenses optimized for astrophotography are very costly. High-end commercial photographic lenses using exotic glass and coatings can easily exceed the cost of an 11-inch SCT. This is a fact of life. As recent as the 1950s, only a few camera manufacturers, such as Leitz, Nikon, and Carl Zeiss, were noted for producing top quality lenses. Today, many independent aftermarket makers like Sigma, Tamron and Vivitar have dramatically increased quality. Some of their lenses are suitable for our purposes. This has come about with the advent of widely available computer-aided design programs that have replaced labor-intensive hand calculations that only relatively few people were capable of.

Today, you can begin your search for a lens with the basic assumption that a group of similar lenses produced by different manufacturers using the same number of lens elements and types of glass should be optically similar. Increasing image quality above today's averages requires more

lens elements and exotic glasses (rare earth and lanthanum) but only results in a moderate increase in image quality. The higher quality lenses
also benefit from greater mechanical precision and quality control during
manufacture. (Yet, even here, computer-controlled machines have significantly increased capabilities while reducing costs). These lenses are tech-
nological marvels and, with few exceptions, can only be justified for
cutting edge scientific work or for commercial imaging where the lens may
be a small part of a much larger budget.

7.2 The Evolution of Camera Lenses

The first useful lenses created in large numbers were spectacle lenses crafted about 1285. It was not until around 1608 that lens making techniques
were applied to create telescopes. For the next 200 years, improvements in
lens design were a trial-and-error process. By the mid-18th century, the
work of people like Chester Moor Hall and John Dollond had led to development of achromatic lenses for telescopes. Up to about 1800, however,
lenses produced only a narrow field-of-view. This was adequate for telescopes and microscopes, but proved inadequate for photography when it
was perfected later in the 19th century. In 1812, W. H. Wollaston discov-
ered that a single meniscus lens with a displaced aperture stop could produce a sharply focused image over a flat surface. This design became
known as the Landscape lens. When the Daguerreotype photographic process was introduced in 1839, the Landscape lens was used to produce a
wide field on the film plane. However, the Landscape lens suffered from
longitudinal chromatic aberration. An achromatic doublet version was
soon perfected to improve the focus of the Daguerreotype camera and
marked the beginning of the development of the modern photographic
lens.

In 1840, Joseph Petzval's mathematical investigation of lens designs
resulted in the $f/3.6$ Petzval Portrait lens, an incredibly fast lens for its day.
Petzval's analysis of field curvature, known as the Petzval sum, is still used
by today's lens designers. In 1856, Ludwig von Seidel mathematically defined the five primary monochromatic optical aberrations that lens designers still strive to minimize. They are:
- spherical aberration,
- coma,
- astigmatism,
- field curvature, and
- distortion.

Today, this list can be expanded to include the two primary chromatic aberrations: longitudinal and lateral chromatic aberration.

The 19th century work of Petzval and Seidel set the course of lens crafting for the next 100 years. It was not until about 1960 that the introduction of computers allowed lens designers to optimize their designs to further reduce aberrations that degrade photographic images.

Most conventional camera lenses trace their origins to the 1817 work of Karl Friedrich Gauss when he perfected an achromatic lens using two air-spaced meniscus elements. This design now bears his name. In 1888, the famed telescope making company Alvan Clark & Sons perfected a wide-angle camera lens by placing two Gauss-type lenses back-to-back with a stop between them. Examples of this design exist today as the classic Metrogon. In 1896, Paul Rudolph modified the basic double-Gauss lens to reduce aberrations and created the $f/4.5$ Planar lens. H. W. Lee further modified this design in 1920, achieving a photographic breakthrough with an $f/2$ lens having a 50-degree field. Today, Lee's double-Gauss lens is the basis for most 50-mm focal length lenses with a speed between $f/1$ and $f/2$.

7.3 A Camera Lens Primer

Lenses used on consumer cameras are characterized by their focal length and focal ratio. The focal ratio is the lens's focal length divided by its aperture, and is usually symbolized by an "f" followed by a forward slash ($/$) and this ratio. Familiar examples are a 50-mm $f/1.4$ lens, or a 35-mm $f/2.8$ lens.

By convention, a camera's "normal" focal length lens, regardless of film or image sensor format, is approximately the same as the diagonal measure of the imaging surface. For the standard 35-mm film format this is approximately 43 mm. The first Leica 35-mm cameras introduced in 1925 followed this convention and produced a 53-degree wide field-of-view. However, a 43-mm focal length when used on the 35-mm film format introduced too many aberrations, so the "normal" focal length was increased to about 50 mm to eliminate them. This slightly longer focal length still closely reproduces a perspective view similar to that seen by the unaided eye. But not all manufacturers adhere rigidly to the 50-mm focal length, and actual focal lengths can vary by a few millimeters. Today, digital cameras do not have a standard sized imaging surface, which results in a wide variation of "normal" digital camera lens focal lengths between models and brands.

Optical designers try to make their lenses perform best at the mid-range apertures where most snapshots are taken. In this compromise, good consumer camera lenses will resolve about 60 lines/mm, really good lenses

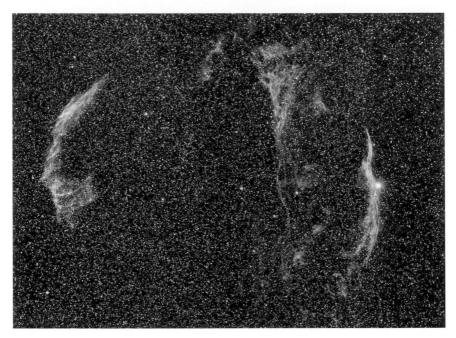

Fig. 7.1 *This four-segment mosaic of the Veil Nebula was imaged through a 4-inch f/5 TMB refractor and a Canon 10D at ISO 1600. Each segment is a combination of 6 five-minute exposures. Photo by Johannes Schedler.*

will resolve around 90 lines/mm, and exceptional lenses will be limited only by the resolving power of fine-grained film. Astrophotographers, however, tend to work with a lens at its widest aperture and are at the mercy of lens aberrations not seen by most photographers.

Manufacturers speak of lens elements and groups of elements when they discuss their products. The number of lens elements refers to the individual pieces of glass of which the optic is comprised. If two or more of the elements are cemented together into a single component, they are called a group. Thus, if a lens is said to have eight elements in seven groups, two of the elements are cemented together. More elements give a designer more latitude to correct aberrations.

There is a point of diminishing returns in adding lens elements. An uncoated glass surface reflects about five percent of the light that strikes it. An eight-element lens with 16 air-to-glass surfaces would thus block 56 percent of the light incident on it, equal to a loss of more than a full *f*-stop of speed. This 56-percent light loss is significant, but less than simple math would indicate (five percent times 16 air-to-glass surfaces = 80 percent). This is because the first air-to-glass surface passes 95 percent of the light, and each succeeding air-to-glass surface passes 95 percent of what it receives from the previous air-to-glass surface. Thus the actual light loss is 0.95 to the 16th power, or about 56 percent. However, the light loss from uncoated optics is not nearly as important as the increase in stray light that causes lens flare and loss of contrast.

Fortunately, since the early 1970s, improvements in lens multicoating technology have reduced reflection losses to about 0.5 percent per air-to-glass surface. Coatings suppress the tendency of a lens surface to act as a partial mirror. However, anti-reflection coatings are not perfect. A coating is just that—a coating. It is not a polished glass surface. Coatings are rough, and while they help prevent reflections, they are rougher than the glass surface and still scatter light to some degree. But coatings nonetheless dramatically improve the performance of modern lenses over their classical counterparts.

Apochromatic elements are becoming popular for today's telephoto lenses just as they are popular for astronomical refractor objectives. The term, often contracted to "APO," means that the lens has such a high degree of correction that it can focus three different colors to the same point. These lenses often contain exotic glass types to achieve such specialized dispersive and refractive performance. Some manufacturers use the term apochromatic and APO rather loosely, so such claims should be examined closely before choosing an expensive APO optic.

The optics term "circle of confusion" refers to the degree of sharpness with which a lens can focus rays of light. When the circle of confusion is small, the images will appear sharp. In most consumer cameras, this is generally regarded as about 0.03 mm. This coincides with the size where small star images will noticeably begin to elongate on unguided exposures of stars. The circle of confusion for an image is formed at the tip of the cone of light converging from the rear of a camera lens. If the lens is not focused or is poorly corrected, the focal plane will intercept the light cone either in front or behind the tip of the cone, and the circle of confusion will be larger, rendering a poor focus.

Another term related to camera lenses is diffraction. This term applies to the fact that when a beam of light, or any other wave phenomenon, passes through an aperture, it has a natural tendency to spread out. The smaller the aperture, the more the light beam will spread. The effects of diffraction are not usually apparent in astrophotography until wide-aperture lenses are stopped down to $f/5.6$ or $f/8$.

There are a number of ways to improve the performance of a lens used in astronomical imaging. The easiest is to close the lens down one or two f-

stops. Experiment with your lens and find how much is needed. Perhaps yours only needs a half-stop closure to reach acceptable performance. After all, the less it is stopped down, the more starlight will reach the imaging plane.

A second way to improve the quality of star images is to narrow the wavelength band of light being photographed. All lenses exhibit chromatic aberration to some extent; that is, they focus each color of light to a slightly different point than others. In modern camera lenses this effect has been greatly reduced with computer-aided design of multi-elements lenses made from exotic glasses, but older lenses can suffer noticeably from this problem. In most astrophotography applications the blue end of the spectrum can be filtered out with a light yellow filter to improve the sharpness of star images. This filter and others are discussed in Sections 11.7 and 11.8.

7.3.1 Several Factors Govern Attainable Star Magnitudes

The faintest stellar magnitude attainable is determined by both the lens' focal length (because the greatest useful exposure time is limited by focal length) and by aperture. However, at less than maximum exposure (such as with most amateur astrophotography) it is the *f*-ratio that matters. Because of the limitations inherent in long-exposure digital astrophotography—primarily noise and dark current in the imaging sensor—digital astrophotography does not employ the lengthy exposures once used with film astrophotography. Thus, we tend to ignore the fact that ultimate exposure length is limited by focal length, but it is mentioned here to allow an understanding of the photographic process.

In practical terms for the amateur astrophotographer, the aperture of the lens will determine the limiting magnitude of stars we can photograph. (Aperture refers to the wide-open diameter of the lens and should not be confused with its *f*-stop.) For example, a wide-open 135-mm *f*/2.8 lens will record more stars with its 48.2-mm aperture than with a 50-mm *f*/2.8 lens with its 17.8-mm aperture, even though both are *f*/2.8 systems. The faintest stellar magnitude recorded is aperture-dependent because, photographically, stars are point sources, not extended sources; and no matter what lens is used, star images cannot be magnified.

Since a difference of one magnitude corresponds to an actual brightness ratio of 2.51, a lens must have 2.5 times the light-gathering area, regardless of *f*-ratio, in order to photograph stars one magnitude fainter for a given exposure length. This is significant because for each additional magnitude reached, the number of stars will approximately triple.

Because the light-collecting area of the lens governs star brightness

Fig. 7.2 *Both the 50-mm f/1.4 and 200-mm f/4 Nikkor set to f/5.6 have a 36-mm aperture and theoretically record stars to the same limiting magnitude. The longer-focus lens will, however, have fewer aberrations, and therefore, sharper focus than the 50-mm lens. Photo by Robert Reeves.*

on the image plane, wide-field star photographs taken through small-aperture wide-angle lenses need longer exposure to reach the same given magnitude than images taken through wider aperture normal and telephoto lenses. On the other hand, two lenses with the same aperture will theoretically record stars to the same limiting magnitude even if they have differing focal lengths. For example, a 200-mm *f*/5.6 lens records the same limiting magnitude as a 50-mm *f*/1.4 lens because both have the same 36-mm aperture. However, another factor comes into play. Longer focal length lenses tend to have fewer aberrations, so the 200-mm lens will produce smaller, better-focused star images than the 50-mm lens over a given area of the sky. The faintest, out-of-focus stars will not be visible in the 50-mm lens exposure, because they will be blurred too much.

7.3.2 Focal Ratio Governs Sky Fog and Dim Nebulae

While star brightness is governed by the lens diameter, another astrophotographic law states that the ability to record both sky fog and dim nebulae are functions of the *f*-ratio and the exposure time.

Natural sky glow is caused by fluorescence of nitrogen and oxygen high in the Earth's atmosphere. If the imaging sensors in digital cameras were noise-free and capable of exposing long enough, natural sky glow would fog the image even if there were no man-made light pollution. Sky fog would eventually limit the faintest magnitude attainable because star images fainter than sky fog would be lost in it. Extended exposure beyond the point where sky fog obscures faint star images simply washes out more stars.

The secret to astrophotography is to match your lens to its target. Wide aperture is needed to record dim stars, while a fast f-ratio is needed to record nebulae. The limiting magnitude of large-aperture high-speed lenses is actually less faint than that of long-focus smaller-aperture lenses because of the rapid buildup of sky fog with the high-speed lens. Fortunately, sky fog can be reduced with the use of certain filters so fast lenses can be utilized to their fullest extent.

Thus, a 3-inch aperture, 30-inch focal length $f/10$ lens would be great for recording stars but be a poor performer on nebulae. On the other hand, a one-inch aperture, two-inch focal length $f/2$ lens would be poor on star fields but could record nebulae, comets, and Milky Way star clouds quite well. These are extended objects, and their brightness on the focal plane is related to the focal ratio of the imaging system.

Let's look at an example illustrating the advantages of longer focal length for recording dimmer stars in astrophotography. Two lenses of equal diameter will record stars at the same rate, regardless of differing focal ratios. A 50-mm $f/2$ lens and a 100-mm $f/4$ lens both have an aperture of 25 mm and will record stars at the same rate. But the 100-mm lens is two f-stops slower and will allow four times the exposure length because the amount of sky fog recorded is related to the f-ratio, not the aperture. Thus, the 100-mm $f/4$ lens will record stars 1.5 magnitudes dimmer than the 50-mm $f/2$ lens because its higher f-ratio suppresses sky fog that obscures dimmer stars, thus allowing longer exposures.

7.4 Lens Considerations for Digital Cameras

The basic types of lenses used in digital cameras are:

- **Fixed focus, non-zoom.** These lenses are usually found on disposable film cameras and inexpensive limited-function digital cameras.
- **Zoom lens with autofocus.** This is the most prevalent lens found on consumer digital cameras. The lens is non-interchangeable, but allows the user to select between wide, normal and telephoto fields-of-view. The camera may or may not allow manual focusing of the lens.
- **Digital zoom lens.** This is not a true optical zoom that magnifies detail, but instead is a software feature that crops out the center of the picture and interpolates the pixels to create a larger image out of the selected pixels. Since this does not actually magnify existing detail, it results in a larger but fuzzy image. This feature is a marketing ploy and is of little actual use in photography.
- **Removable lens.** Digital SLR cameras have the ability to change lenses allowing the use of different focal lengths or the ability to attach the cam-

era body directly to another optical device, such as a telescope or microscope.

In Section 6.3 we covered digital point-and-shoot cameras that have fixed lenses. By design, all of these have lenses that matched the format of their image sensor. However, here we will deal with cameras that have removable lenses—more specifically, those that closely resemble a 35-mm film single lens reflex camera (SLR). Today's film SLRs are a trial-and-error consequence of 70+ years of market forces and represent a proven balance between size, cost and complexity. From the very beginning, digital camera makers sought the flexibility of an SLR, and used adaptations of film camera bodies purchased from Nikon and others. Kodak and Fuji DSLRs are examples. It did not hurt that most of the pioneers in this market were not optical houses and lost nothing by allowing photographers to use their current inventory of lenses; in fact, it made the transition financially easier. As a result there are numerous digital models that now use lenses originally intended for 35-mm film cameras. The fundamental differences between film and electronic imaging sensors have led to the propagation of several myths about the use of "standard" camera lenses on digital imaging bodies. They are:

1. Older film camera lenses do not have enough resolution for digital imaging.

2. Digital imaging sensors in DSLR cameras exaggerate chromatic aberration.

As we shall see in the following discussions, these myths are based on partial truths. But it is a fact that many 35-mm film lenses do work quite well with some DSLR cameras. Indeed, I have achieved good results using 35-year old Nikkors on a Canon 10D with a Novoflex Nikon-to-EOS adapter. On the other hand, three of my lenses (20-mm f/3.5, 35-mm f/2.8 and 45-mm f/2.8 GN) will not fit the digital body because of mechanical interference. Further, even if some older wide-angle lenses do fit, they may give poor results if the light cone exiting the rear of the lens exceeds a critical angle.

Despite the fact that there has been endless discussion about what constitutes a "digital lens," there is no commonly accepted definition. However, from an understanding of how a digital imaging sensor functions, we should be able to dispassionately arrive at the general characteristics:

1. A retrofocus design with a long image "throw" behind the lens.

2. Exit pupil distance far from the focal plane.

3. A reduced image circle size at the focal plane.

Fig. 7.3 *Where is the aperture ring on the 50-mm (left) and 50- to 200-mm zoom (right) Olympus Zuiko Digital lenses? They have none. The camera body electronically controls the aperture function in digital lenses. Photo courtesy of Olympus.*

4. Reduced chromatic aberration.

5. Increased resolution.

6. A focal length selection that produces image scales similar to 35-mm film photography.

7. The ability to use existing common-size filters.

8. Sufficient internal baffling to reduce flare and ghosting and increase contrast, especially on telephoto lenses.

7.5 Digital-Friendly Lenses

To match a lens' capability to the camera's needs, designers are creating a new generation of digital-friendly lenses with a reduced image circle size. Lenses that form an image much larger than needed are both costly and larger than necessary, so sensors that need a smaller image circle can also use physically smaller lenses. The beauty of a camera lens is that it can be scaled up or down in size with relative ease to cover different imaging formats. The disadvantage is that as the lens is scaled up from a 45-mm focal length for use with 35-mm film to a 420-mm focal length for use with an 8 x 10 view camera, the lens weight also balloons by a factor of eight. Unfortunately, aberrations are also scaled up. However, for smaller imaging formats, such as digital sensors less than half the size of 35-mm film, the lens becomes smaller and lighter.

However, with the advantage of a smaller lens for use with smaller imaging sensors there comes a penalty: the exit pupil gets closer to the focal plane as the lens gets smaller. This problem is correctable by adding lens elements that give the design more back-focus.

Fig. 7.4 *If you already own a number of Nikon film camera lenses but desire to use a Canon DSLR body, Novoflex makes the NIK EOS adapter that will allow the use of Nikkors on an EOS body. Photo by Robert Reeves.*

Lens designers have already addressed the exit pupil issue with the new popular consumer lenses. Nikon now has a 17- to 35-mm $f/2.8$ zoom lens that moves the exit pupil to 98 mm from the focal plane at the 17-mm setting to 78 mm at the 35-mm zoom setting. Sigma has also released lenses with similar exit pupil distances of about 80 mm, with a resulting negligible eight-degree maximum divergence on 1.6 crop factor sensors, and only 6 degrees with 2.0 crop factor sensors. This means DSLR cameras using these lenses will be essentially free of vignetting.

Camera and lens companies advertise digital-friendly lenses with extreme exit pupil distance as being "telecentric" lenses. This term is used loosely as a true telecentric lens has an exit pupil that is mathematically located at infinity. True telecentric lenses are costly; however, light does arrive perpendicularly at the sensor. Near-telecentric lenses with exit pupils in the 80-mm to 90-mm range are more the norm. An 80-mm exit pupil distance will work well for most digital sensors with a crop factor of 1.6 or 1.7, and cameras with a crop factor of 2.0 are essentially vignette-free.

Fortunately, it is possible with clever lens design to have the exit pupil seemingly be placed beyond the physical front of the camera lens. This "retrofocus" concept has been used for the past 50 years. SLR cameras have a reflex mirror that must swing up out of the way so that the incoming light can reach the film. Retrofocus lenses provided the space for the mirror movement. This distance is often paradoxically longer than the focal

Fig. 7.5 *The use of digital sensors that are smaller than a 35-mm film frame increases the apparent focal length of a lens, with a resulting loss of wide-angle capability when used with traditional "wide-angle" camera lenses. A 1.6 image sensor crop factor results in a 28-mm wide-angle lens performing like a 44-mm lens, producing a "normal" lens field-of-view. Photo by Robert Reeves.*

length of the lens itself. With Nikon film cameras, for instance, 38 mm of back-focus, often called back focal length or back clearance, is provided to allow mirror movement. This clearance is greater than the focal length of wide-angle lenses, but has the effect of forcing the light rays exiting the lens to be more nearly parallel. Still, even with this back-focus, the light will strike the extreme edges of the 35-mm frame at a maximum of 19 degrees. This exceeds the 12-degree limit of the popular Kodak sensors. However, other factors work in favor of the photographer.

For decades, designers have crafted lenses specifically to cover the 43-mm diagonal measurement of a 35-mm film frame. The lens thus had to produce an image circle on the focal plane at least, and preferably slightly larger than, the diagonal measure of the film surface. However, digital sensors in most consumer cameras are smaller than a 35-mm film frame, often having only one quarter its surface area. This reduced size results pri-

Fig. 7.6 *The majority of lenses designed for DSLR cameras are zoom lenses. These "digital-friendly" lenses are longer in physical length than their short focal length would normally indicate because of the digital sensor's need for a lens with a high exit pupil. A disadvantage for astrophotographers is the fact that most digital zooms feature a small aperture. Photos courtesy of Nikon and Pentax.*

marily because making full-size sensors would be costly and increase the camera's price beyond the reach of the consumer market. Because the sensor is much smaller than classical film size, a digital parameter known as "crop factor" arose to provide a ready comparison to 35-mm photography. Most DSLR cameras currently have a crop factor around 1.6. Thus a given focal length will act as if it is 1.6 times longer than if used with 35-mm film. This means that the divergent light angle from a 50-mm exit pupil that was an excessive 19 degrees at the edge of a full 35-mm film frame is now an acceptable 12 degrees at the edge of a sensor having a crop factor of 1.6. Further helping the situation is that the "four-thirds" standard discussed in Section 3.11 will have a crop factor of 2.0, reducing the incident angle to a maximum of about 10 degrees.

Some users might be concerned that the shorter focal lengths used with smaller digital sensors to maintain "normal" focal length will create distorted image perspective like that found with very wide-angle lenses used with 35-mm film photography. This perspective distortion is evident when foreground objects appear larger than they really are and background objects seem father away than they really are. This will not be the case with digital photography as long as the shorter focal length "normal" lens is about the same focal length as the diagonal measure of the image sensor. But, even if this perspective distortion were present it would be of little consequence in astrophotography because all celestial targets are at an apparently infinite distance.

To achieve high image quality, lenses destined for DSLRs with high crop factors need higher resolution than their film camera counterparts. Fortunately, some lenses with superior resolution already exist: the Canon 135-mm $f/2.0$ L, Nikon's 45-mm $f/2.8$ or 85-mm $f/1.4$. However, future

Fig. 7.7 *Fixed-focal length "digital-friendly" lenses also retain the trait of being physically longer than their same focal length film camera lenses. This Pentax 14-mm lens is as long as some popular 85-mm lenses used on film cameras. The physical length is dictated in part by the need for greater exit pupil to focal plane distance in order for light rays exiting the lens to be as parallel as possible to properly illuminate the digital imaging sensor. Photo courtesy of Pentax.*

digital lenses will have to achieve similar performance because of the smaller imaging sensor. Manufacturers are already developing new generations of digital lenses: Canon has the EF-S series, Olympus has their E-Series, Nikon has the DX series, Sigma offers the DG and DC series, and Tamron is producing their Di series of lenses dedicated to the digital imaging market.

While some 35-mm film camera lenses fit the "prescription" for digital lenses, they are of focal lengths that result in non-standard fields-of-view. For example, the general-purpose zoom lens with a range of 28 to 80 mm is considered standard for event coverage. It provides a wide to short telephoto range in one convenient lens. With 1.6 to 2.0 crop factor image sensors, however, these focal lengths aren't convenient anymore. The wide angle disappears altogether while the short telephoto becomes a true telephoto. This is why newer digital camera zoom lenses have odd-sounding focal lengths like the Canon 18- to 55-mm, the Nikon 16- to 55-mm and the Olympus 14- to 56-mm. After considering the crop factor, these are the new "standard" wide-angle to short telephoto lenses in the DSLR era.

The bottom line is that with many "film" camera lenses performing well with reduced size "crop factor" sensors, do we really need digital lenses for DSLRs? The answer depends on the needs of the user. Existing lenses may perform well, but at higher effective focal lengths due to the crop factor. This leaves a void in the wide-angle area to be filled by digital designs. Additionally, the exaggerated effects of chromatic aberration with wide-angle lenses on digital applications call for attention to that design detail.

An interchangeable camera lens is rarely thought of as dedicated to a

Fig. 7.8 *NGC 7635, the Bubble Nebula near the star cluster M52 in Cassiopeia, was imaged using 20 five-minute exposures through a Canon 10D at ISO 800 and a 4-inch TMB refractor at f/5. Photo by Johannes Schedler.*

single camera body. It must be assumed that any new lens bought today will eventually be used on future, more advanced camera bodies featuring a higher-resolution imaging sensor. A lens used on a sensor with a 1.6 crop factor is going to have to be 1.6 times as sharp as a lens used with film in order to retain the same image detail. Some would argue that six-megapixel sensors do not achieve the same image detail as do fine-grained films. However, there are two factors that make that a moot point. First, cameras in the near future will have higher pixel counts and will achieve true "film-equivalent resolution." Second, prints made from digital images are enlarged more than their film counterparts. In the era of the digital darkroom, it is common to print 5- by 7-inch images when before, we accepted 3.5- by 5-inch machine prints from the photo lab, and 13- by 19-inch digital prints are common. Between crop factor and larger prints, new DSLR lenses need to have near diffraction-limited optics.

7.6 Zoom Lenses the Standard for DSLR Cameras

The DSLR camera is aimed at the advanced and professional user and is designed to allow maximum versatility in a wide range of shooting situations. For the consumer, this means zoom lenses are now the normal lens offered with a camera body. While this may mean added convenience for consumer use, it limits us to smaller apertures and higher *f*/ratios. The astrophotographer has traditionally avoided zoom lenses because early models were inferior to fixed focal length lenses. However, this is usually not the case anymore. Zooms can deliver adequate sharpness but the astrophotographer has to accept that their design reduces maximum aperture. Until fairly recently "normal" lenses sported *f*/1.4 or *f*/1.8 optics. These wide aperture lenses were needed, in part, to provide a bright image for manual focusing. The majority of new digital lenses designed for DSLR cameras feature motorized autofocusing capability and thus for a zoom lens, *f*/2.8 is a "large" aperture. Zooms are also inherently slower to control image aberrations. A typical zoom can have 12, 15, or even more, elements and lens designers must limit the speed of the lens to keep aberrations at a manageable level. Gone are the days of the inexpensive, bright, fast *f*/1.4 optic. For the best results in astrophotography, try to stay with fixed focal length lenses, as they will usually produce better star images.

7.7 Dealing with Common Lens Defects

There are many reasons why astrophotographic images suffer degradation, but they can be summed up by six root causes. The first is errors in the design process. Residual design errors can be reduced, but they can never be completely eliminated, even in the best lenses. The second reason is errors in the mechanical fabrication and optical alignment of the lens elements. Third, diffraction is a function of the wave nature of light and cannot be eliminated. It therefore creates a limit to how sharp an image can be, even with a perfect lens. Atmospheric turbulence, or "seeing," also blurs the image. Image motion, caused by movement of either the camera or the target, is the fifth cause of image smearing. Finally, stray light scattered or reflected within the lens assembly will reduce image contrast. Of these six causes of image degradation, four are completely out of the photographer's control for a particular lens. Only image motion can be completely controlled. If atmospheric turbulence is too severe, our only option is to try again on another night.

There are a number of optical defects and aberrations inherent in most lens designs that conspire to degrade the quality of celestial images—and especially pinpoint subjects like stars. The nature of targets in celestial photography quickly exposes the following defects and aberrations:

- **Coma** makes the star images near the edge of the field look like little comets or parachutes, and is visually the most disturbing lens defect. It is caused by variation in the magnification with increased pupil distance

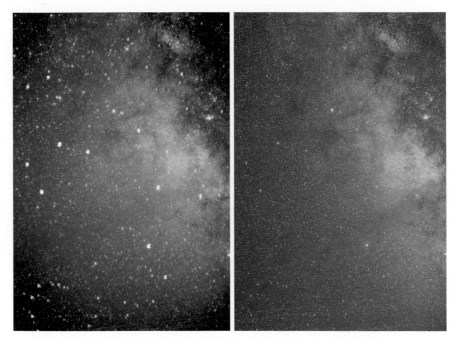

Fig. 7.9 *The improvement in star image quality gained by stopping the lens down two f/stops is apparent in these two images of Sagittarius taken with a 50-mm lens. On the left, taken at f/1.4, the stars are mushy and bloated. On the right, taken at f/2.8, the stars are sharp and compact. Photos by Robert Reeves.*

from the focal plane. Stopping down the lens can control it.

- While most lenses have been corrected for **primary longitudinal chromatic aberration**, most will display some **secondary longitudinal chromatic aberration**, meaning that they do not focus all colors onto the same focal plane. This defect is caused by varying focal length for a given lens at different wavelengths of light and makes star images on the optical axis seem swollen. The only optical systems completely free of chromatic aberration are all-mirror systems. In lens-based optical designs, these lens aberrations can be essentially eliminated by filtering out wavelengths that do not focus at the same plane as the desired colors of the astronomical target.

- **Primary lateral chromatic aberration** is caused by varying magnification at different wavelengths, and appears in images off the optical axis. Most lenses are corrected for this aberration but may display some secondary lateral chromatic aberration. Also called lateral color, this happens because the blue wavelengths are refracted more than the red. Thus, depending upon the construction of the lens elements, blue light will focus either closer or further away from the optical axis than the red wavelengths. This causes stars to progressively smear radially from the center of the field and appear as small spectra. Lateral color can be controlled by using different glass types in multiple element lens designs. In astrophotography, filtering out undesired wavelengths can further reduce it. Most fixed-lens consumer digital cameras display chromatic aberration, often called "purple fringing" by digital users. These bluish fringes show up around bright objects like stars.

- **Field curvature** is often mistaken for coma because it is usually mixed with astigmatism. It appears when a poor lens focuses light not onto a flat plane, but onto a shallow spherical surface instead.

- **Spherical aberration** appears when light rays passing through the edge of the lens are not focused at the same distance from the lens as the light rays passing through the center of the lens. It can be reduced by stopping down the lens. A variation of this defect called spherochromatism is displayed in an optic that suffers divergent degrees of spherical aberration for differing wavelengths of incoming light.

- **Astigmatism** occurs when a lens has differing optical power in its tangential and sagittal planes, and thus forms an image on both the tangential and sagittal focal surfaces. The plane through the target and the lens axis is the tangential plane, while the plane at right angles to the tangential plane is called the sagittal plane. Point sources focused on the tangential surface appear blurred into lines tangent to circles centered on the optical axis. Point sources focused on the sagittal plane appear as lines pointing radially from the center of the optical axis. Focusing can reduce the blur in either the tangential or radial direction, but not both simultaneously, so star images will appear elliptical instead of round.

- **Distortion** is caused when lens magnification varies with increased distance from the optical axis. This effect causes straight lines to either curve out from or to pinch in toward the center of the field and is often called barrel or pincushion distortion, respectively. Barrel distortion causes images to be spherized at their center. This is most pronounced with extreme wide-angle lenses and will show up to some extent at maximum-wide setting on zoom lenses. Most consumer digital cameras will display about one percent barrel distortion. Some cameras like the Nikon Coolpix 950/990 suffer from a large amount of barrel distortion. Pincushion distortion occurs at the telephoto end of the zoom lens scale, and in most consumer digital cameras, vertical and horizontal straight lines will pinch inward about 0.6 percent. Panorama Tools makes a freeware Photoshop plug-in that will digitally rectify the effects of barrel and pincushion distortion.

- **Vignetting**, or image dimming near the edge of the field (which is not an aberration), appears when the image is larger than the area the lens can evenly illuminate. This problem is difficult to avoid with wide-angle lenses. It can occur as the result of mechanical beam-clipping where some of the light rays entering off-axis encounter a foreshortened aperture and are cut off. There is also cosine-fourth vignetting (fall off) where three things happen to darken the edges of the image field. First, the entrance pupil is foreshortened off-axis and becomes elliptical when viewed at an angle. Second, the distance from the entrance pupil to the focal plane is greater at the edge of the field so light intensity is less. Third, light rays from the edge of the lens are focused onto the focal plane at a shallow angle and, therefore, are spread out and become less intense. Light rays impinging at 30 degrees from the optical axis are 44 percent dimmer because of this effect. It can be compensated for in lens design, but at the expense of a more complex and costly optic. Though photographers complain about vignetting being a defect, it is used by lens designers to improve overall sharpness of the image because clipping the off-axis rays is a creative way to eliminate aberrated rays that would otherwise blur the image.

Basically all lenses have aberrations. Stopping the lens down can reduce the effects of all aberrations, except distortion and lateral color.

7.8 Other Optical Limitations

A ghost center spot can occur if the target has a dark center and a bright object near the edge of the field. This problem is common to both film and digital imaging. It is caused when light rays reflect off the imaging surface back onto the rear lens element and back onto the imaging surface. The ghost image often shows the shape and pattern of the variable lens stop.

While a digital sensor is not as reflective as photographic film, it is more sensitive to ghost images. This is because a digital sensor reflection is always specular (the reflections are not scattered in all directions); they are light rays that bounce directly back in the direction from which they came. Some sensors are more prone to specular reflection ghosts than others.

Environmental factors can also affect image quality. Two major atmospheric phenomena that affect astrophotography are "seeing" and "transparency." Seeing simply refers to the relative steadiness of the air. Atmospheric eddies and currents driven by convection cause air turbulence, which creates an effect that can be equated to "heat waves," causing stars to twinkle to the eye and ripple and dance when seen through a telescope. In wide-field astrophotography, the resolution of an image is determined primarily by the lens focal length. The low magnifications used generally mean that poor seeing will reduce the overall quality of a shot by just a slight amount. But with high-magnification telescopic astrophotography, bad seeing will noticeably degrade the image.

Atmospheric transparency is affected by airborne dust particles, by water vapor, and even by the molecules of air itself. Dust and water vapor in the air reflect outdoor lighting, causing increased sky glow that degrades an astrophoto.

7.9 Testing a Lens for Astrophotography

Camera magazines have long rated the resolution performance of lenses by the number of lines per millimeter the optics can resolve at the focal plane. This figure is derived by analyzing images taken of a standard test target such as the U. S. Air Force 1951 Three-Bar Resolution Test Target. This test target has clusters of progressively smaller white and black lines, which, when viewed at the proper distance, will provide between 0.25 line per millimeter and 225 lines per millimeter. While a lens may be said to resolve a certain number of lines per millimeter using this target, the true terminology is that the lens is resolving a certain number of line pairs per millimeter because each line consists of a bright line and its adjacent dark line. Clusters of test lines are arranged at right angles to each other to test both tangential and sagittal resolution, which will reveal astigmatism and other lens defects. Such targets can be purchased from stores that cater to the professional photographer, or Sine Patterns LLC (see Appendix A).

Another useful test target is the PIMA/ISO12233 test chart that has targets similar to the Air Force test chart, but many are at angles instead of simply vertical and horizontal. This is useful in comparing one lens to another.

There is more to evaluating a lens than just testing its resolving power. A lens must be in mechanically good shape as well. Most brand name models will be adequate for astrophotography, but it is possible to get one that has become defective from wear or abuse. A new or used lens destined for astrophotography should be purchased with the condition that if it is not up to par, it can be returned.

When evaluating used optics, make sure the focusing barrel rotates smoothly. Progressively set the aperture to smaller *f*-stops and cycle the aperture blades to be sure they close to the same pattern at each *f*-stop setting.

Examine the lens mount and try it on the camera body you intend to use. All DSLR camera bodies, except those using the ⁴⁄₃ format, employ one of two types of bayonet lens mount—either the Canon EOS style, or the Nikon style lens mount. The older Pentax-style screw mounts or Canon breech-lock mounts were more secure in that they automatically compensated for wear and always held the lens firmly. Bayonet mounts cannot compensate for wear and can develop play in the mount that allows the lens

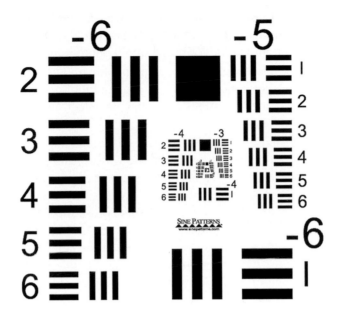

USAF - XL (40" x 40" on photographic paper)

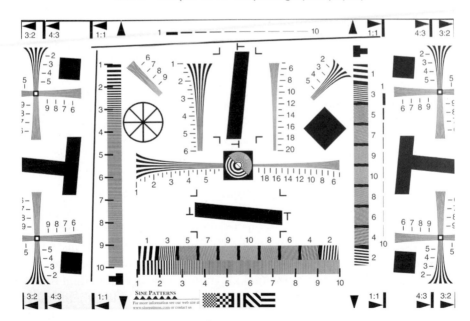

Fig. 7.10 *Lens resolution test charts such as the 1951 USAF target (above) or the PIMA/ ISO12233 target (bottom), both available from Sine Patterns Applied Image, will reveal if a given film camera lens will work well with a digital camera. Charts courtesy of Sine Patterns Applied Image.*

to remain slightly loose. In spite of this drawback, the bayonet mount is now universally used because lenses can be quickly and conveniently changed, an important feature for professional photographers.

Examine the interior of the lens elements for fungus growth, or fogging caused by evaporating lubricants or separation of cemented lens elements. A single scratch on a lens often has no affect on the image. If it does, fill it in with black ink.

To sky-test a lens, make three separate 15-second unguided exposures of the area around Polaris. Shoot first at the widest *f*-ratio, and then close the lens down one, then two *f*-stops on subsequent exposures. Examine the images at high magnification in an image processing program. If the lens is perfect it will show the stars as pinpoints across the field of all three shots. However, it is rare to find a really good lens at less than a premium price. Accept no lens that does not focus well at the center of the field. If the field edges at the widest aperture are on the soft side, or the stars flare badly near the edge of the field, you have to decide if the lens's performance is adequate when closed down one or two *f*-stops. If further stopping down is required to improve performance, reject the lens. If possible, arrange the purchase of the lens to be contingent upon passing this star test.

If you are "blessed" with the usual extended period of cloudiness that follows the purchase of anything destined for astronomical use, another method of lens testing involves shooting city lights that are at least a kilometer away. Again, close the lens down one additional *f*-stop with each succeeding exposure. Using city lights that dimly illuminate buildings will also allow examination of how the improved contrast at smaller apertures effects extended objects.

7.9.1 Variations in Focus

Never assume a lens's infinity mark is true infinity. Many aftermarket lenses work very well for astrophotography, but focus past infinity, as do all of my Nikon telephotos. Moreover, some new telephoto lenses, particularly models destined for autofocus cameras, do not have an infinity stop. More plastic and aluminum are being used in today's cameras instead of brass and steel, and they permit more room for play in focus mechanisms. Further, manufacturers are now making more allowances for slight differences among camera bodies. Indeed, the longer the focal length, the more variation manufacturers are allowing for focus shifts caused by temperature changes and the expanding and contracting of metal and lens components. This is not a defect, but a design feature to allow the lens to work under all climatic conditions. For example, the Olympus 350-mm *f*/2.8 sports telephoto is so sensitive to temperature variations that some photographers make temperature versus focus position charts for each separate filter they use with that lens.

It is simply not possible to look through the viewfinder and tell if a lens is well-focused at infinity. Most do not focus well at full aperture, so it is fortunate that few pictures in the daylight terrestrial world are taken with the lens wide open. A maximum aperture of $f/1.4$ looks impressive, but it is of little use in daylight beyond making the viewfinder brighter during focusing.

Even if you can clearly see star images on the view screen through a focusing magnifier, it is still best to determine the lens's true infinity focus photographically using a Hartmann mask as shown in Figure 7.11. This is because the dark-adapted eye best focuses at about 520 nanometers, in the green region of the spectrum, while bright emission nebulae shine by hydrogen-alpha emission at 656 nanometers, in the red. Most lenses have slightly different focus for the two wavelengths, so a lens focused by the eye will not be focused for the red wavelengths.

Inaccuracy in the placement of the camera's focus screen in the viewfinder further complicates visually determining the precise point of infinity focus. The wider the aperture of the lens, the more critical this placement is. If the camera is attached to an 8-inch $f/10$ telescope, the focusing tolerance is plus or minus 0.25 mm, for a total of 0.50 mm. With an $f/1.8$ "normal" camera lens, the total tolerance is reduced to just 0.1 mm. It is even less for an $f/1.4$ lens. For a modest fee compared to the camera's purchase price, a reputable repair facility can insure that the focus screen in a camera is placed optimally.

7.10 Finding Precise Focus

Even if a particular lens does focus exactly on the infinity mark, that point will change when a filter is used. Each filter will slightly shift the focus of the lens because most lenses focus different wavelengths to slightly different points. The result of this is that all lenses in the telephoto range should be individually focus-tested on stars both with and without any filter. Fortunately, this is not a complicated process.

The easiest way to test the focus of a lens is to make a series of short, fixed-tripod exposures with different focus settings through what is known as a Hartmann mask (also called a Scheiner disk because it was invented by Christoph Scheiner in 1619). Basically, a Hartmann mask is an opaque covering with two small holes cut in it placed over the lens. Each opening should be about ⅓ the lens's full aperture and separated by a distance equal to their diameter. Be sure the two apertures are completely in front of the

Fig. 7.11 *A dual-aperture Hartmann mask can be used to verify the exact infinity focus of a lens (upper left). An auxiliary focus scale is attached to the lens to monitor the lens focus setting (upper right). A series of short, unguided star exposures is taken at different focus settings. Out-of-focus images will show stars as dual streaks, while those taken at the proper infinity focus setting will record as a single streak (bottom). Photos by Robert Reeves.*

lens element itself, and not partially blocked by the metal rim in front of the element.

To test a lens's focus with the mask, the exact position of the focusing ring must be known. This is best accomplished by taping a finely graduated paper scale on the lens barrel next to the focus mark because the normal rangefinder marks on the lens are too crude for this purpose. I made a scale for my lens using the DOS computer program CIRCLE.BAS featured on page 304 of the March 1988 issue of *Sky and Telescope* magazine. This program will print out a linear 0-to-360 degree scale of any length for use as altazimuth telescope mount setting circles. In my case, I set the program to print the most compact scale possible with my printer, a series of graduated lines each about two millimeters apart.

After attaching the auxiliary focus scales to my own lenses, I subsequently found some of the more modern computers would not run old DOS programs like CIRCLE.BAS. My solution was to use the text tool within Photoshop to create a graphic with a row of vertical marks, and then resize the graphic to small proportions so my printer created a graduated scale with the marks about two millimeters apart. To create such a graphic, select the Ariel Black font and type a series of vertical bars found as the upper case selection key just below the backspace key (just above the Enter or Return key).

To perform the focus test, set the camera on its highest resolution and the ISO at 200. Aim the camera at a first magnitude star near the celestial equator and align the Hartmann mask so the cutouts are perpendicular to the direction of the target star's drift. The test will not work if the holes are aligned parallel with the star's drift. Open the lens to its widest aperture. Turn the focus away from the infinity point, and then advance it back to several scale gradations short of the infinity mark. Record the location on the scale and expose the target star for 45 seconds. Do not worry about sky brightness overexposing the image. The small apertures of the Hartmann mask allow longer exposure even if the Moon is bright. After the exposure, advance the focus ring half a scale mark closer to infinity. Expose again for 45 seconds, then advance the focus another half mark toward infinity. Continue the process until the infinity stop is reached, or if applicable, the lens is past infinity.

Using an image processing program, open all the images in separate windows. Zoom in on the target star in each image and compare the sequence of images. They will show slightly trailed stars, but where the lens was out of focus, the star will appear as a parallel double trail. As the lens focus approaches true infinity, the parallel trails will get closer together. At the point where the lens was truly focused on infinity the star trail will be

Fig. 7.12 *Wide-converters (left), that thread into the front of a camera lens, or teleconverters (right) that fit between the camera body and lens, are popular ways to gain either a wider field-of-view or greater focal length with a given lens. Neither are recommended for astrophotography because they introduce additional aberrations and the image quality suffers. Photos courtesy of Canon and Olympus.*

a single line as shown in Figure 7.11. Refer to your notes to see which scale marking the lens was set on for that image. That mark is this lens's true infinity mark.

Repeat the Hartmann mask test for each individual lens as well as each filter you plan to use with that lens. With my telephoto Nikkors, I was surprised to find the true infinity mark was 1 mm short of the mechanical infinity stop in each lens. One millimeter may seem like splitting hairs, but at wide-open aperture, it is the difference between mediocre star images and great star images.

After the infinity focus of a lens, or lens-and-filter combination, has been determined, always turn the focus ring in the same direction to achieve focus so the play in the focus mechanism always acts in the same direction.

7.11 Focal Reducers and Teleconverters for Digital Lenses

Focal reducers, when placed at the focus of a telescope, effectively decrease the focal length of the instrument and give it a faster effective focal ratio. These devices enable wide-angle views not normally possible in a given telescope. The same optical principle can be applied to camera lenses to allow wider fields-of-view with a given lens. We will photographically refer to these devices as "wide-converters" to avoid confusion with established astronomical telescope terminology. Wide-converters have not been popular in the past because it is simpler and more convenient to just use a dedicated wide-angle lens on a camera. Today, with the crop factor affecting the field-of-view of digital cameras, the wide-converter may see resurgence in popularity as photographers try to use existing lenses and avoid the expense of purchasing new digital-friendly wide-angle lenses.

Although a wide-converter introduces more lens elements into the optical path, light reduction is not a problem. Assuming a generous two percent transmission loss per lens element, a four-element converter will block 15 percent of the light coming from the attached lens. However, when this is compared to the 112 percent gain in light density achieved by the 0.66x converter, the light loss is negligible and there is still a one *f*-stop speed gain. However, this gain in speed comes with a penalty; any lens used with a converter will produce inferior images compared to the lens without a converter. For a converter to work well enough to not degrade the image, all lens aberrations must be corrected and the converter itself must also be aberration-free. For these conditions to be met, the optics by necessity will be beyond consumer affordability.

Teleconverters used with telephoto lenses also introduce aberrations. They may work reasonably well on terrestrial subjects, but stars are notoriously difficult targets for a lens to properly image. While a teleconverter may provide larger image scale, they will degrade the image. Additionally, since a DSLR with less than a full 35-mm frame size sensor introduces a crop factor of 1.6, 1.7, or even 2.0, a telephoto lens is already producing an image using only the central portion of the lens' field-of-view. The crop factor is like already using a teleconverter with the lens because the image field-of-view is only using half the lens' field-of-view as compared to a 35-mm film frame. For the most part, the lenses used with crop-factor sensors are already using all their resolution. So, adding a teleconverter on top of that will only result in empty magnification. The image may be bigger; but there is no additional detail in it.

The bottom line with wide-converters and teleconverters is that to achieve the best astrophotographic results, do not use them. Any lens used with an additional converter will produce inferior results. The user will be much happier with the results gained from using a separate lens with the desired focal length.

7.12 Cleaning Lenses

Eventually all lenses need cleaning to restore maximum performance. However, it is amazing how much "trash" can accumulate on a lens without harming its performance. Cleaning a lens the wrong way may actually damage it and cause more degradation than the dirt. The best way to prevent an optic from becoming dirty is to keep a skylight or UV filter on it at all times. It is far cheaper to replace a skylight filter than a damaged lens!

The first step in cleaning a lens is to remove large dust particles that could cause damage if ground into the lens with a cloth. Blow them off with a commercial canned air blower. Take care not to tip the can, or liquid may be sprayed onto the lens.

Next, soak natural cotton balls (not the artificial-fiber kind) in either a 50/50 mix of Windex and distilled water, or a few drops of pure liquid soap and distilled water. Soaps containing no coloring or perfume are best and can be found at health food stores. Partially squeeze the solution out of the cotton ball so it is not dripping. Lightly swab the lens using only the weight of the cotton ball. Then rinse with distilled water and dry by blotting with a dry cotton ball. If necessary, repeat the process.

Use distilled, not tap, water for lens cleaning because the latter may leave hard-water mineral deposits on the glass. Also, never use alcohol to clean lenses because it can dissolve the glue between cemented lens elements.

Chapter 8
Accessories for
Astrophotography

There is more to digital imaging than just a camera and telescope; and in this chapter we deal with a number of seemingly mundane subjects that will, if you anticipate and deal with them, make your imaging experience more enjoyable and productive.

8.1 Telescope Covers Reduce Heat

If your instrument is going to be set up in the field during the day, cover it with a reflective aluminized Mylar "space blanket" or "survival blanket" to limit the heat build-up.

8.2 A Ground Cover Tarp Saves Frustration

When you set up on grass or dirt, place a heavy plastic or canvas tarp about 8-feet square under the telescope mount. Pull the tarp tight and stake it down; set up the telescope in the middle. A ground cover will reduce dust from foot traffic near the telescope, and will make it much easier to find small items dropped while you go about your business adjusting and operating equipment. Anyone who has suffered the frustration of searching for small set-screws or an Allen wrench dropped in the grass knows what a lifesaver a ground tarp can be.

Don't spend a lot of money on the tarp; it is expendable. Eventually the mount will puncture or tear it, and it will need replacing. Be sure to use short stakes at the corners of the tarp to prevent tripping over them in the dark. Nine-volt red LED "blinkie" lights are perfect for marking the corners, and also help prevent trips.

8.3 Photo Tripods

A solid tripod is essential for fixed-camera night-sky exposures. A simple tabletop or collapsible travel tripod will not reliably support a camera during long exposures in a wind. If you already have a lightweight tripod that is otherwise stable but lacks the mass to remain steady in the wind, you can increase its mass by hanging a suitable weight from its apex, perhaps a camera bag.

Your tripod is a long-term investment that should outlive many cameras. Look for strong legs with no more than three collapsing sections and spike tips for good ground contact. To increase its versatility it should have rubber feet that screw downward and over the spike tips so it can be used indoors without damaging the floor. Make sure that it centers the load over the apex of the leg junction and not on an offset beam or rail. Pay particular attention to the controls. Can they be operated in the dark with gloved hands in the winter? An astronomical tripod must be capable of aiming straight up; many do not. Large, stable video models cannot be aimed at the zenith. My heavy tripod can only move $45°$ in altitude; to get a full travel of $90°$ I built a $45°$ wedge adapter.

8.4 Lens Shades Necessary for Piggyback Photography

Interchangeable camera lenses generally have accessory lens hoods or shades that either attach to the lens barrel filter thread, or are built-in and nested inside or outside the lens barrel from where they can be pulled out into position. Most cameras with fixed lenses also have a provision for a shade. Shades are designed for a specific lens field-of-view. One made for a telephoto lens will vignette a normal lens, even if the shade will screw into the filter thread. Conversely, a wide-angle shade provides little or no protection for a telephoto lens.

Lens shades reduce or delay dewing and prevent stray light from flooding the image. Stray light striking a camera lens from just outside its field-of-view can cause contrast-robbing flare. In astronomical photography, lengthy exposures record light so dim the human eye cannot detect it so even very dim stray light can flood a lens while you are imaging. The proper lens shade will prevent this from happening. (See Figure 8.1)

If you have a large filter mounted on a lens with a step-up ring you cannot use its "regular" lens shade. In this instance, an inexpensive collapsible rubber lens hood should be fitted to the filter.

8.5 Afocal-Projection Couplers

As soon as consumer digital cameras appeared on the market, people sought ways to attach them to all kinds of optical instruments. It did not

 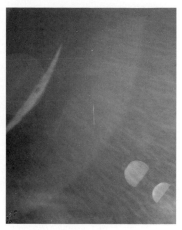

Fig. 8.1 *(Left) If the filter being used on a telephoto that has a built-in lens shade is too large to allow the shade to extend, a simple solution is to make a lens shade out of black construction paper and tape it in place. The shade should extend 1.5 times the clear aperture of the lens. (Right) The need for a lens shade, even at night, is illustrated by this image of an Iridium satellite flare. If a lens shade had been used, a nearby light would not have ruined the image by causing reflections and light-scatter within the camera lens elements. Photos by Robert Reeves.*

take long for many sources of coupling devices to develop. Using the digital camera's LCD preview screen, accurate framing and focusing can be accomplished through binoculars, spotting scopes, microscopes, and telescopes. This has resulted in a large range of options for the astro imager. There probably are two or three ways to adapt virtually any kind of consumer fixed-lens digital camera to any kind of optical device. Sorting out which one to use is a challenge, and an excellent resource to find the adapter that works best for your setup can be found at a web site maintained by Simon Szykman and Gregory Pruden (see Appendix B).

I personally have a Scopetronix coupler, so I'll use various couplers from this manufacturer as generalized examples for discussion purposes. There are many other manufacturers making quality products that have similar features.

Coupling a compact lightweight camera to a telescope eyepiece. Coupling a camera to an eyepiece for afocal imaging is usually done with a two-piece assembly that costs slightly less than $100. One piece fits over the eyepiece and mates to the second, that in turn, attaches to the camera. The eyepiece adapter is the part for which you will find the most variation between manufacturers. The Scopetronix Digi-T attaches only to eyepieces that have a removable rubber eyeguard. When this eyeguard is removed, it exposes a machined groove. Scopetronix calls its eyepiece adapter a Digi-T ring, and it mounts over the eyepiece and is held there by three set-

Fig. 8.2 *The Scopetronix Digi-T system will adapt compact digital cameras to most 1¼-inch eyepieces that have a removable rubber eyeguard. The Digi-T-adapter is attached to the machined rim of the eyepiece with three setscrews (left). A step ring and lens adapter collar is then threaded onto the Digi-T-adapter (center), and the entire eyepiece and adapter assembly then threads into the camera lens or camera body (right). Photos by Robert Reeves.*

screws that tighten into the machined groove. This eyepiece/adapter assembly is then attached to the camera via another coupler called a step ring. See Figure 8.2.

Because this adapter is low-profile, it is suitable for telescopes that have limited back-focus. However, with this and similar adapters there are two issues that cannot be generalized: suitable eyepieces and vignetting. While an adapter may fit the majority of 1.25-inch eyepieces that have a rubber eyeguard, it will not fit them all; you must check for compatibility. And just because the adapter fits an eyepiece, there is no guarantee that it will work well; vignetting can be a problem. Your supplier will usually have a web site and/or provide telephone support to deal with these issues.

Adapters for larger eyepieces. Another convenient way to couple a compact camera to the telescope is with the Digadapt adapter. At first glance, it looks much like a conventional variable eyepiece projection adapter. The difference is that it is larger so that it can hold bigger eyepieces, and it is designed to collapse (not extend) so the camera lens can be placed as close as possible to the eyepiece eye lens. It is "T" threaded so another adapter is needed to couple it to a camera. If the eyepiece can fit through the T-thread, it will couple close to the camera lens. It is not recommended for Newtonians or refractors where back-focus is often an issue, but it does work with Schmidt-Cassegrain telescopes.

Adapters for even larger eyepieces and small camera lenses. The TeleVue brand 2-inch eyepiece adapter is designed for eyepieces that cannot project through a T-thread adapter and allow close spacing between camera lens and eyepiece lens. To avoid vignetting, this adapter should be

Fig. 8.3 *Binoculars or other optics with odd-sized eyepieces can be attached to compact digital cameras using the AdaptaView coupler, and the same step ring and lens collar used with the Digi-T system. The three set-screws on the AdaptaView will fit a wide variety of eyepieces. Photos by Robert Reeves.*

Fig. 8.4 *Any camera capable of accepting a T-mount can be coupled to odd-sized eyepieces on binoculars or spotting scopes using the Scopetronix Uni-T-adapter. A standard T-adapter threads onto the Uni-T while the nylon set-screws clamp the unit onto any eyepiece between ¼- and 1½-inches in outside diameter. Photos by Robert Reeves.*

used only with long eye relief eyepieces and cameras that have small lenses.

Adapters for small pocket cameras that do not have a threaded lens. If your compact camera cannot attach to the eyepiece with the previously mentioned devices, the EZ-Pix adapter clamps around the telescope eyepiece and attaches to the camera's tripod socket hole. The bracket has a wide range of adjustments to properly hold the camera over the eyepiece. This adapter is somewhat inconvenient to use because it has to be readjusted and accurately aimed through the eyepiece each time it is set up. For this reason, it should be considered a last resort option for use only when other adapters will not work. A plus is that it does not require additional back-focus and is suitable for use with all Newtonians and refractors.

When using a small camera with the EZ-Pix adapter, it may be necessary to add a spacer between the camera and adapter if the distance from the center of the lens to the adapter bracket is less than 1.75 inches. With-

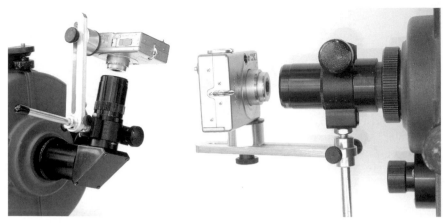

Fig. 8.5 *Ultra-compact cameras that lack filter threads on the lens or body can be attached for afocal photography by using a bracket like the Scopetronix EZ Pix to support the camera at the eyepiece. The EZ Pix can clamp onto the star diagonal (left) or directly to the focuser and eyepiece (right). The EZ Pix comes with clamps to attach to both 1¼- and 2-inch eyepieces and has the spacers needed to position small cameras over the eyepiece. Photos by Robert Reeves.*

out the spacer, some compact pocket cameras with a low-mounted lens may not be able to reach the eyepiece when mounted on the adapter.

Adapters designed to fit any eyepiece with an upper outside diameter between 1- and 2-inches. The Uni-Adapt firmly attaches to a wide variety of optics other than astronomical telescopes, such as spotting scopes, microscopes, and night vision scopes, so long as they have a fixed eyepiece. If the eyepiece can fit through the T-thread, it can couple the camera as closely as the Digi-T. Like the previous adapters, this is also "T" threaded so additional components are needed to couple it to a camera.

Another adapter capable of attaching a camera to any fixed eyepiece is the Uni-T-adapter. This device has six nylon thumbscrews that can be tightened onto optics between 0.25- and 1.5-inches in outside diameter. Like the Uni-Adapt, this one is "T" threaded and needs additional components to couple it to a camera, and it is compatible with the Digi-T system parts. A larger 2.5-inch Uni-T-adapter is also available. Neither version requires extra back focus.

Projection adapters. The MaxView 40 by Scopetronix is a proprietary-design wide-field eyepiece with a built-in adjustable projection adapter. It can be visually used as a 40-mm eyepiece with an adjustable eyeguard for comfortable viewing. The eyeguard can be unthreaded to expose an integral thread that allows a camera adapter to be attached directly to the eyepiece. The adjustable collar can vary the distance between the camera lens and the MaxView 40 flush-mounted eyepiece lens, allowing

Fig. 8.6 *The Scopetronix MaxView 40 is a wide-field 1¼ inch 40-mm eyepiece with a built-in variable projection adapter. Compact digital cameras can be coupled to the eyepiece using the Digi-T-adapter system. The eyepiece's built in projection collar can then be adjusted to provide the best view with the camera being used. Photos by Robert Reeves.*

the user to tweak the most usable image area out of cameras with large lenses. The MaxView often provides 10 to 40 percent more usable field than other eyepieces. The correct adapter components are needed to couple the MaxView to a specific camera; and because it does not require additional back-focus, it can be used with any Newtonian or refractor. The MaxView II is a 2-inch version of this design for use with cameras having large lens openings.

The ScopeTronix MaxView 14 mm to 18 mm wide-angle oculars for Nikon cameras are new eyepieces designed to work best with the early model Nikon Coolpix cameras possessing 28 mm threaded lens barrels. Visually, they offer a wide 66° apparent field-of-view. Removing the eye-guard exposes a 28-mm thread that allows them to screw directly onto the Nikon cameras. The design allows use of the full zoom range to vary the power with little or no vignetting. No modifications to the back-focus are required when using this eyepiece with the Coolpix cameras.

Variable projection adapters for DSLR cameras. The MaxView DSLR adapter has the rigidity to firmly support large cameras such as the Canon 10D and Nikon D70. Sizes are available to fit 1.25- and 2-inch focusers. It has a removable telenegative lens for applications where limited back-focus is an issue. With the telenegative lens removed, it couples camera and telescope for prime-focus imaging. This particular variation is quite large; it can be a projection adapter for the largest Nagler eyepieces, even the massive 31-mm Nagler and 41-mm Panoptic.

Fig. 8.7 *The MaxView II 40-mm 2-inch eyepiece allows compact digital cameras to be coupled to 2-inch focusers and achieve a wider field-of-view than available with the standard 1¼-inch MaxView 40. The rubber eyeguard adapter is unthreaded from the top of the eyepiece and a step ring attached in its place (left). The same lens collar used with the Digi-T system is then attached to couple the camera to the eyepiece (center). Photos by Robert Reeves.*

Fig. 8.8 *The Scopetronix MaxView DSLR Variable Projection Adapter is a dual-purpose unit that can be used as either a prime focus adapter (left) or an eyepiece projection adapter. If used with an eyepiece installed, the adapter body can be slid in or out (right) to achieve the desired projection magnification. Photos by Robert Reeves.*

Fig. 8.9 *For use as an eyepiece projection adapter, the included telenegative lens is removed from the MaxView DSLR adapter (left), and an eyepiece is installed inside (center). The unit is then attached to a camera using a standard T-adapter (right). Photos by Robert Reeves.*

Fig. 8.10 *Left) Electronic cameras need electronic remote shutter releases. The Canon TC80-N3 for the EOS 10D is typical and is also capable of controlling a sequence of preprogrammed time exposures. Photo by Robert Reeves. (Right) Many compact digital cameras are capable of remote shutter operation using a wireless infrared remote switch. The Olympus Camedia series uses the RM-1 wireless remote shown above. Photo courtesy of Olympus.*

8.6 Remote Shutter Releases

A shutter release cable—or more commonly, a cable release—allows shake-free actuation of a camera shutter. A cable release is an absolute necessity for any camera mounted on a telescope. Most digital cameras have an electronic release: either a wireless infrared remote controller or a wired push-button hand switch. Both the wired and infrared remotes will control the camera from 10 to 15 feet away. A major advantage of wireless remote shutter controls is that nothing is hanging from the camera that can snag you in the dark. Some fixed-lens point-and-shoot digital cameras, like my Olympus Camedia 3020, do not have any form of remote shutter capability. Should that be the case with your camera, a special fixture can be fabricated to enable a standard mechanical cable release to press on the camera's shutter button. Details of how I constructed one are in Section 12.7.

DSLR cameras usually have both electric single-shot remotes and timers to hold the shutter open for "bulb" exposures, or time exposures longer than the camera's longest programmed exposure time. Also, interval meters that can be programmed to automatically take a series of timed exposures are available for most DSLR cameras. This feature is particularly useful for meteor patrol work where the camera is often left unattended. The interval meter allows the creation of series star-trail images where a series of short exposures are digitally combined with an image processing program to create stunning results that are impossible with film cameras.

Details about this extended star-trail technique are in Section 14.12. Remote operation is also possible using a computer since almost all digital cameras come with a computer utility that allows it to be controlled via a USB cable.

Canon offers the RS-80N3 remote shutter control switch for the 30D, 60D, 10D and 20D models that features a built-in programmable interval-ometer. With this accessory, a series of time exposures can be programmed to be made without further attention from the photographer. When combined with an autoguided telescope, imaging essentially becomes a "begin-and-forget" operation. An alarm clock may be a needed accessory simply to remind the photographer that the exposure sequence is complete.

Nikon Coolpix 880 and 990 series cameras, which use the MC-EU1 remote for sequential time exposure photography, have the option of using the Harbortronics DigiSnap 2200 that has a starting price of about $125. The DigiSnap allows control of most Nikon Coolpix camera functions and makes automated long exposures easy. The DigiSnap operates on AAA batteries instead of the Nikon's expensive lithium button batteries. The device must be programmed in advance using a computer, but in operation, the computer is not needed.

8.7 Power Supplies in the Field

Most deep-sky imaging has to be done at dark sites that are often far away from convenient sources of commercial power. Today's imager has a lot of capability—electronic cameras, computers, clock drives, displays, autoguiders, anti-dew heaters, and possibly even internal GPS receivers. All of this capability adds up to a significant amount of power consumption that requires batteries and/or a generator. You should not try to run your system off your car battery. It is no fun discovering, after a night of imaging, that you can't start your car!

While Honda and other manufacturers make generators in the 1KW to 4KW range that produce exceptionally clean (and quiet) 120VAC (the larger ones up 240VAC) and 12VDC, the vast majority of astrophotographers will use batteries, so I will concentrate on that power option.

8.8 High Capacity, Lead Acid Battery Types

High capacity, lead acid batteries can be divided into 4 distinct categories

1. **Car Batteries** are designed to provide a brief burst of high current to start an engine; they are not suitable for deep discharge applications. The vehicle's alternator supplies the power to run the vehicle and recharge

Fig. 8.11 *This wide-field image shows the Andromeda Galaxy shining through the foreground stars of our own galaxy. Three five-minute exposures with a Canon 10D at ISO 800 using a 50-mm lens at f/2.8 were combined for this image. Photo by Johannes Schedler.*

the battery. The 'Cold Cranking Amps' rating does not indicate the battery's storage capacity. The thin plates in these batteries are good for a quick release of energy, but deep discharging will cause rapid deterioration.

2. **RV and Marine Batteries** are a compromise between car batteries and deep discharge batteries. The plates are similar to the engine starting batteries and will not withstand repeated deep discharging. No matter the claim, they are not deep discharge batteries.

3. **Motive Power Batteries** are deep-discharge batteries used with electric vehicles such as golf carts and forklifts. They have thick plates that will withstand many deep discharge cycles. These batteries are durable and have good storage capacity.

4. **Stationary Batteries** are commonly identified as utility backup cells for telephone companies and computers. Most are not designed for deep cycling.

8.8.1 Rating a High Capacity, Lead Acid Battery

Battery ratings will help you select the proper battery to meet your needs. An automobile battery's "amp/hour" rating and its advertised "cold-cranking amps" are not the same thing. Cranking amps relates to a battery's ability to deliver large amounts of power in a short time (to start an engine) while the amp-hour rate relates to the battery's overall power capacity and the rate at which it can be delivered (discharged).

As power is drawn from a battery, the output voltage slowly drops. The rated amp/hour capacity for a 12-volt battery is referenced to that point where the voltage drops to 10.5 volts. Bear in mind that some astronomical accessories, such as the ST-4 autoguider, will not function below 11 volts. For this reason, amp-hour ratings should not be used as the sole guide to determine suitability for voltage-sensitive devices. The amp-hour rating does not remain constant as the battery ages. Old batteries, while functional, do not deliver the same amount of total power as new ones of similar rated capacity.

Some battery manufacturers no longer print the amp/hour rating on the battery, so the user will have to test it to insure it has the needed capacity. A relatively easy way to determine the useful capacity of a 12-volt battery is to discharge it under controlled conditions and measure how long it takes to reach the minimum acceptable voltage. To perform this test, fully charge the battery then bridge the terminals with a wire-wound 12-ohm/20watt resistor and time how long it takes the battery to discharge to the minimum acceptable voltage. For a 12-volt battery, the 12-ohm resistor will draw one amp per hour. While it is true that the power draw will slowly change as the voltage drops, a fully charged "12-volt" battery starts at about one volt more than 12 volts and is considered to be discharged at a little less than 11 volts. The average voltage during the test will be around 12 and the calculated amp-hour rating will be close enough for our needs. If your battery takes 30 hours to discharge below the point of minimum useful voltage, you can count on 30 amp-hours of power once it is fully charged. You should conduct this test at approximately the temperature at which you expect to use it. As the temperatures drops so does battery capacity. On the other hand, too much heat destroys batteries.

8.8.2 Charging a High Capacity, Lead Acid Battery

More batteries die from overcharging than from old age. A battery that will not be used for long periods of time should be continuously connected to a "smart" or float charger that is properly matched to the battery type. Do not be misled by so-called trickle chargers; they can overcharge a battery and destroy it. A smart charger slowly recharges the battery when it drops below its 80 percent "state of charge." The "state of charge" for a deep cycle battery can be determined with a digital voltmeter with a 12-volts DC scale divided into 0.1-volt increments. This test is conducted at room tem-

Table 8.1 Deep Cycle Battery State of Charge

Battery Voltage	State of Charge
12.7–13.0	100
12.5–12.6	80
12.3–12.4	60
12.1–12.2	40
11.9–12.0	20

perature, and the battery must not have been charged or discharged for several hours. See Table 8.1

It is important not to recharge lead-acid batteries at a rate greater than one-tenth of the total rated amp-hour capacity. For example, recharge a 30 amp-hour battery with no more than a three amp-hour charge rate. Charge rates above this level will heat the battery and damage it. Most heavy-duty car battery chargers exceed this rate. These devices are designed for a one-time emergency quick charge to get your car started. Repeated overcharging will eventually damage a battery by overheating. For deep cycle batteries, avoid discharges of less than 10 percent or more than 80 percent. A deep cycle battery's service life will be extended if it is connected to a low-voltage disconnect device set at 80 percent depth-of-discharge.

8.8.3 Battery Safety Issues

Maintain a lead-acid battery in an enclosed plastic carrying case. Do not "top off" a battery with electrolyte (except to replace spills) and always recharge batteries within 24 hours of each use. If the plates are not covered with electrolyte, add distilled water. Wear eye protection when adding either electrolyte or water. The plates should not be allowed to run dry; it kills them.

Storage batteries hold a great deal of instantaneous energy. Never allow the positive and negative leads to come into direct contact with each other. Shorted wires can instantly overheat, burn you, and severely damage equipment.

8.9 Gel Cell Batteries

Rechargeable gel cells are popular lightweight 12-volt telescope power sources. Because they are leak-proof and spill-proof, they can be operated in virtually any position except upside-down. Gel cells can also be recharged, without adverse effects, at any time during the discharge cycle; but they must be recharged slowly. A Sears DieHard wheelchair gel cell or a computer backup power supply gel cell will work well as a telescope power source.

A gel cell is a high capacity lead acid battery that is pressurized and sealed using special vents, and thus it should never be opened. It is sensitive to both overcharge and prolonged undercharging, both of which will shorten its life. Gel cells cannot be recharged using a high-amperage car battery charger or even an automobile alternator unless there is provision for regulating the voltage and current. If gel cells are recharged too quickly, they can be damaged by swelling and bursting their case, or possibly exploding.

Gel cells need more care than standard maintenance-free lead acid batteries. It is best to store them fully charged, and always recharge them immediately after use. Do not leave them discharged for prolonged periods or they will fail. Conversely, continuously overcharging a gel cell will dry out the electrolyte and shorten the battery's life. Use only a temperature-sensitive trickle charger with a gel cell option.

Here are some tips for gel cell care and use:

1. A 12-volt cell has six 2-volt cells.
2. The maximum charging voltage for a 12-volt model is 2.4 volts per cell, or a maximum of 14.4 volts.
3. If the charging voltage is 2.4 volts per cell, the charger must be turned off when the battery is fully charged. Conversely, if the charging voltage is 2.15 volts per cell, the charger can be left on indefinitely.
4. The maximum charge rate is the amp-hour rating divided by 5.
5. It takes about 40% of the total charging time to recharge the final 10% of a battery's capacity. The final charging current is about 1/200 the amp-hour capacity of the cell.
6. Gel cells must be recharged every six months if not used.
7. Beware of flea market batteries that have been sitting around for a year or more.
8. There is a direct relationship between the depth of discharge and the life of a gel cell in terms of recharge cycles. This is summarized in Table 8.2.

8.10 Other Battery Notes

A majority of today's fixed-lens cameras use AA batteries or proprietary lithium-ion battery packs. The power provided by brand-name alkaline batteries varies according to battery size. Their capacity is rated in amp-hours, just like gel cells or a car battery. For instance, a D-cell produces one amp-hour for 14 hours, two for seven hours or 14 for one hour. In spite

Table 8.2 Depth of Discharge vs. Gel Cell Cycle Life

Average Discharge	Battery Life in Cycles
100%	200
50%	400
30%	1300
10%	2000

of the television commercials claiming one or the other is the more powerful battery, the fact is that in low-drain applications they are virtually identical in performance. Let price be the guide.

Never mix different brands, different types, or different capacity batteries in the same application.

As good as alkaline batteries are, they should be considered only as emergency replacement batteries; they only power a camera for about an hour before they depleted; they cannot be recharged. Rechargeable nickle cadmiun (NiCad), nickel metal hydride (NiMH) or lithium-ion (LiON) batteries are the best choice for economic reasons, and each has advantages over the other.

8.11 Nicad Batteries

Rechargeable NiCad batteries can be recharged hundreds of times. However, most NiCad only have about one-tenth the power capacity of similar size alkaline cells and they often prematurely run out of power. Their voltage is typically 1.3 volts for a AA cell versus 1.5 volts for a disposable alkaline battery. NiCad batteries are extremely sensitive to how they are charged; if they are recharged prior to being completely discharged, they will achieve only a partial recharge. To deal with this problem some battery chargers have a "recondition" feature that completely discharges the battery prior to recharging. The best NiCad chargers negatively pulse the battery to condition it before recharging.

8.12 Nickel Metal Hydride Batteries

Today, the rechargeable battery of choice is the nickel metal hydride (NiMH). These batteries store about 40 percent more power than a similar size alkaline battery.

The specifications for AA-sized NiMH batteries may cause needless concern because each cell is rated at 1.2 volts instead of the customary 1.5 volts for older battery types. A fully-charged NiMH battery starts off at 1.4

volts, but quickly drops to 1.2 volts where it will stay for 90 percent of its discharge cycle at which point it quickly loses all voltage. In use alkaline batteries are no better; they quickly lose voltage and drop below 1.2 volts for most of their life. I have never seen a device designed for an alkaline battery fail to function properly with NiMH batteries, and they run far longer.

NiMH batteries must be charged and discharged three to five times before they provide full capacity. One charge cycle constitutes fully charging the batteries and allowing them to cool after charging, discharging them through normal use, and then recharging them again. They do not retain a charge "memory" and can be recharged at any point in the discharge cycle. They can be recharged between 500 and 1000 times—it is best to use a microprocessor-controlled charger designed specifically for these cells. If the charger has a cover, leave it open because these cells get hot, as high as 130°F.

It must be understood that the charge shelf life of NiMH batteries is less than alkaline cells. At room temperature a fully charged NiMH battery will lose about 30 percent of its charge in 30 days, and continue to discharge about one percent a day thereafter. This is normal. If stored at higher temperatures (like 100°F), they will lose 70 percent of their charge in 30 days. By the time it has been through 500 charge cycles, it will have lost 20 percent of its original capacity. Remember that battery packs in a camera are composed of multiple batteries that are connected in series; it only takes one bad battery to render the whole cluster inoperative; so do not condemn all the batteries without first individually testing them.

Rayovac produces the I-C^3 (In Cell Charger Control) NiMH battery that is rated at 2000 mAh in the AA size. These batteries are sold by Radio Shack who bundles them with a special charger that can recharge them in just 15 minutes. A four-cell charger, with two batteries included, is offered as Radio Shack item #23-039. Additional AA cells are available as item #23-531 for a two-pack and item #23-532 for a four-pack of I-C^3 batteries.

8.13 Lithium-Ion Batteries

Lithium-ion batteries offer about twice the power contained in a NiMH battery, but they are not yet available as individual AA-size cells. Most camera manufacturers offer a proprietary lithium-ion battery pack for their cameras. Those batteries are more expensive, but for many applications their performance justifies their high cost.

Fig. 8.12 *Inexpensive widely available 12-volt inverters manufactured by companies such as Xantrex can power low-wattage accessories such as 120-volt powered battery eliminators for digital cameras. Photo courtesy of Xantrex.*

8.14 Power Inverters

Devices that need only 12-volts DC can be hooked directly to a battery, but those that run on 115-volts AC require a DC/AC inverter. Hardware stores now stock many brands of inverters that will run off a 12-volt battery and produce high quality 120-volt AC. An important point to remember is that many laptop computers create electrical noise. This noise can affect other items attached to the same power source; and CCD cameras, such as those used in autoguiders, are sensitive to voltage fluctuations and noise. It is a good idea to power a CCD autoguider with its own battery to prevent problems, or power the computer with its own battery as shown in Section 12.2.

If you do not need large quantities of 12- or 115-volt power, you should consider a self-contained battery-powered inverter like the Portawattz Power PAC distributed by Kendrick Astro Instruments. This is an inverter with an 18 amp-hr battery delivering either 12-volts DC at 15 amps or 2 amps (300 watts/500watts surge) at 115VAC. It will allow you to simultaneously draw a combination of both voltages, but at reduced maximum capacity from each. The battery can be damaged by complete discharge but has an audible alarm that warns you when this is about to happen. It delivers a reliable one amp at 12VDC for 12 hours. Its capacity

Fig. 8.13 *Low-wattage 120-volt accessories can be powered at remote locations using a 12-to-120 volt inverter that plugs into a cigarette-lighter socket. Units such as the one here are manufactured by Xantrex and marketed by popular outlets such as Radio Shack. Photo courtesy of Xantrex.*

for 115VAC devices depends on the wattage (volts times amps). Field tests by Dennis di Cicco found that it powered a low-draw 10-watt device for 20 hours, but a higher wattage draw depleted the battery almost twice as fast as the consumption ratio between the two devices would suggest. Should a backup battery be needed, you can bypass the internal battery and hook it to a regular car battery with a cable and heavy-duty clamps. The device comes with a wall transformer outputting a ½-amp trickle charge. An interesting feature is the ability to charge the battery in the field using a solar panel (sold as a separate accessory). The panel can recharge the unit's battery in 10 to 12 hours on a sunny day.

If you choose to build your power supply using batteries and an inverter, be aware that some inverters are better than others; and their output wattage alone should not be used as a buyer's guide. Most 120-volt devices are designed to use sine-wave alternating current—in the United States this is 60 Hertz (Hz). Some inverters may claim that they are outputting 120-volts 60 Hz AC, but in reality they are producing noisy square-wave output. Some electronic devices do not run properly with square-wave AC. Costco sells the Xantrex X-Power 1000-watt inverter for about $70 that produces modified sine-wave current and is fully compatible with electronic devices. Refer to Figure 12.3 to see a self-contained battery box that I built for this inverter and two GFCI-protected outlets.

Another power-to-go option is a 140-watt 12-volt to 120-volt inverter, made by Xantrex, that plugs into an automobile cigarette lighter and is distributed by Radio Shack (#22-148). Be careful when you power equipment from your car's battery; a discharged car battery makes for a long walk home.

8.15 Foreign Country Line Converters

If your imaging adventures take you abroad, you almost certainly will en-

Fig. 8.14 *A common AA-battery Mini Mag flashlight converts to a great astronomical flashlight by inserting red cellophane between the bulb and the lens. Photo by Robert Reeves.*

counter public electrical systems with different voltages and line frequencies than your equipment was designed for. Voltage Valet makes various transformers that can convert alternative power systems for use with domestic devices. Radio Shack also offers several devices to convert various foreign voltages, line frequencies and plug-pin configurations to allow compatibility with U. S. standard devices. For non-electronic devices of up to 1600 watts, Radio Shack offers their power converter item #273-1413. For electronic devices up to 50 watts (such as a laptop computer), Radio Shack offers item #271-1410. Their item #273-1405 is a set of four separate wall outlet adapters, one of which will work virtually anywhere in the world. Note that these wall outlet adapters and voltage converters do not convert 50 Hz to 60 Hz. A separate converter is needed to change the power frequency.

8.16 Flashlights

There is an old astronomical proverb that says an astronomy flashlight should taste good, because it is going to end up clutched between your teeth at some point during the night. There are many LED and bulb-illuminated "astronomy" flashlights. But when it comes down to universal availability of batteries, accessories and replacement bulbs, few flashlights can beat the utility of an ordinary Mini Mag with AA batteries. These flashlights can be converted for astronomical use by inserting two layers of red cellophane behind the front lens.

Mini Mag flashlights have a number of accessories that are useful when working around the telescope. A lanyard that clips to the base of the flashlight allows it to hang around your neck while the Bite-a-Lite accessory slips onto its end to cushion your teeth as they act as your third hand.

To protect night vision (and avoid the wrath of those around you), the light should be relatively dim. The Mini Mag can project a diffuse wide pattern, or zoom into a tight spot when greater illumination is needed. A red flashlight should be truly red. An unofficial test for this is to shine it on the cover of a *Sky and Telescope* magazine in a dark room. If the magazine's white name is indistinguishable from the surrounding red border, the flashlight passes.

8.17 Accessory Carrying Cases

A compartmentalized accessory carrying case will make working in the field easier. To keep all of my telescope accessories organized, and to insure nothing is left at home or in the field by accident, I constructed a wooden four-compartment carrying case for all my telescope equipment. The lid has storage for star charts and books while the body accommodates two lift-out trays over its bottom compartment. Small optical items like eyepieces, Barlows, diagonals and adapters go in the foam-lined top tray. Tools, flashlights and camera mounting gear go in the lower tray, while bulky items like the drive corrector, counter weights, and dew blower go in the bottom compartment.

Camera gear is also easier to keep track of in the dark and protected from damage if it is in foam-lined carrying cases instead of a camera bag, but these cases are expensive. A cheaper option is a foam-lined plastic carrying case that is often found for sale in a sporting goods store. The cases I use for my photo gear were designed for competition pistol shooters. They come with a foam filler that you carve with an Exacto knife to make it snugly fit your equipment.

8.18 Laptop Computers for Astrophotography

Over the past decade, laptop computers have become common observing aids. For the digital imager, a laptop is an absolute requirement since so much relies upon having one: telescope control, sky mapping, focusing, camera control, and image storage. Fortunately, the cost of these devices has steadily declined, and even basic models have the processing power to perform the tasks needed. A laptop by the telescope is now the perfect "helper" to accomplish many tasks. Some examples are:

Fig. 8.15 *(Left) Inexpensive aluminum cases are excellent for storing astrophoto accessories. Units like this one can be purchased at Lowe's Home Improvement Warehouse. Photo by Paul Hyndman. (Right) Plastic accessory cases can be purchased at sporting goods stores where they are sold as pistol cases. The foam liners are cut with an Exacto knife to custom-fit the contents. Photo by Robert Reeves.*

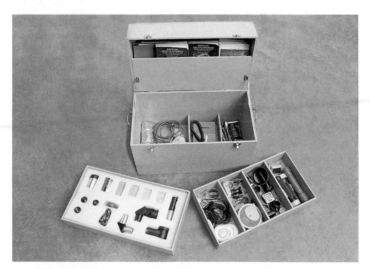

Fig. 8.16 *The author built this four-level custom accessory case from plywood. The foam-lined top tray contains delicate optics, the middle tray contains lightweight accessories while heavy items are stored in the bottom section. The compartment in the lid houses books and charts. Photo by Robert Reeves.*

1. By loading MegaStar Star Atlas on a laptop, the need for large star charts at the telescope is reduced. Large atlases are invaluable planning aids, but under the night sky, MegaStar can control your telescope, show extreme close-ups of star fields, provide framing fields, etc.

2. A laptop simplifies record keeping for your observing runs.

3. A laptop can be used for digital camera control and image acquisition, freeing the user from the limitations of finite memory card size.

4. The ability of having a cell-phone connected modem or observing from

a Wi-Fi hot spot can allow email and instant messaging chat with other observers anywhere in the world via an Internet connection.

Laptop display screens can be objectionably bright at night and wreak havoc on your night vision. Bright computer screens also make the owner unpopular at star parties. Some astronomical programs feature a "night vision" mode where the display is rendered a deep red, but the intensity of the screen brightness may still be objectionable. Some models of Dell laptops have a feature to control screen brightness. If needed, a full-screen filter made from tinted Plexiglas, rubylith foil or deep red cellophane will go a long way toward making a laptop more astronomer-friendly in the night.

If you already have a laptop, you probably do not need another for digital astrophotography. Assuming it has a Pentium processor and a reasonable amount of RAM and disk storage space, it will probably perform well for your digital imaging needs. If it lacks a CD-RW drive to archive your digital images, it may be more cost effective to set up a wireless network with an existing desktop PC that does have this capability instead of buying a whole new laptop or an external CD-RW drive.

Laptops used for astronomy duty need to be protected from dew and nighttime low temperatures. Even southern states, such as in south Texas where I live, occasionally experience freezing cold snaps. Power saving options that turn off hard drives to save battery power should be disabled in order to keep the moving mechanisms inside the laptop warm. Should moisture from the air condense on the hard drive and then freeze, it will fail to function. An insulated carrying case with auxiliary power supply to maintain the laptop on cold nights can be modeled after the one shown in Figure 12.4.

A laptop should be considered part of the imaging system, but not the final destination of images destined for work in an image processing program. For the finishing touches in any image processing, do not use the laptop's display screen alone. The liquid crystal display used with even the best laptops still lacks the basic image colors and dynamic range that a standard desktop cathode ray tube (CRT) computer monitor offers. Initial image processing may be done on a laptop, but the final corrections in color and shades should be done using a CRT display in order to see the full range of the image. If your laptop has the proper graphics driver, it is often possible to plug in an external CRT monitor and use the laptop for the processing while viewing the results on the CRT.

If you are buying a new or used laptop for your digital imaging efforts, consider the following points:

1. Get as much hard drive capacity as you can afford. It is amazing how

quickly images, and various processed versions of images, accumulate on your hard drive. Remember, each download of a common 256-Mb camera memory card is adding another one-quarter of a gigabyte of data to your hard drive, and each 5-Mb RAW file from a camera such as a 10D or D70 will expand to a 36-Mb 16-bit TIF for processing.

2. Get a minimum of 512 Mb of RAM. The added cost of RAM upgrades at the time of purchase is minimal, but provides a significant increase in computer performance; this is especially true for image processing programs.

3. Purchase a laptop with the highest screen resolution and smallest dot pitch that you can afford. The minimum desired screen resolution should be XGA, or 1024 x 768. A low-resolution monitor makes working with large images very difficult.

4. Get a CD-RW or DVD-RW optical drive to archive your images for safe-keeping. Remember, a CD full of digital images is the same as a plastic negative or slide-file page in film photography.

5. Consider the keyboard configuration when choosing a laptop. A standard desktop keyboard has a 19-mm key pitch, and the space bar is centered under the "b" key. This is ¾-inch between the centers of adjacent keys. Some notebooks maintain this standard pitch while other shrink it to downsize the keyboard for ultra-portable machines. Most users can use an 18.5-mm pitch with little problem, but keyboards with 18- or 17-mm pitch are just too small for comfort.

6. Most laptops have at least two USB 2.0 ports, but accessories like mice, camera control and interface cables, accessory keyboards and external drive all compete for USB interfaces. Get at least four USB ports if possible.

7. Consider buying a spare battery to provide an increased flexibility in the field.

Refurbished laptops can be a bargain. Check Compaq, Dell, or Gateway's web sites for current offerings on fully warranted used laptops at significantly lower prices than new items.

Be sure to accessorize your laptop with a mouse and DC to AC power inverter.

If it is to be extensively used in the field, consider purchasing an extended warranty. For the extra $50 to $100 you have peace of mind that the nighttime elements will not permanently kill off your laptop.

Many astrophotographers use a high-end laptop as their only computer. Powerful models may have good graphics cards in them, but compared to standard CRT displays, their flat-panel display screens are limited in image dynamic range. The solution is to plug in an external CRT monitor to allow viewing and processing images on a larger screen possessing a higher dynamic range. It is tempting to substitute a space-saving larger flat-panel display in place of a desk hog CRT, but as discussed below this is not advised.

8.19 Flat-Panel Computer Monitors

Advances in the manufacturing technology of flat-panel computer displays have dramatically lowered the price of these space-saving units. However, flat-panel displays still lag considerably behind their CRT counterparts in their ability to show the widest possible dynamic range of grayscale. The problem is inability to fully show all shadow detail, a factor that is important in displaying astronomical photos that typically have large dimly-lit areas.

Shadow detail falls into the display area known as "black level." The term black level is seemingly a misnomer because there is only one level of black, zero luminance; but it applies in principle to the dark area of the display's grayscale. Flat-screen displays still tend to compress the grayscale, slightly clipping the high and low end and creating a higher contrast image with reduced dynamic range.

The problem is that each display pixel on a flat-panel screen is never completely "off." On a CRT display, the screen phosphor will be black if there is no electron beam projected on it. But both types of display always have some level of signal applied to each pixel to prevent image burn-in, and black levels on a flat panel are inherently brighter than the phosphor on a CRT screen. The result is that a good CRT such as a Prinston Display-Mate can achieve contrast ratios of 440:1 while a plasma display will achieve only 60:1. Careful measurement of the darkest grayscale levels on each display reveals the CRT registers 0.2 nits. A nit is an industry-standard computer display luminance measurement. Similar measurement of the flat-panel display registers 3.6 nits; therefore, the flat panel display shows "black" 18 times brighter than the CRT. The net result is that on a flat-panel display showing a 16-step grayscale ranging from pure white to completely black, the two blackest steps blend together, limiting the dynamic range of any image shown.

With careful attention the human eye can perceive wide contrast ranges but in everyday viewing conditions, it is limited to a contrast ratio of about 100:1. Under such circumstances a flat panel monitor looks quite good. They are flicker-free, have bright colors, and recently have achieved wider viewing angles, making them very attractive. However, they still lack the full dynamic range of a CRT monitor, which is an important factor for image processing.

Fig. 8.17 *One of the most convenient accessories a digital photographer can have is a memory card reader. Not only does a reader speed up downloading of images, it also acts as an external memory module and allows images to be moved from one computer to another. Photo by Robert Reeves.*

8.20 Software for Planning an Observing Session

Out under a wide-open sky at a dark location, a pre-planned imaging session that moves from object to object in an efficient sequence will make for a productive evening, and it is much more relaxing than hopping helter-skelter around the sky. This is especially true even for city observers. Here, buildings, trees, and possibly offending streetlights are all obstacles to be overcome. By knowing the relative location and height of each of these obstacles you can consult a computer star chart program like MegaStar and see if, or when, a particular target will be conveniently visible. In my case, a large pecan tree blocks the eastern sky up to an altitude of about 70° while an oak tree blocks the north up to about 35°, and a neighbor's tree blocks the southwest up to about 25°. By consulting MegaStar and entering the expected time I plan to do my photography, I can click on the designated target and instantly see what its altitude and azimuth will be at a desired time. This information will tell me if the object has yet cleared those obstructions to my east, and about how long it will remain above western obstructions.

8.21 Memory Card Reader

If you have a camera with a large capacity memory card, or have the need to continue to use the camera with a second memory card while the first one is downloaded to a computer, you should purchase a memory card reader. These devices traditionally plug into the computer's USB port, operate as if they were another disk drive, and have their own drive number on the computer's desktop or Explorer window. A USB card reader is up to twice as fast as the camera's own USB connection to the computer. Card readers that accept up to ten different styles of memory card are fairly inexpensive and should be considered a necessity for every digital imager. Not only do they download images up to 100 times faster than old serial cable camera connections, it is far safer and more convenient to simply remove the memory card from the camera instead of removing the entire camera from the telescope each time that images need to be downloaded. A computer will write files to a memory card as well as read them; therefore, a card reader can also be used to transport images between a laptop and a desktop computer.

8.22 Applying Digital Techniques to Old Film Astrophotos

Digital astrophotography, by its very nature, requires computer image processing in order to achieve the desired image. As the user works with the digital medium, he or she will gain experience with the basics of image processing and hopefully explore more advanced techniques to further improve their images. This image processing experience can also prove useful to the veteran film astrophotographer who has many images taken on 35-mm film and slides. By scanning old film and slides and applying these digital processing techniques, better and more highly detailed prints can be extracted from old photos.

A high quality film scanner is needed to adequately digitize 35-mm astrophotos because of their inherent dark—and often black—backgrounds. Such scanners have become more affordable in recent years but are still costly enough to deter the casual astrophotographer from purchasing one. Fortunately, owners of DSLR cameras already have a device that can produce outstanding scans of 35-mm films; the imaging sensor in their camera. By attaching an inexpensive slide copier to the DSLR with a T-adapter, the camera becomes a high-quality film scanner. Making the situation even better is the fact that inexpensive film camera accessories are becoming widely available on the used equipment market as more people switch to digital imaging.

The slide copier shown in Figure 8.18 features an enlargement function that can zoom in as small as one-quarter of a 35-mm film frame. The slide stage behind the illumination diffuser is movable on its vertical and horizontal axes so that any area of the slide or film can be enlarged. 35-mm negatives can be copied by sandwiching them in an unsealed cardboard slide mount.

Using a DSLR is faster and more versatile than using a standard slide scanner. By attaching the camera to a computer by its USB cable and starting the camera with no memory card installed, it will automatically down-

Fig. 8.18 *Those who wish to digitally process their old film astrophotos can make high quality digital copies of them by using a slide copier attachment on their digital camera. Photo by Robert Reeves.*

load images directly to the computer. Results of the "scans" will be seen immediately and any exposure, lighting, or white balance adjustments can be made on the spot.

An inexpensive slide copier coupled to a DSLR is the perfect tool to bridge the gap between film and digital astrophotography. By applying new digital processing techniques to old film images, hidden gems will be revealed that were not apparent through standard film processing.

8.23 Odds and Ends

In order to maximize night vision, you need to shield your eyes against bright sunlight before nighttime astronomical observations. It only takes a little daytime solar ultraviolet radiation to reduce the acuity of one's night vision that evening. Sunglasses should thus be considered an astronomical accessory to protect your eyes before an evening under the stars.

Many headaches can be avoided by preparing a checklist of your gear and taping it to the inside lid of your telescope or accessory case. Be sure to review it before you leave home—it's surprising how an important item can be forgotten while packing for a trip to a remote observing site.

Most DSLR cameras allow adjusting the brightness of their LCD pre-

Fig. 8.19 *A camera's LCD display can disturb the night vision of nearby observers. A magnifying hood by Hoodman (left) will shield the LCD while allowing the photographer to view a magnified view of the screen (right). Photos courtesy of Hoodman.*

view screens; but even at their dimmest settings, the glow can be annoying. A solution is to use a bellows hood to shield the screen. A hood such as the one shown in Figure 8.19 is available from Hoodman and contains a 2x magnifier to allow the LCD screen to be viewed when needed, but remain shielded so it will not create night-vision problems.

Setting up for an observing run has become a major logistical undertaking. You can easily expect to move the equivalent of a small observatory. To start with, it is not uncommon to have a 12- or 14-inch SCT or a 5-inch refractor that, because of its size, is not truly portable. Then there is the auxiliary equipment: mounts and tripods, Schmidt-Cassegrain wedges, guidescopes, power supplies, and creature comfort items like chairs, chart table, and an ice chest. Unless you find a manageable way to deal with moving this equipment, you will soon come to the conclusion that it is much easier to remain indoors with your stamp collection.

My partial solution to the transportation problem was to build two separate telescope mounts as shown in Section 12.6. They are made from common steel and hardware store parts. They are essentially identical, except one has heavy casters on which to roll it out of the garage onto the driveway, while the other remains at my dark-sky location where it rests on solid jackscrews. These mounts are robust and are not affected by weather. The remote site mount weighs several hundred pounds, so the livestock that share their pasture with me do not disturb it. Of course, this option will work only if you can leave the remote mount on private property.

Even moving from the garage onto the driveway can be a challenge. Many observers use the rolling tool chests commonly used by mechanics to solve this problem. A three-drawer cabinet can be purchased from Sears

Fig. 8.20 *The Milky Way from Sagittarius to Cassiopeia is shown in this mosaic of images taken with a Canon 10D at ISO 400 using a 20-mm lens at f/3.5. Each mosaic segment is a combination of 3 five-minute exposures. Photo by Johannes Schedler.*

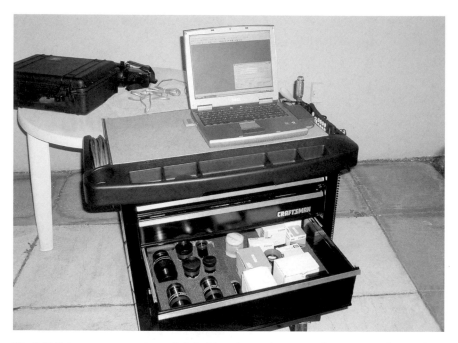

Fig. 8.21 *Telescope accessories and digital imaging equipment can be conveniently stored and rolled to the telescope in a small multi-drawer mechanic's toolbox. This unit is available from Sears. Photo by José Suro.*

for about $100. The tool chest top is a perfect for a laptop computer, and its outer edge has recessed trays that are handy for keeping track of small items in the dark.

Carrying cases that come with telescopes and accessories may be fine for casual use, but they are often inadequate for extensive traveling. Pelican brand cases are foam lined, waterproof, airtight, and extremely strong. They make excellent travel cases for accessories. Telescopes themselves, especially today's long refractors, usually do not fit into aftermarket carrying cases. ScopeGuard has a line of carrying cases for various telescopes, however.

There are many different ways to time long exposures. I have seen stopwatches, oven timers, alarm clocks, digital countdown timers, even an hourglass pressed into service. Today, just about everyone owns a digital wristwatch with a countdown timer option. Many digital cameras have electronic shutters allowing exposures up to 30 seconds. Digital cameras with "bulb" capability also have optional remote shutter release mechanisms with built-in timers. Some cameras are so silent that you will need a timer just to alert you when the exposure has automatically finished. Since a laptop computer will likely control a telescope-mounted DSLR camera,

Fig. 8.22 *This image is a sequence of 6 two-minute exposures aimed at the area of the sky where two Iridium flares were predicted to occur. During the 12 minutes of exposure, five flares were actually recorded using a Canon 10D at ISO 200 through a 50-mm lens at f/2.5. Photo by Johannes Schedler.*

a convenient way to keep track of timing is to open the computer's "Adjust Time and Date" window from the bottom desktop toolbar. The Windows clock has an easy-to-see analog face.

A black plastic drinking cup can hardly be considered an astrophoto accessory item, but it is. If approaching car lights or a passing airplane threaten to fog or streak a long celestial exposure, briefly cover the lens with the plastic cup to interrupt the exposure until the light source is shut off or moves away.

Long telephoto or zoom lenses are heavy enough to flex a single-point camera mount like a camera attached piggyback to a telescope with the tripod mounting socket hole. Small wedges made from wood or dense foam can be used to support front-heavy lenses against the telescope tube.

Flexible exercise weights make good temporary counterweights for odd telescope balance problems. They come in sizes from 0.5 to 5 pounds and are easily held in place with bungee cords.

Among the don't-leave-home-without-it items in digital photography is an extra memory card for your camera. You never know what adventure will await you on a trip to a remote dark observing site and unexpected photo opportunities can arise.

Chapter 9
Polar Alignment and Guiding

With a stationary telescope, the Moon and Sun can be imaged with an exposure lasting only a fraction of a second. However, you will quickly find that the constantly moving sky makes it necessary to shift the telescope every few seconds to keep the target centered in the field-of-view. It is futile to attempt long exposures through a telescope that is not both clock-driven and accurately polar-aligned. Accurate polar alignment is essential for success with long exposure astrophotography. Fortunately, achieving good alignment is not difficult; it just takes time.

Many of today's telescope drives are capable of accurate unguided tracking for up to a minute provided they are accurately aligned with the pole. This capability has added a new phrase to the imager's vocabulary: "tracked, but not guided." Accurate tracking allows stacking of successive short exposures to simulate a much longer exposure, but the mount must be well aligned to prevent field rotation. Each successive exposure must be identically framed. This is accomplished by insuring that the same guide star is centered in the crosshair at the beginning of each exposure. A key advantage of stacking multiple short exposures is that if any one is bad—due to a tracking error, a bump to the telescope, wind shake, etc.—it can be discarded. Removing a minute of poor tracking is not possible with a long exposure film astrophoto image.

9.1 Principles of Polar Alignment

A permanently mounted telescope only needs to be polar aligned once. As long as the mount is not moved on its pier, it should retain the alignment. However, a portable telescope needs polar alignment every time it is set up. Most of the alignment procedure can be accomplished in twilight before it gets dark enough to begin imaging.

Polar aligning a telescope brings the telescope's right ascension (polar axis) exactly parallel to the Earth's rotational axis. Once aligned, the telescope's drive motor turns the telescope around its polar axis at a rate that exactly matches the apparent movement of the sky due to Earth's ro-

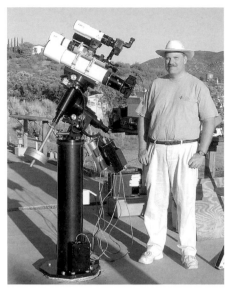

Fig. 9.1 *This handsome rig owned by Jim Windlinger typifies the massive, accurate-tracking mounts used by today's digital astrophotographer. Windlinger's homemade mount carries both 3- and 4-inch Borg refractors using a combination of film and digital cameras. Photo by Dave Kodama.*

tation. The easiest way to visualize this is to think of the mount's right ascension axis as aiming at Polaris (the "Pole Star" in the northern hemisphere). However, Polaris is not the true pole; it is currently less than a degree from that spot. A long star-trail exposure will show Polaris tracing a tiny arc around the true pole. The right ascension axis must, therefore, be properly offset from Polaris to be aligned with the true celestial pole.

The drive rate for a telescope tracking the stars is not the familiar 24-hour mean solar day. Earth's diurnal motion causes a star to appear to make a complete trip around the pole every 23 hours, 56 minutes, and 4 seconds, which is a sidereal day. The four minute difference between a mean solar day and a sidereal day is due to the effect of the Earth's orbital motion around the Sun. As the Earth moves along its orbit, the Sun seems to move west-to-east relative to the stars. Thus, every 23 hours, 56 minutes, and 4 seconds the stars have returned to their starting positions, but the Sun appears a little farther east among them. It takes an additional four minutes for the Sun to appear to return to its original starting position (usually taken as local noon). Because of this daily four-minute displacement, the stars apparently shift position in the sky at the rate of about one degree per day, presenting different constellations to us as the seasons change.

9.2 Telescope Mechanical Checks

There are several reasonably quick ways to align a telescope mount that will allow piggyback astrophotography; however, they only work if the telescope itself is in good mechanical order.

First, if you are using a star diagonal, check to see if the star field you see through the diagonal is the same you see without it—many do not! Sight a bright star through the telescope, then lock it so the drive tracks the star. Insert the diagonal and verify that the star has not moved. If it has, the diagonal is at fault. Few star diagonals are adjustable, so plan on purchasing another.

Next, check that the telescope's declination axis is precisely perpendicular to the polar axis. Start by using the setting circles as a reference to adjust the declination to 90 degrees. Next, lower the mount's latitude adjustment to zero degrees (horizontal) and sight an object on the horizon; this should be at least a quarter-mile away for a fork mount and a mile away for a German equatorial mount. Rotate the tube around the polar axis and observe the target in the eyepiece; it should stay centered as the tube rotates. If the target makes an arc in the eyepiece field, then the declination setting circle may not read correctly, or the declination axis may not be perpendicular to the polar axis. Make sure the optics are properly collimated. If the image shift occurs in a Schmidt-Cassegrain telescope, insure that "mirror flop," (caused by the primary mirror moving on its focusing collar), is not the cause of the shift. If the declination and polar axes are not mutually perpendicular, either the telescope tube and/or the declination axis will have to be shimmed. Hardware stores carry thin brass shim stock, and in a pinch, an ordinary soda can may be cut with heavy scissors to make shim stock. If thinner shims are needed try strips of paper.

9.3 Basic Polar Alignment

A quick polar alignment can be done by roughly aiming an instruments' axis toward Polaris. This will perform adequately for casual visual observing, but it falls far short of the precision needed for even basic imaging. This is especially true the closer the target is to the pole where misalignment will cause field rotation, even if the object at the center of the image is tracked perfectly. When this happens the telescope/camera system and the sky are rotating on two divergent axes. For instance, if the polar axis is just one degree below the true pole, a star on the celestial equator within two hours of the meridian will drift eight seconds of arc per minute. Within three minutes, this will cause noticeable trailing even on an image taken with just a 200-mm lens.

Some German equatorial mounts have a small sighting telescope built into their polar axis that is equipped with a special reticle for setting on the true pole by offsetting from Polaris. These polar finders are convenient,

Fig. 9.2 *A rough polar alignment can be done in the daytime by using an inclinometer to adjust the polar axis elevation and a compass to point the polar axis toward north. Photo by Robert Reeves.*

and some are set up to even work in the southern hemisphere. But they must be carefully checked for accuracy if they are being used for the first time. Not every reticle is adjusted properly at the factory, and even if they were, rough shipping or wear and tear might have caused a problem. Use the drift alignment method (explained in Section 9.5) to verify that it is positioned correctly. Once the mount is properly aligned, Polaris should appear to stay aligned in the reticle as the mount is rotated. If it does not, then the reticle is misaligned. Be sure to use the proper Standard or Daylight Savings Time when setting up the polar finder.

If your German equatorial mount is not equipped with such a finder, an alternative is to rest a finder scope directly on the polar axis using a machinist's v-block. The true pole can then be found using any of the popular star charts or star charting software that show the dim stars that are closer to the polar than Polaris.

For equatorial mounts not covered by the above examples, use the following method. First, point the polar axis towards Polaris. This step can be done while it is still daylight by using an inclinometer to set the polar axis elevation to the local latitude (See Figure 9.2); then use a good magnetic compass to adjust the mount's azimuth to within a few degrees of north. Next, aim the telescope toward a bright star of known right ascension near the celestial equator. Set the right ascension setting circles to the current year's R. A. for that star. (You can generate a list using MegaStar or similar sky-mapping programs.) Next, position the telescope using the setting circles to aim at 2 hours, 32 minutes right ascension and +89 degrees 16 minutes (or the current year coordinates for Polaris). Leave the setting circles alone and move the entire mount in azimuth and elevation to sight Polaris.

Repeat the process by sighting on a bright star near the celestial equator and proceeding as before. If you can shift between the equatorial star and Polaris using setting circles alone, no further alignment is necessary. However, be aware that this process sets the mount's polar axis on the refracted, not the true, celestial pole. This method will also work for the southern hemisphere by using the star Sigma Octanis at 21 hours, 09 minutes right ascension and –88 degrees, 57 minutes declination (or the current year coordinates for Sigma Octanis).

If you perform polar alignment in twilight, beware of the star Kochab. This star in the Little Dipper's bowl is about 15 degrees from Polaris and is almost exactly as bright. Ordinarily there is no mistaking this star for Polaris, but when in unfamiliar territory one's sense of true north can be off a number of degrees. At certain times of the year, such as early evening in April or November, Kochab is at the same elevation as Polaris, so it is easy to confuse the two, particularly if you are hurried. The mistake will soon become obvious, but valuable twilight setup time will be lost.

9.4 Other Factors in Polar Alignment

Star-drift caused by misalignment can be corrected using the mount's slow-motion controls, but field rotation is a tougher problem. Rotation cannot be corrected by just using the telescope's directional controls. To be free of field rotation on a full 35-mm frame-size imaging sensor using a 300-mm lens after an accumulated one hour of exposure near the celestial equator, the polar axis must be aligned within 22.8 arcminutes, or about a third of a degree, of the true pole. This tolerance is reduced to 11.4 arcminutes for a field at 45 degrees declination and is just 1.28 arcminutes at 85 degrees declination.

Three things affect both declination drift and field rotation:

1. the location of the true pole,
2. the polar alignment point in the sky, and
3. the celestial location of the target.

Michael Covington, in *Astrophotography for the Amateur,* explains that if the polar alignment error is no more than a few degrees, both declination drift and field rotation are linearly proportional to the amount of alignment error and are geometrically interrelated; if the true pole, the alignment point, and the target form a right triangle, then the declination drift is greatest and field rotation is zero; but if the true pole, the alignment point, and the target lie in a straight line, then declination drift is zero and field rotation is the greatest.

This analysis shows that the declination of the target itself has much to do with which factor—declination drift or field rotation—comes into play. Declination drift is fairly independent of the target's own declination and quickly diminishes for targets above 80 degrees. On the other hand, field rotation increases with higher declination; at 60 degrees, it is twice its value at the celestial equator, and it increases dramatically very close to the poles.

It is mathematically possible to predict the greatest amount of declination drift and field rotation for a given set of circumstances. In many cases the actual drift or rotation will be less than calculated. The maximum declination drift is

$$15.7 \, \text{arcseconds} \times \text{Exposure Time (in minutes)} \times \text{Alignment Error} \, (°)$$

and the maximum field rotation is

$$0.01(15.7) \, \text{arcseconds} \times \text{Exposure Time (in minutes)} \times \text{Alignment Error} \, (°)$$

Generally field rotation will not be visible unless it exceeds 0.1 degree.

9.5 Star-drift Polar Alignment

Permanent telescope installations, or any prime-focus astrophotography, require high-precision polar alignment. The drift method of polar aligning should be used here. Brad Wallis and Richard Proven popularized the following procedure, that I have augmented with information provided by Chuck Vaughn. Stars bright enough for the drift alignment procedure can usually be seen about 15 minutes after sunset. If you begin then, alignment can usually be completed by the time it is dark enough to begin imaging.

1. Roughly aim the polar axis toward Polaris and level the tripod.
2. Place the crosshairs of an illuminated-reticle eyepiece (about 200 power) on a star that is near the meridian and within five degrees of the celestial equator. Align the crosshairs so the star moves up and down the declination crosshair and left and right on the right ascension crosshair.
3. The star should noticeably display drifting in declination within 30 seconds. (Ignore any east-west drift for now.)
4. If the star drifts northward, the polar axis is too far west of the pole. To

correct, rotate the mount in azimuth so the star moves to the east. During this process, beware of confusion arising from the fact that eyepieces and star diagonals can invert and/or reverse apparent motions.

5. If the star drifts southward, the polar axis is too far east of the pole. To correct, rotate the mount in azimuth so the star moves to the west.

6. If the star drifts noticeably within five seconds, the polar alignment is off by at least 10 eyepiece field diameters. If it takes the star at least 30 seconds to drift, the setting is probably not more than one or two eyepiece fields off.

7. Once the north-south drift is minimized, place the crosshairs on a star near the celestial equator and about 15 degrees above the eastern horizon. If the eastern horizon is blocked, go to a star near the western horizon and reverse "above" and "below" in the next step.

8. If the star drifts northward, the polar axis is above the pole; adjust the elevation of the polar axis to move the star toward the bottom of the field. If the star drifts southward, the polar axis is below the pole; adjust the elevation of the polar axis to move the star toward the top of the field.

Once the corrections have been made for polar axis position, repeat the procedure because each subsequent correction will slightly upset the previous adjustment. When properly aligned, and assuming the drive motor is turning at the proper speed, there should be no noticeable star drift for five minutes.

9.6 Star-drift Alignment Variations

If you can't find an adequate star along the celestial equator for drift alignment, stars at higher declinations will still work. A star at 10 degrees declination, for instance, will suffice because the cosine of 10 degrees is 0.985; thus a star 10 degrees from the equator will, in a given amount of time, still move 98.5 percent the angular distance of a star on the celestial equator. Indeed, you can easily drift-align on a star 40 degrees from the equator. The cosine of 40 is 0.766; thus, a star at 40 degrees declination will still move 76.6 percent the angular distance in a given time that a star along the equator would. In practical terms, this means that if you let a star on the equator drift for three minutes, a star at 40 degrees declination should be allowed to drift for four minutes to achieve the same accuracy.

Thomas Krajci devised a quick drift method technique for determining how much to correct the polar axis's azimuth and elevation. Tom assumes the mount is already in rough alignment with the pole. You will need an eyepiece with a readable scale, such as Celestron's Micro Guide; or Edmund Scientific markets transparent scales that can be inserted into an inexpensive 12-mm Kellner eyepiece.

You begin by measuring star drift at two locations just like the traditional drift alignment method—the meridian and near the horizon. Krajci orients the eyepiece reticle and monitors the drift in declination. After five minutes he has quantified the amount of drift and can calculate how much to move the polar axis to correct the alignment. An important point to remember with this technique is that you measure the north-south drift on the meridian, but you must move the telescope to a star 90 degrees away (±6 hours from the meridian) when making the polar axis adjustment. Here are the steps involved:

1. Roughly polar align to within a degree using the finderscope.

2. Find a star on the meridian, near the celestial equator, and measure its north-south drift in reticle divisions for five minutes.

3. Multiply this drift value by 46. This is how many reticle divisions you need to move the mount in azimuth (but don't move it yet).

4. Point the telescope to a star on the celestial equator that is also near the horizon.

5. Now move the telescope in azimuth to shift the star by the number of reticle divisions calculated in step 3.

6. Remain on the star near the horizon and measure the north-south drift with the reticle for five minutes.

7. Multiply this value by 46. This is how much you need to move the mount's polar axis in elevation (but don't move it yet).

8. Move back to the star along the meridian near the equator.

9. Now move the mount in elevation to shift the star by the number of reticle divisions calculated in step 7.

Repeat the above steps if needed to further refine the polar alignment.

Krajci's method is independent of focal length. The accuracy will be limited in Schmidt-Cassegrain optics that can shift; you must eliminate this mirror movement before the 46-to-1 rule will work satisfactorily. How to go about locking a Schmidt-Cassegrain mirror is explained in Section 12.12.

9.7 Guiding Corrects Alignment Errors

Guiding keeps the target stationary on the camera's focal plane by incrementally adjusting the telescope's right ascension and declination control during an exposure. The need for guiding arises from inaccuracies in the telescope's polar alignment, its mechanical drive components, electrical power fluctuations, and atmospheric turbulence. Done properly, the image will be sharp with stars recorded as pinpoints instead of streaks and trails.

A telescope can be guided manually or electronically. The manual method relies upon the observer visually sighting a guide star with an illuminated-crosshair eyepiece and manually correcting the telescope's aim. Today this is done with a hand controller connected to an electronic drive corrector that slightly speeds up or slows down the telescope's movement in right ascension and/or declination. This method places no significant burdens on the operator for exposures of several minutes, but the unforgiving need for constant attention to the guide star makes lengthy exposures an uncomfortable challenge. Improvements in film astrophotography techniques over the past several decades have allowed exposures to reach into the multi-hour range, severely taxing the endurance and skill of the operator. For this reason, electronic autoguiders that relieve the imager of the burden of guiding have become very popular, and the use of old-fashioned manual guiding has waned. Autoguiders automatically detect the need for corrections in right ascension and declination and issue the proper commands to the mount's drive motors to compensate for tracking errors. However, this convenience comes at a price—they are expensive, ranging in some cases up to $2000.

The story is much better for digital astrophotography, which inherently limits exposures to just a few minutes (for now anyway) before digital noise becomes objectionable. This reduces the guiding burden both financially and physically. Instead of the challenging task of manually guiding lengthy and uninterruptible half- to one-hour exposures, the digital imager has an easier job of taking a series of short exposures that are later digitally combined to create the longer ones. Even better, if there are guiding errors in mid-series, you just toss the image and go on; you have not ruined the entire exposure. The shorter exposures possible with digital imaging have revived interest in lower-cost manual guiding techniques.

9.8 Causes of Guiding Errors

A major source of guiding error can be traced to the telescope's clock drive. Moderately well-aligned telescopes will need only occasional corrections in declination, usually always in the same direction, but the right ascension drive is prone to tracking errors caused by a number of reasons. Let's examine the cause of these errors:

- **Imperfect polar alignment.** In order for an equatorial mount to properly follow the stars, its right ascension axis must be perfectly aligned with the Earth's axis of rotation. As we have seen in this chapter, there is no easy and absolute way to do this, but there are several procedures that can

Fig. 9.3 *Camera mount flexure over a three-hour period trailed and ruined the image at the top. The advantage of stacking the inherently shorter exposures taken with a digital camera is shown in the bottom image. Each of the 45 separate three-minute exposures was too short to suffer from flexure. When all images were stacked and aligned the effects of any flexure were eliminated. Photos by Robert Reeves.*

be followed to achieve an adequate polar alignment.

- **Periodic error.** The heart of a motor-driven equatorial mount is the worm and gear (or spur and gear depending on the design) assembly that actually turns the mount in order to follow the rotation of the stars. Tiny manufacturing errors in the worm and gear's machined mating surfaces cause recurring errors as the worm or spur rotates. These errors, called periodic error, cause the mount to slightly speed up, then slow down over a cyclic period lasting several minutes.

- **Incorrect drive rate.** Drive rate is usually not a problem with today's electronically-controlled drive correctors, but the drive rate must match the rotation of the sky in order for the mount to track correctly.

- **Flexure.** The mounting between the telescope and a piggyback-mounted camera or separate guidescope must be absolutely rigid. Any mechanical play or bending between the two as the telescope moves, or changes of position due to gravity, will cause the optical axes of the two instruments to diverge from each other. This will cause stars to trail on the image because the telescope or camera will no longer be aimed at the same spot in the sky as that seen through the guidescope. Even low-magnification piggyback cameras with long telephoto lenses should be supported under the lens barrel, in addition to the mount holding the camera body. This is important with lenses longer than 135-mm focal length because the ¼-inch tripod socket in the base of the camera body is too small to hold the heavy camera and lens combination steady as the telescope changes position. With an unsupported long lens, even short exposures will suffer from flexure. For example, even if the individual frames do not show trailing from flexure, a series of images taken over an hour's span will show star drift between the first and last image (see Figure 9.3). When this series is digitally stacked to create a deeper exposure, the stars will be out of registration, making it necessary to perform additional processing steps to realign the images.

- **Atmospheric refraction.** Refraction, or the bending of starlight, is greater the closer the star is to the horizon. If a star is descending toward the horizon, it will seem to slow down as refraction makes its apparent position seem higher. Conversely, if a star is rising from the horizon, it will seem to speed up as it approaches higher elevations. Shooting only above about 30 degrees elevation, if possible, can minimize the effects of refraction.

- **Field rotation.** The wider the field-of-view of an image, or the lower the magnification, the more leeway you have in guiding precision. However, if a telescope or tracking platform is not well polar aligned, field rotation will occur, as discussed in Section 9.4.

9.9 Dealing with Tracking Problems

No telescope drive is so accurate that it can track the stars for long periods of time. Even if the drive motor tracked perfectly and everything was properly aligned, atmospheric scintillation and refraction would conspire to slightly shift the image. However, many of the mechanical and electrical guiding problems that were encountered in telescopes a decade ago have been solved by today's high-end mounts. Accuracy comes at a price that ranges from several thousand dollars to more than twelve thousand. Mounts in this class are made by Losmandy, Mountain Instruments, Takahashi and Software Bisque. Of course, all of these mounts require good polar alignment to produce good results.

However, I suspect that the majority of digital images will, at least initially, use mounts that do not track as well as these high-precision models. There are a number of things that can be done to improve tracking accuracy or make known errors more manageable. The most common cause of small tracking errors is random manufacturing errors in the drive gear. Each tooth in a telescope's drive gear is slightly different than the next. Consequently, it is not good to have the right ascension axis of a telescope perfectly balanced; a slight imbalance toward the east will cause the drive to "pull" the telescope from east to west and better compensate for irregularities in the gear teeth. Minimizing play in the mesh between the drive worm and gear will reduce backlash and prevent the telescope from shifting if buffeted by the wind.

Periodic error is a machining defect in the drive gear worm that will show up as a cyclic back and forth motion in right ascension and causes the drive to alternately move the telescope too fast then too slow. Not all drives have the same periodic error. Mounts benefiting from modern machining techniques may display a cyclical periodic error of only a few arc-seconds while older mounts may have up to several arc-minutes of periodic error. Mounts with severe periodic error will not work with electronic autoguiders as the star's apparent motion will overwhelm the autoguider. Periodic error receives the lion's share of blame for tracking problems, but in reality, only the good drive gears have true repeatable periodic error. The rest have random errors that are erratic and much harder to predict. Some mounts with electronic controls incorporate Periodic Error Correction (PEC) that can be programmed to anticipate true periodic error and issue tracking commands to significantly reduce its amplitude.

To test your mount for tracking error—periodic, random or both— take a long unguided high-magnification star exposure with the polar axis deliberately aimed about 20 degrees east of the pole. This will require the use of a film camera, as the exposure has to be continuous and about 20 minutes long to record several rotations of the drive's worm gear. By deliberately skewing the polar alignment, the stars will drift southward as they cycle back and forth from periodic error. If the resulting star trails demonstrate a smooth sine wave appearance, then the telescope drive has a true periodic error. However, if the trails have random zigzags or bumps, the worm and drive gears need aligning, the telescope is grossly out of balance, or the gear set is poorly machined.

Nearly all mounts now have a pushbutton controller for fine adjustments in tracking. Older telescopes require a separate drive corrector to vary the frequency of the right ascension drive motor voltage, and thus its tracking speed. Declination adjustments with older mounts can be made either manually or with a remote switch controlling a small motor on the declination knob. With four-button controllers that govern the right ascension and declination adjustments, orient the controller so that the buttons coincide with the respective motions they control, so there is no confusion over which to push. If a direction is reversed on the controller and there is no function reverse switch, turn it upside down and push the buttons from underneath.

9.10 Choosing a Guidescope

A piggyback camera is mounted directly atop a telescope which is to function as its guidescope. The camera needs to be mounted as rigidly as possible so the camera will not shift position as the telescope moves through different inclinations. If the camera is used to photograph through the telescope, it is no longer possible to guide with the telescope because the camera blocks the eyepiece. There are two ways to guide when the camera is shooting through the scope: a separate piggyback-mounted guidescope, or a device called an off-axis guider.

A separate guidescope is usually a piggyback-mounted refractor that is smaller and lighter than the main telescope. Since a camera is at the focus of the main instrument, a guidescope's sole purpose is to track a guide star and thus "steer" the main telescope to accurately follow the stars. It is usually attached to the main scope by a pair of adjustable mounting rings that allow it to be aimed at a guide star located within a degree or so of the main target. To work properly, a guidescope and its mounting hardware must be light enough to not overpower the drive mount, yet strong enough to prevent the guidescope from shifting, drooping, or flexing in any way during the exposure. Any relative movement between the main and guidescope will cause oval, trailed, or hook-shaped stars on the image, even if the guiding is otherwise perfect.

Refractors are the usual choice for guidescopes because once assembled correctly, they tend to stay in collimation whereas reflecting telescopes are notorious for going out of collimation. The user should be especially wary of Schmidt-Cassegrain models unless a means has been added to solidly lock the primary mirror as shown in Section 12.12. Schmidt-Cassegrains focus by sliding the primary mirror up or down the central light baffle tube. Mechanical play in the mirror mounting assembly is the root cause of the notorious SCT "mirror flop" phenomenon that rears its ugly head as the telescope slews across the sky. The image shift is slight and not a problem for visual observers, but to have the guide star jump during an exposure is disastrous for an astrophoto!

By combining an eyepiece with a Barlow lens, it is usually possible

Fig. 9.4 *An off-axis guider eliminates differential flexure between the guide scope and the main imaging telescope by allowing the photographer to both photograph and guide on stars from the same telescopic field-of-view. Photo by Robert Reeves.*

to achieve a magnification that is four times the magnification of the image on the camera's sensor. And by magnifying the guide star (and hence any tracking errors) much more than the image on the camera sensor, guiding errors can be detected and corrected before they've had a chance to degrade the image. An electronic autoguider requires less magnification because it can detect very small variations and can respond and correct them much more quickly that a human.

9.11 Off-Axis Guider

An off-axis guider, usually called a OAG, may be the only option when a telescope is too small to carry a piggyback guidescope, or if the optical configuration of the main instrument cannot maintain good collimation. An off-axis guider is placed at the telescope's focus and uses a small prism or mirror to "pick off" a portion of the image and divert it to a guiding eyepiece or autoguider. This pick-off prism lies just outside the imaging sensor's field-of-view, so it does not appear in the image. With this device therefore, both guiding and imaging take place within the same optical system. The advantages of an OAG are:

- It is lighter and cheaper than a separate guide scope.
- It does not have separate optical components that can shift and flex relative to the main telescope.
- A focal reducer can be installed between the OAG and the camera to provide a wider, brighter field-of-view for the camera, while the guiding

eyepiece uses the telescope's full focal length, making it easier to detect and correct drive errors.

The camera is focused normally when using an off-axis guider; either with an auxiliary knife-edge focuser or by examining sample images downloaded to a monitor or computer. The operator then turns his attention to the guiding eyepiece and focuses it visually by sliding it in or out of its holders.

Finding a suitable guide star is the next step. Usually there are no suitable stars near the guiding eyepiece reticle. The following conditions may be present:

- The desired target is small, and there is maneuvering room in the camera's field-of-view. Move the telescope to a suitable star within the eyepiece field-of-view. Here, the desired image will be offset somewhat in the final image frame.

- There is no room to shift the telescope and capture the target. A projection reticle can be inserted between the telescope and eyepiece. This has several advantages; it projects a virtual reticle onto the field-of-view of any eyepiece allowing the selection of different magnifications or better eye relief for eyeglass wearers. It also allows the reticle to be moved to any star in the field-of-view.

- No suitable star is visible in OAG's field-of-view. Hunt for one by rotating the entire guider assembly around the optical axis of the telescope. This is a tedious but necessary step. If the guide star is found at a point where OGA is rotated into a difficult viewing position, then a star diagonal may allow more convenient access to the eyepiece.

Separate guidescopes have the advantage of presenting a sharp bright star to guide on. This reduces visual fatigue or presents a crisper image for an autoguider to lock onto. Off-axis guiders, on the other hand, use the edge of the telescope's light cone where field curvature, coma, and possibly other aberrations render the star image dimmer, distorted and blurry. The usual view is of fan-shaped stars. OAGs present challenges to the user, but they do allow guided astrophotography with optical systems that otherwise could not be used because of weight limitations.

9.12 Illuminated Reticle Eyepieces

There are two types of illuminated reticle eyepiece; one has the reticle permanently built into it; and the other is a device called a projection reticle. The latter is a device that installs between the telescope and the eyepiece and projects a virtual reticle onto the focal plane of the regular eyepiece. This allows any ocular to become a guiding eyepiece. With interchangeable eyepieces you can select various magnifications and degrees of eye

Fig. 9.5 *Existing wire-powered illuminated guiding eyepieces can be converted to wireless operation by installing an aftermarket battery and illuminator such as the PulseGuide offered by Rigel Systems. The two knobs adjust brightness, and control adjustable-duration blinking illumination. Photo by Robert Reeves.*

relief. Most projection reticles are also adjustable; that is, you can move the location of the reticle relative to the eyepiece's field-of-view. This is a major advantage because the camera's aim and subject framing, often restricted by the size and celestial orientation of the target, sometimes cannot be moved without losing part of the target.

Guiding eyepieces with built-in reticles come in a variety of styles; single crosshair, dual crosshair resembling a tic-tac-toe pattern, or a grid for estimating angular separations. These devices also have a wide range of features: self-contained batteries, wire-connected power supplies, and blinking options. The blink option causes the reticle to wink on and off to allow viewing extremely dim guide stars that would ordinarily be obscured by a continuously illuminated reticle.

Most guiding eyepieces have a standard 1.25-inch diameter barrel. The 12.5-mm focal length, dual-crosshair model is most popular and is available from almost all manufacturers. Meade also makes a 9-mm Super Plössl that has the advantage of wireless operation, higher magnification, and an adjustable reticle. Celestron offers the 12.5-mm Micro Guide eyepiece that has a reticle with various angle, size and position scales as well as a guiding bull's-eye and crosshair. These scales may be useful for double star measurements and other aspects of visual astronomy, but most astrophotographers find the view through the Micro Guide to be too "busy" and cluttered for convenient guiding.

An important thing to check when purchasing one of these units is that the reticle and guide star lie on the same focal plane. You make this check by shifting the eye from side to side; the guide star should not move relative to the reticle. If it does change, it will create phantom movements of the guide star when the eye inevitably moves slightly during the course of the exposure. To correct this problem, most dedicated guiding eyepieces also have a diopter adjustment so the focus of the illuminated reticle can be set for your particular vision.

I recommend the dual-crosshair variety of guiding eyepiece because the "box" formed at the intersection of the pair of crosshairs provides a reference not available with single crosshairs. This box is useful to estimate the acceptable amount of guide star drift before the image begins to trail. I have found that for most piggyback imaging situations when the guide star is confined within this box, images will be very well tracked. I align the crosshairs so the right ascension and declination movement of the telescope will move the star in the same direction as the appropriate crosshair. As we will explain in Section 9.15, there is usually considerable tolerance for guiding error in piggyback wide-field imaging.

Do not choose a guiding eyepiece that does not allow the illumination to be dimmed. Nearly all have this capability. It is needed when a bright star cannot be located near enough to the target field. I personally like to use a guiding eyepiece with a blinking option for dim stars. It is fatiguing to follow very dim guide stars for any period of time, especially late at night. Setting the reticle to blink off and on at a time cycle comfortable for you makes the job a lot easier. It allows the eye to briefly rest from the glow of the reticle and stay more sensitive to dim stars. The guiding process becomes a series of updates, as first the star is very visible by itself, then the reticle briefly overlays it for reference, and blinks off again. This system is admittedly not for everyone, and some photographers find it annoying.

Another technique for very dim stars is to turn the crosshairs 45 degrees relative to the movement of the right ascension axes. This way the guide star will not become lost to sight as it drifts a distance behind the crosshair.

Most illuminated guiding eyepieces use a red LED for illumination. After staring at bright red crosshairs for a time, the eye can become blind to them and not perceive needed corrections. Some eyepieces allow switching to green illumination to rest the eye. If you suffer eye fatigue from lengthy use of red LED illumination, the red LED in most eyepieces can be changed out for a green LED obtained from Radio Shack. Just insure the polarity is correct when installing the new LED. If the guiding eyepiece you have does not have a pulsing option, it may be retrofitted with a pulsing illuminator with a battery package offered by Rigel Systems.

Guiding eyepieces cost from $50 to $200. Older units usually have a 12-mm dual-crosshair eyepiece while newer ones usually are a 9-mm Plössl for better eye relief. Excellent examples are often available on the used astronomy equipment market.

9.13 Manual Guiding Tips

- Wherever possible, use a refractor for polar alignment and guiding to avoid mirror shift from changing the true aim of the instrument.
- Do not attempt to polar align a telescope that is set up on grass or soft dirt. Place a concrete stepping stone or a one-foot square piece of ¾-inch plywood under each tripod leg to prevent it from sinking into the soft ground.
- Make sure the finder and telescope optical paths are parallel. Use a straight-through eyepiece on both when polar aligning. It makes for simplicity (image orientation, etc.) and avoids possible star diagonal aimpoint errors.
- Do not add or remove accessories after performing a critical polar alignment. Changing the weight carried by a mount, particularly a fork mount, will alter the "droop," and thus the aim of the polar axis.
- When using a Schmidt-Cassegrain telescope for guiding, be sure to reach final focus while turning the focus counterclockwise. This leaves the focus adjusting screw pushed against the mirror and prevents the mirror from settling backwards due to play in the focus mechanism.
- Make sure the declination setting circle is aligned. On fork-mounted SCTs they can slip, but can be adjusted by loosening the bolt at the center of the declination circle.
- The drift alignment method is time consuming, but it will align the telescope's mechanical axis, regardless of its optical alignment.
- Frequently switch from one eye to the other when guiding to prevent strain. Long periods of guide star monitoring can cause your eye to see non-existent crosshairs.

9.14 Manual Guiding Techniques

Piggyback guiding tolerances are far greater than those required for prime focus imaging. However, spending a night of piggyback guiding will instill in you a great deal of respect for those who manually guide high-power prime focus astrophotos at ten times the precision. But, as I repeatedly state in this book, digital imaging is fundamentally different. We get a deep exposure by stacking a series of short exposures that are set just below the point at which electronic noise becomes objectionable. So, while taking a series of short exposures, the imager gets a brief, but restorative, break while each one is downloaded to a laptop or written to the camera's memory card.

There are many different ways to manually guide. You will eventually develop a personal style.

Get comfortable. If the overall exposure is going to be lengthy, make sure you are seated so the eyepiece will be accessible throughout the shot. The eyepiece can rise or fall quite a bit as the telescope tracks an object across the sky during a lengthy exposure. Practice guiding for a minute with the shutter closed. See which way the controls move the telescope in response to the guide star's movement.

Align the pushbuttons on the hand controller to correspond with the directions the star moved along the crosshairs.

Expect that with even good polar alignment, the guide star will slowly and gently rock back and forth in right ascension because of the telescope drive's periodic error.

Unless an extremely accurate polar alignment has been achieved, the guide star will usually drift in one direction and require a correction every few minutes. This drift is usually constant and can be predicted after the first couple of corrections.

Do not defocus the guide star to make it more visible against the crosshairs, because it will introduce parallax that can shift the guide star's true position relative to the crosshair. This may not show up with low-magnification piggyback work, but it can lead to erratic guiding with high-power prime focus photography.

Too much backlash creates a "dead zone." When a dead zone is present, the telescope will not track properly and guiding corrections will not be effective. Almost all telescope drives have some backlash. In right ascension, backlash is usually taken up by the drive motor, that is continuously pulling the telescope toward the west. In declination, backlash slack can cause delay in corrections in the reverse direction, as the reversed motor has to take up the backlash slack before it can move the telescope. Excess slack can be minimized by slightly unbalancing the telescope in the north-south direction. If this does not work, deliberately move the polar axis slightly east or west of the true pole. This will cause the guide star to drift in only one direction, thus removing the backlash.

I recommend that if you are a novice at guided astrophotography and are planning to use an autoguider, you should learn to manually guide at first in order to have an understanding of what an autoguider is doing once you start to use one. Even if you don't pursue manual guiding for long, the illuminated reticle eyepiece will be a handy accessory to help achieve good polar alignment.

9.15 Guiding Tolerances

When you move to high-resolution prime focus astrophotography, there will be one rule about guiding tolerances… there aren't any. If the guide star does not stay tucked into the intersection of the eyepiece crosshairs, the image will exhibit trailing. This stringent need for precision is intimidating to beginners and a constant source of frustration for even experienced imagers. Simply said, it is hard work to manually guide a prime focus astrophoto with such precision. Advanced imagers are happy to pay for an autoguider that will relieve them of the task, even if digital imaging has inherent exposure time limits.

Piggyback wide-field astrophotography is far more forgiving of guiding errors. When 35-mm film was the primary means of doing astrophotography, it was easy to calculate the allowable error for a particular camera lens. Digital cameras have no similar standard because of widely differing sensor sizes. For the most part, long exposure piggyback imaging is going to be done with digital SLR cameras. This somewhat simplifies the tolerance calculations because the image scale is either equal to a full 35-mm film frame, or to one of various "crop factors" governed by the sensor size. Depending on camera brand, the crop factor can multiply a lens's apparent focal length by 1, 1.3, 1.5, 1.6, 1.7 or 2 times. Manufacturer's web sites, and comprehensive camera review sites will show the crop factor for all DSLR cameras.

The illustrative guiding tolerances we are going to calculate in this section assume a full 35-mm frame sized imaging sensor. If your camera has a crop factor, you should divide the calculation by the crop factor. For instance, using a 12-mm double-crosshair eyepiece with a crosshair spacing of 0.2 mm on an 8-inch *f*/10 Schmidt-Cassegrain telescope produces an eyepiece image scale where the box at the center of the intersecting dual crosshairs is about 20 arcseconds in diameter. With a full 35-mm film sized imaging sensor, like the Canon EOS 1D, we know that when using a 50-mm lens, a star can move about 120 arcseconds before trailing is apparent on the imaging sensor. This corresponds to six tracking box diameters in the guiding eyepiece—and provides a very wide tolerance before the image is degraded. On the other hand, if the camera has a 1.6 crop factor, such as found with Canon 60D, 10D, 20D and 300D cameras, the 50-mm lens resembles an 80-mm lens, and we must divide the 120 arcsecond tracking tolerance by 1.6 to arrive at the actual allowable error—75 arcseconds, which is equivalent to about 3.5 tracking box diameters.

For telescopes other than the 2000-mm focal length Schmidt-Cassegrain in the above example, the size of the tracking box will be different. A

Fig. 9.6 *The guiding tolerances for piggyback astrophotography are surprisingly lenient. If a 200-mm or longer focal length is used, the guide star should never leave the dual crosshair "box" (left); but if a normal or short telephoto lens is used, the guide star can wander a few box diameters (center) without affecting the image. If a wide-angle lens is used, the guide star can drift four or five box diameters (right) without affecting the image. Photos by Robert Reeves.*

rule-of-thumb method to determine the box size is to compare it to an object of known size, such as the planet Jupiter, that averages approximately 40 arcseconds in diameter. For more precision, we can use a series of formulae published by Barry Gordon to calculate the field size of the guiding box in a 12-mm dual-crosshair eyepiece on other telescopes.

The first of Gordon's formulae is

$$A = \frac{41{,}250}{F}$$

where

    A = guide box size in arcseconds

    F = Guide scope focal length in mm (assuming a guide box reticle size of 0.2 mm).

The degree of guiding accuracy is dependent upon the magnification of the image on the imaging sensor. Extreme wide-angle views are naturally more tolerant of guiding and tracking errors than a narrow telephoto view. We can calculate the amount of error a particular photographic system will tolerate—in other words, how big a mistake we can get away with—by using another of Gordon's formulae:

$$T = \frac{DG}{BF}$$

where

    T = tracking tolerance in box widths

    B = guide box size in mm

    D = allowable image drift in mm

    G = guidescope focal length in mm

F = camera lens focal length in mm

Since we want to keep star drift under 0.03 mm on the imaging sensor, and we know that the guide box size in the 12-mm eyepiece is 0.2 mm; the above formulae can be simplified to

$$T = 0.15\left(\frac{G}{F}\right)$$

Thus, with a full 35-mm film sized imaging sensor and a 50-mm lens, the guiding tolerance works out to six times the tracking box width. Switching to a 200-mm lens reduces the guiding tolerance to just 1.5 tracking box diameters.

If a DSLR camera with a smaller imaging sensor is used, the results of the calculations must be divided by the camera's appropriate crop factor. Table 9.1 summarizes the guiding tolerances in tracking box diameters for popular lenses and various crop factors using the same 12-mm eyepiece and 8-inch f/10 telescope. As can be seen, the guiding tolerances for cameras with higher crop factors are tighter, but wide-angle views are still quite forgiving of guiding errors. In many cases with wide-angle lenses the amount of allowable error exceeds the periodic error of the telescope mount, making guiding a simple task.

9.16 Electronic Autoguiders

An autoguider is essentially a small CCD camera used in place of the illuminated-crosshair eyepiece. This device for all practical purposes instantly images the guide star and compares successive before and after images to detect when a star moves on the image sensor. Circuitry within the autoguider then determines the direction of movement and automatically issues electronic commands to move the telescope in order to re-center the guide star on the proper pixel. Although an autoguider is more convenient and less fatiguing for the astrophotographer, it is far more expensive than a simple illuminated eyepiece. The device also requires an interface between the guider and telescope drive motors to convert the guider's signals into voltage that is sufficient to run the drive motors. Additionally, the telescope mount must have electric controls on both the right ascension and declination axes, and inherently have an accurate drive rate free from excessive periodic error.

The trend in high-end CCD cameras has been to include a self-guiding function (circuitry and logic) to detect star drift and issue compensating commands. Less expensive, stand-alone autoguiders, are also

Table 9.1 Tolerances for Various Lenses and Crop Factors

Lens (mm)	Magnification	Tracking Tolerance (arcseconds)	Box Diameters
Full 35-mm Film Size Imaging Sensor			
18	0.36	333	16.7
28	0.56	207	10.3
35	0.7	171	8.6
50	1.0	120	6.0
85	1.7	71	3.5
135	2.7	44	2.2
200	4.0	30	1.5
300	6.0	20	1.0
1.3 Crop Factor			
18	0.49	256	12.8
28	0.73	159	7.9
35	0.91	132	6.6
50	1.3	92	4.6
85	2.21	55	2.8
135	3.51	34	1.7
200	5.2	23	1.2
300	7.8	15	0.8
1.6 Crop Factor			
18	0.58	208	10.4
28	0.89	129	6.5
35	1.12	107	5.4
50	1.6	75	3.8
85	2.72	44	2.2
135	4.32	27	1.4
200	6.4	19	1.0
300	9.6	13	0.6
2.0 Crop Factor			
18	0.72	167	8.3
28	1.12	103	5.2
35	1.4	86	4.3
50	2.0	60	3.0
85	3.4	36	1.8
135	5.4	22	1.1
200	8.0	15	0.8
300	12.0	10	0.5

Fig. 9.7 *The development of electronic autoguiders such as the Santa Barbara Instrument Group ST-4 has allowed the astrophotographer to achieve flawless long exposures without fatiguing and tedious manual guiding. Photo by Robert Reeves.*

available. These include the discontinued Santa Barbara ST-4 and Meade 201XT, 208XT and 216XT autoguiders. Meade units are simpler than the ST-4, but the 201XT lacks sensitivity, which limits it to wide-field imaging where the chances of finding a bright guide star are greatest. Meade autoguiders range from $400 to $700. The discontinued ST-4 is often available used for around $500 and is still, in my mind, is the best all around autoguider.

9.16.1 Setting Up a Typical Autoguider

The ST-4 is controlled with software running in a separate control box. For consistent results, the ST-4's array must be aligned with the east-west (right ascension) and north-south (declination) directions of the telescope's field-of-view. This is done by aligning the unit's serial number label with the proper direction of star movement as seen through a temporary eyepiece. The exposure and gain are roughly set for the guide star, then fine tuned by a trial-and-error process. The exposure and gain set-up must be repeated for each new guide star because the star and seeing conditions are never exactly alike, and then a dark frame is taken every time the exposure and gain setting are changed.

Focusing is accomplished by sliding the unit in or out of its holder until the star's readout on the ST-4's display shows maximum brightness. A good focus is critical to the proper operation of the unit. Since focus rarely changes significantly, quicker setup is possible if you mark the ST-4's barrel at the point of focus. Do not allow the brightness value to exceed 90 because this usually indicates a saturated star image, leading to poor guiding. The best brightness readings are between 10 and 60 in "normal mode."

Lower readings, between 8 and 16, are acceptable in "faint mode" and in fact, the ST-4 guides better with faint stars. Brightness is adjusted using a combination of exposure time and gain. The ST-4 cannot take images while it is issuing corrections, so the exposure time will be limited by the frequency of needed drive corrections. For a typical Schmidt-Cassegrain, this is usually every 2 to 3 seconds. Shorter-focus telescopes with brighter star images allow shorter exposures and correction commands at between a half- to one-second intervals.

Once the ST-4 is set up, a test should be run to confirm that it is properly aligned with the equatorial mount's cardinal directions by slewing in one direction and viewing the *x* and *y* axis error count. One count should not change while the other changes significantly, meaning the star is slewing properly along a row of pixels in the imaging head. Allow the ST-4 to properly calibrate itself to the telescope's response to tracking commands before beginning an exposure. The scintillation adjustment on the ST-4 allows the unit to dynamically adjust the tracking calibration if a star is bloated or dancing about because of scintillation (bad seeing). The effect of bad seeing will cause the autoguider corrections to overshoot. Setting the scintillation adjustment to a higher number will allow the ST-4 to change the calibration of its tracking response so as to remove overshoot.

9.16.2 STV: The Next Generation Autoguider

The Santa Barbara STV replaced the ST-4 and retails for just under $2000. In addition to autoguiding, it is also a cross between an entry-level prime focus electronic camera and an electronic amplifying eyepiece. The autoguider functions now have increased sensitivity, a video display, and software that eliminates the need for special orientation of the guider head. It does not require a separate computer. The system is comprised of a one-pound camera head that fits into a 1¼-inch focuser, and an 11½- by 9½-inch control box that runs on 12V DC and draws 2 amps. It has options that map telescope tracking errors in RA and Dec and outputs the results as either numerical data or as a graph on its LCD screen to help fine-tune polar alignment. Another function measures the size of a star's image to help focusing.

The STV is so versatile that a separate guide scope is not required. If the optional focal reducer lens is installed as an imaging objective, the STV can be mounted piggyback on a telescope. The unit is so sensitive that it eliminates the need to search for guide stars; many are always visible to the unit no matter where it's pointed in the sky. Enabling the unit's Auto mode sets in motion a sequence of events to automatically calibrate, calculate exposure time for the stars it "sees," and then select the brightest of these stars to guide on. Pushing the Calibrate and Track buttons then sets the STV to guide with arcsecond accuracy.

Chapter 10
Using a Digital Camera for Astrophotography

10.1 Introduction

In this chapter, we discuss how to set up your camera to image celestial objects. If you have not already done so, please read Chapter 4 , taking special note of the various forms of astrophotography discussed there. If you are doing piggyback or prime focus deep-sky astrophotography and are an experienced film astrophotographer, you should rapidly transition to digital imaging.

Regardless of what kind of celestial imaging you are attempting, the following points are critical:

1. The camera must be securely mounted. Lengthy exposures are not possible with a handheld camera or cameras mounted on a tripod independently of the telescope.

2. Do not activate the shutter manually. Vibration will degrade any telescopic exposure more than a few hundredths of a second in length. Use a remote shutter release or self-timer for hands-off operation.

3. Leave the LCD display off whenever possible. The heat from the LCD display is the biggest source of preventable noise in a digital camera.

4. Control the exposure yourself; do not let the camera's autoexposure control the shutter. Use the camera's manual settings in astronomical applications.

10.2 Built-In Camera Flash

The very first thing to do, if your camera is so equipped, is turn off the flash. While this may seem obvious, even the most experienced imagers have been known to accidentally "flash" their first celestial shot of the evening. This is even more likely since the camera can easily have been used for snapshots during the day.

10.3 Lens Focus Setting

For either prime focus or for eyepiece projection imaging, DSLR cameras are mounted with their lenses removed. The telescope focuses the image directly onto the image sensor. Fixed-lens cameras have to be set up with their lenses in place and properly focused for afocal photography. So then, how is the camera focused? When we look through a telescope, the eye is not focusing on the lens of the eyepiece; it is focused on a distant image produced by the telescope. The afocal-projection camera works the same way. The camera focuses an astronomical object at infinity. If we tried to focus the camera closer, we would simply be imaging the inside of the telescope tube. For deep-sky and planetary photography, the targets will be too dim for a fixed-lens camera's autofocus to work properly, so the camera must be set for manual focus with the lens set at infinity.

If the target is a bright object like the Moon, the camera's autofocus mode may be able to achieve the proper focus. However, be aware that it takes a second or so for the camera to adjust the focus after releasing the shutter so there will be a delay before the actual image is taken. Also, beware that not all autofocus cameras are analyzing the actual image; some focus using a sensor on the camera body, and thus will attempt to focus whatever is within the sensor's line-of-sight and not the image at infinity as seen through the telescope.

Although it is not a good idea to use a camera's digital zoom for astrophotography—the reasons will be discussed later in this chapter—this feature can be used along with the camera's LCD preview screen as a focusing aid—focus on a bright star using maximum zoom, and then refine the focus until the star is as small as possible on the LCD screen. However, most people find it difficult to achieve good focus this way. If your camera is equipped with a video output jack, an external 12-volt monitor can be used to provide a larger, easier to see image. A small automotive LCD video display like the InterAct Mobile Monitor works well. Some cameras have provisions to download images directly to a computer where they can be inspected for proper focus.

Another aid to focusing is to temporarily set your camera to its highest ISO rating that will give you a brighter image. After you have achieved focus, return the camera's settings to a lower ISO for imaging.

Solar photography presents a unique set of challenges. Here the problem is not too little light, but too much. High ambient light makes viewing the camera's LCD preview screen difficult, particularly when it is attached to the back of the telescope and the screen cannot be adequately shielded from glare. In cases like this, cameras that allow focusing with an external monitor or immediate downloading of focus test images to a computer are a great help. However, even with an external monitor, some sort of shade may be needed to shield the monitor from direct sunlight to adequately view the image.

It is important to know how your digital camera achieves focus. Not all cameras focus and preview the image with the lens set at its widest aperture. If you are shooting at a wide-open aperture setting, the camera may not actually be set to that aperture until the shutter button is partially depressed—that would mean that you may be trying to focus and frame with a dimmer preview that was taken with the lens stopped down.

Software can sometimes be a problem. You need to keep your camera's software up to date. Some early Nikon Coolpix models had a software flaw that prevented focus at infinity at full zoom. This was fixed by updating the camera's built-in software. Early models of Olympus fixed-lens cameras also had infinity focus problems. These cameras achieved "infinity" at a setting that was closer than true infinity, regardless of zoom setting. This was also addressed with changes in the camera's software. Consult your manufacturer's web site frequently to check for updates.

10.4 ISO Speed Setting

The ISO settings on digital and film cameras are essentially the same. The higher the number, the more sensitive the camera will be to light. With dim targets, it is tempting to use the highest possible ISO sensor speed, but there is a point of diminishing returns. Film cameras achieve higher ISO speed by using larger-grained emulsions to intercept more photons so the photochemical reaction occurs faster for a given brightness of illumination. Digital cameras, however, use the same sensor with the same number of pixels regardless of the ISO speed setting. Sensitivity to light is increased by electronically amplifying the gain on the sensor. This amplifies the signal created when incoming light strikes the sensor, but it also amplifies the electronic noise present in the sensor. Under normal sub-second daylight, or flash "snapshot" usage, this is not usually a problem; but on dim astronomical objects, the exposures span seconds or minutes, and any noise present will accumulate and become a factor in overall image quality.

To minimize electronic noise, it is generally best to use the lowest possible ISO setting that will image your target. You need to find a balance between sensitivity and unacceptable image noise; this balancing act is not unique to digital imaging. Astrophotographers have always sought a balance between having the finest film grain for high detail, but the fastest possible film for reasonable exposure length. No matter how you image, film or digitally, there will be a compromise between speed and image quality.

10.5 Exposure Setting

The inherent limitation to digital photography is noise that places an upper limit on how long a deep-sky exposure can be before sensor noise degrades the image, see Figure 10.1. Scientific-grade CCD cameras achieve long exposures with sub-zero cooling systems to reduce the electronic noise generated within the sensor. For the most part, an advanced consumer DSLR camera (uncooled) will be able to expose five minutes without being too noisy. However, do not despair about the short exposure limitations. These cameras' exposures can be "built up" by digitally stacking successive exposures of the same object using an image processing program. How "deep" the overall exposure will be is determined by how many successive images are stacked. The actual individual exposure time for each image is determined by trial-and-error to minimize the effects of noise. Fortunately, the instant feedback from digital cameras makes this relatively easy and quick to determine.

Planetary photography usually does not rely on a long exposure, so once the target is framed and focused, it is quicker and easier to simply try several test exposures to find the correct one. The results can be evaluated visually on the camera's preview screen faster than the correct exposure can be mathematically calculated or looked up in an astrophoto exposure reference such as the *LFK Exposure Guide*.

10.6 Resolution

In celestial photography, we usually strive for the greatest amount of resolution in our images. In film photography, we increase image resolution by using finer-grained film. In digital photography, we don't have this option; but by selecting various menu options in the camera's operating system, we can control the image resolution by choosing an image size with the proper amount of pixels. The greater the image resolution, the greater the number of pixels, and thus the image file size is increased. Typically, a camera will allow selecting between small, medium, and large image sizes with a range of file compression levels that can be applied to the image. If your camera has the option, selecting the RAW file output option will automatically set the image size to your camera's maximum resolution without any form of compression.

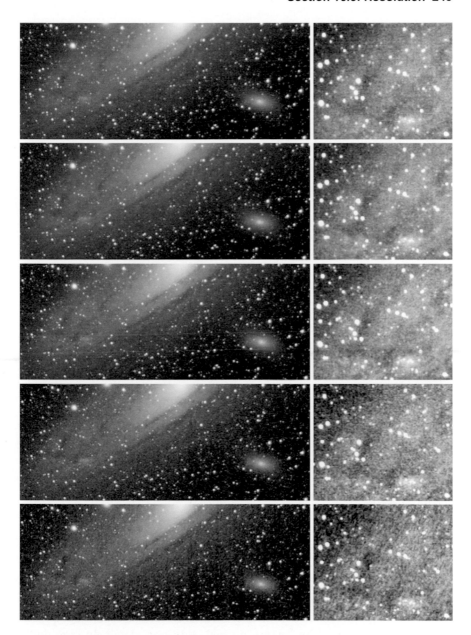

Fig. 10.1 *This comparative series of images taken with a Canon 10D demonstrates that the higher the ISO setting, the noisier and "grainier" the image will be. The top image is a combination of 16 one-minute exposures at ISO 100 with each lower image being half the number of exposures at twice the ISO. The lower image is a one-minute exposure at ISO 1600. Note the lack of difference in the overall sensitivity between the 16 exposures at ISO 100 and the single exposure at ISO 1600. There is, however, a marked difference in image noise at higher ISOs as shown in the enlarged frames. Photos by Phil Hart.*

By selecting image sizes smaller than the maximum, your camera "bins" the image. Binning is the process of combining the information from a group of pixels to create new "larger" pixels. Typically, the information from groups of pixels in a 2x2 array is combined to create a single pixel. This technique has the advantage of reducing the image download time that will speed up achieving initial focus through a telescope. However, this has the disadvantage of creating smaller images with less resolution; so once a reasonable focus is achieved, the camera should be reset back to high resolution to fine-tune the focus and for actual imaging.

The term resolution can mean different concepts beyond just the total number of pixels produced by the camera. It can mean the degree of sharpness as seen on a computer screen or on a printed page. It can also mean the number of bits per pixel. As applied to the displayed image, resolution can have two meanings; it can refer to the total image size, as in a 1200 by 1500 pixel image, or it can refer to the pixels per inch displayed in the print, as in 300 pixels per inch, which for our 1200 by 1500 pixel image will make a 4- by 5-inch print.

10.7 Optical and Digital Zoom

A fixed-lens camera with optical zoom capability is a good match for afocal imaging. Zooming in on the image projected by the eyepiece significantly reduces vignetting and magnifies the detail. However, the digital zoom feature should be avoided. It is generally agreed that the digital zoom is useful as a focusing aid, but basically useless for astrophotography.

While optical zooming spreads detail over larger areas of the imaging sensor, digital zoom simply takes the detail already on the central portion of the imaging sensor and uses software to expand that portion of the image to cover the entire area of the sensor. This is, in effect, simply enlarging a cropped part of the original image—you gain no additional detail. Specifically, with digital zoom, whatever detail there is to begin with is resampled from a small area and expanded to a larger area through software interpolation that fills the void between each "real pixel" with a "virtual pixel." The software usually does this by estimating the brightness values of the virtual pixels needed to fill the voids between the real image pixels. This results in a larger image, but no real additional detail. In fact, under critical examination, the detail is less sharp.

10.8 White Balance Setting

For the veteran film photographer, the white balance setting can be thought of as the choice between daylight and tungsten film. Daylight film is for-

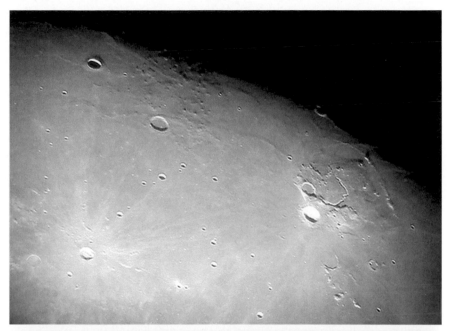

Fig. 10.2 *The use of digital zoom should be avoided because it provides no additional image detail. Using the camera's digital zoom feature on the Marius Hills shown in the lunar view in the top image results in a larger image (bottom) but no additional image detail. Photos by Robert Reeves.*

mulated to color balance bluish daylight illumination. If redder incandescent light illuminates the subject, tungsten-balanced film will compensate for the redder illumination to create an image that the eye perceives to be natural. Traditionally, astronomical imaging is done with daylight color-balanced film, and this is also true for digital astrophotography where the white balance is set on 5700K.

Most digital cameras have a white balance setting that achieves color balance under varying lighting conditions. Simple point-and-shoot cameras will have a selection between automatic and manual mode. More advanced DSLR models offer the ability to set the white point to a particular color temperature. For astrophotography, when saving images as TIF or JPG files, it is best to set the white point control to manual and leave it locked on the outdoor daytime setting of 5700K. When saving images as RAW files, the white balance setting does not matter because it is adjusted later during image processing. If multiple images are being taken of an object for stacking/averaging or dark frame subtraction, then it is very important that the white point be locked in a fixed manual setting. Stacking and dark frame subtraction require consistent images. If the white point is left on automatic, it can vary between exposures and lead to poor results.

10.9 Black and White Option

The Moon is a relatively colorless target. It appears to be monochromatic through a telescope with all detail appearing to be varying shades of gray. There are actually some areas of the Moon that record as dark brown in a color image, but the shade is subtle and often not noticed. Because the Moon appears to be a black and white object, it is tempting to use a digital camera's black and white imaging mode to image the Moon. However, this is only a pseudo-black and white! The camera's imaging sensor records all scenes in color, including those recorded as "black and white." While the image visually appears to lack color, it is, however, recorded as red, blue and green components that are assembled in the final image. The camera's software does this by changing the color image to render the scene in varying shades of gray, but the image is still a color picture. However, in the process of converting the image to this false black and white, some image information may be lost. It is far better to record lunar images in the camera's natural color mode. If true black and white images are desired, use an image processing program to convert them to true grayscale during processing. This way, no original image information is inadvertently lost during the interpolation process where the camera's software "guesses" at a shade of gray to render the color image.

10.10 Remote Operation

The ability to remotely trigger the shutter of any astro-camera without having to touch it is an absolute necessity for sharp images. It may be possible to achieve sharp hand-triggered images of the Moon using high shutter speeds, but planetary and deep-sky photography will require shake-free shutter operation. Fortunately, the majority of today's digital cameras have remote shutter triggers. Most have an infrared remote trigger that operates much like a small remote car alarm control. Other cameras may have a remote switch that attaches via a plug-in cable. Some models even offer remote controls with built-in intervalometers that allow programming a hands-free sequence of exposures of any given length. Computer programs are available for certain cameras that allow a laptop to control the exposure and other camera functions remotely via a cable link to the camera. These programs also allow downloading the image directly to the computer hard drive, freeing the photographer from the memory card capacity limitations.

If your camera has no provision for electronic remote operation, like my own Olympus 3020 Zoom, you have to trigger the shutter by hand while it is attached to the telescope. This is because electronic cameras do not accept the mechanical cable release used by standard film cameras. With any exposure faster than a high-speed lunar shot, this induces motion and vibrations that can degrade the image. A mechanical release can be used to press on the shutter button by fabricating a homemade adapter as shown in Section 12.7. However, if you are without any means of remote hands-off operation, use the camera's self-timer to make a delayed exposure. In the eight- to ten-second delay after the shutter release is pushed, the telescope motion will damp out by the time the shutter does trip. Another option is to use multiple exposures burst mode or sequence mode to make repeated undisturbed hands-off exposures after the first image is taken. This function varies from camera to camera and depends upon the size of each individual picture and the size of the camera's memory buffer. The smaller the resulting image files, the more images the buffer can hold. In some cameras, this option may be bundled into a menu that forces the choice between multiple exposures and some other option you are already using. However, if it is available, it is a convenient option for making repeated exposures of the Moon and planets.

10.11 Use of Filters

Filters have been used in film astrophotography for decades to either enhance certain types of celestial objects or block unwanted natural or man-made stray light that degrades the image. For the most part, the same filters that are useful for film astrophotography will also be useful in digital astrophotography. However, the infrared blocking filter placed in front of the imaging sensor of most digital cameras reduces their spectral sensitivity to red and thus reduces their effectiveness with objects that emit primarily the deep red hydrogen-alpha light. For this reason, red filters will have a diminished advantage with digital cameras.

The common light pollution rejection filters are effective and generally allow an exposure to be two to three times longer. The noise limitations of digital cameras may not allow light pollution filters to be used to the camera's longest possible exposure, but the resulting images will be "cleaner" with less light pollution. Shorter, cleaner, images can be stacked with an image processing program to increase the effective overall exposure.

A #15 yellow filter can help reduce the effects of chromatic aberration by reducing the excess blue wavelengths that produce an unwanted halo around bright stars. A #44 blue filter can be useful for photographing star clusters by filtering out the red nebulosity that surrounds some of them. A #44 filter is also effective for blue objects, such as some galaxies, and will reject enough light pollution to allow exposures of galaxies to be doubled. The benefits of the blue filter come at the expense of rendering the entire image blue. This situation is remedied by using an image processing program to convert the image to grayscale and print it as black and white. Details about the characteristics and use of various filters for astrophotography are discussed in Section 11.7.

10.12 Care of Light Pollution Filters

Light pollution rejection filters are more than just colored glass. They consist of layered deposits of various atomized metals and chemical compounds that are deposited on the glass filter base. For best optical quality, these layered deposits cannot be sandwiched under protective coatings, so they are "soft" and very vulnerable to damage during cleaning. For this reason, the manufacturer is specific about how to clean them if they get smudged or dirty. Lumicon recommends the following for their deep-sky filter:

1. First, wash your hands thoroughly.
2. Using compressed air or a blower brush, carefully blow away loose dirt.
3. Remove the filter element by unthreading the retainer ring using a jeweler's screwdriver in each of its notches.
4. Fold an unscented, plain facial tissue in half, lengthwise, and cut into approximately three equal pieces.

5. Pour a small amount of pure methanol (and no other type of cleaner) into a small container.

6. Soak the lower half of one of the tissue strips in methanol. Using only the weight of the strip itself, slowly draw it across the filter surface to remove any particulate matter.

7. Repeat step 6 up to two more times, each time with a fresh strip. If necessary, "breathe" a fog onto the glass along, with using the methanol swipe, to remove organic compounds. Be careful not to let any spray from your mouth deposit on the glass since it contains calcium carbonate, that, if allowed to dry, is nearly impossible to remove without damaging the sensitive coatings on the filter.

If you use a different brand of light pollution filter, check with the manufacturer for their specific cleaning instructions.

10.13 Electronic Image Noise

The warmer the imaging sensor is, the greater the amount of electronic noise it produces. Noise is the greatest enemy of long-exposure digital imaging and doubles for each six degrees Centigrade increase in sensor temperature.

A digital camera has numerous heat sources. Internal camera batteries warm-up significantly with constant use; so if possible, use an external power source so as to move battery heat away from the camera. Leave the LCD preview screen off as much as possible; turn it off after focusing and framing. The LCD screen, most often placed directly behind the imaging sensor, is a ready-made source of heat—and night-time astronomical imaging needs the coolest sensor possible. If the preview must be used, set it at its dimmest. Normal brightness settings are usually excessive for dark-adapted vision, so dimming the viewer will improve the user's comfort.

Another way to combat noise is to use the lowest ISO speed setting possible that still allows you to get the desired image. The Moon and most planets are bright enough that ISO 100 and 200 will allow short noise-free exposures. Deep-sky and piggyback imaging may require stacking a number of noise-free, low ISO exposures instead of attempting a higher ISO single exposure.

Every imaging sensor, whether it is film or electronic, detects both signal and noise. We define the signal to be the theoretical noise-free image. Noise is the background fog that dilutes or degrades the signal; in film photography, this can be light pollution, or it can be chemically-induced fog that builds in the film base during development. Noise in electronic imaging comes from a number of sources, such as:

Fig. 10.3 *Electronic image noise is an inevitable problem on long exposures with uncooled consumer digital cameras. As shown in these progressive dark frame images of a 10-minute exposure, noise builds during the exposure until it becomes objectionable and degrades the image. Photos by Paul Hyndman.*

- **Thermal noise**—the heat noise caused by electron activity in the sensor itself or the associated amplifier circuits. This is worsened by external heat sources, such as high ambient temperature, or warm components in the camera itself, such as the LCD screen.
- **Reception variation**—another way of saying something got in the way of the target, such as clouds, haze or dew.
- **Read noise**—caused by errors in the camera's amplifier circuit.
- **Quantization noise**—noise inadvertently added to the image during analog-to-digital conversion of the data.
- **Sensitivity variations**—caused by variations in the sensitivity of different pixels across the sensor.

Non-thermal sources of noise that are essentially built into the sensor and are present before an exposure is taken are called bias. When using CCD sensors in specialized astronomical cameras, the bias signal can be removed by subtracting what is called a bias frame, or zero second exposure, from the final image. This step is usually not required unless critical photometry is being performed. In digital cameras, there is no special function for eliminating a sensor's internal bias, and it is ignored as it rarely effects the image.

Tests of the Canon 10D by Roger Clark demonstrate that you can minimize noise and maximize sensitivity by carefully selecting the exposure time and ISO setting. For instance, if a total of 10 minutes exposure is desired, we know that a single exposure is going to have an excess amount of noise, rendering the image difficult to process at best and possibly ruining it. Roger's test, performed at room temperature and looking at signals at the limit of detection, showed that highest signal-to-noise is achieved by combining a series of 10 one-minute exposures taken at ISO 1600. However, these ISO 1600 exposures have a higher peak noise, that will increase visible speckles on the final image when compared to a lower ISO image. A combination of 5 separate two-minute exposures at ISO 400 showed less noise while retaining 90 percent of the signal-to-noise ratio displayed by the ISO 1600 combination. The sensitivity is statistically almost the same between the two combinations, but the ISO 400 combination will have less noise and produce a more pleasing image.

At near freezing temperatures, Roger's tests show that two- to six-minute exposures at both ISO 400 and 1600 both produce good results because the sensor is significantly cooler. Remember that Roger's tests were performed to detect the faintest possible signal. With bright objects, such as the Orion Nebula, a lower ISO setting can be used to record the brighter signal with even lower sensor noise.

The trend in digital photography has been to increase pixel count. For a variety of reasons, this is usually done without a corresponding increase in the sensor's physical dimensions. This means that fewer photons strike each photodiode. Every imaging sensor has an inherent amount of noise. If there are fewer photons striking the photodiodes, but the background noise remains the same, the signal-to-noise ratio is decreased.

When consumer digital cameras entered the two-megapixel range, noise became the hot topic even for snapshot pictures. Some experts predicted that three megapixels would be a barrier that could not be crossed without significantly increasing the size (and cost) of the imaging sensor. Today, thanks to improvements in the efficiency of electronics we have six- and eight-megapixel consumer cameras using relatively small imaging sensors that have acceptable noise levels for consumer applications—but noise does remain a limitation for long-exposure photography.

Unlike film cameras that use a common brand and type of film, no two digital cameras will respond exactly the same when making long exposures. For optimal performance, you will have to perform tests to find the point where noise overcomes the added sensitivity of higher ISO settings.

10.14 Auto Bracketing

Exposure bracketing can be useful in certain circumstances. This function automatically repeats the exposure at several slightly slower and faster speeds or aperture settings around the one selected. This can help when photographing the lunar terminator where the Sun illuminates the landscape at an extreme angle. Under these conditions, it is difficult to get a perfect exposure of both the fully-illuminated Moon and the dark shadowed area where the Sun is just rising or setting. Auto bracketing may or may not hit the proper balance of exposure for such an image, but automating this function into a multiple exposure burst is easier than manually altering the settings.

10.15 Auto Exposure Lock

Auto exposure lock is useful when taking multiple exposures of the Moon or planets if the camera has only automatic exposure control. This option allows the camera to retain the same exposure parameters when using aperture priority or shutter priority modes. In both of these modes, you set either the aperture or the shutter speed, (usually the aperture is set to wide open when using afocal photography) and the camera controls the other setting to produce an acceptable exposure. (Ignore this option if your camera allows full manual control of both the shutter and aperture.) Auto exposure lock is useful when imaging multiple areas of the Moon to assemble a high-resolution mosaic. Once the camera selects the optimal exposure for a typical area of the Moon, engaging the auto exposure lock retains this same exposure for the rest of the images regardless of the changing illumination across the Moon's surface. This simplifies matching the different areas when assembling the mosaic.

Auto exposure lock can also be used to make "light frames" which are used to subtract background light from the image. This differs from the more familiar dark frame technique where we use the dark frame to subtract the camera's internal sensor noise from the image. Light frames are exposures of a blank part of the sky near the desired target. They pick up light pollution or Moon glow. Auto exposure lock will prevent the camera from changing exposure settings between the target and sky background when applying the light frame technique.

10.16 Memory Card

Cameras vary greatly in their image storage capacity. With anything involving computers, memory is an important consideration and digital cameras are no different. A camera's memory is determined by its removable memory card that can be solid-state flash memory or a miniature hard disk. The capacity of the memory card is analogous to how long a roll of film is in a film camera: the longer the roll, the more pictures. The greater the memory of a digital camera, the more pictures it will hold. But unlike film, memory cards can be erased and used again once their contents have been

Fig. 10.4 *The high-capacity CompactFlash memory card (right) is commonly used with DSLR cameras. Newer fixed-lens models are using the small Secure Digital memory cards (center) that have capacities up to one gigabyte. Older Smart Media cards (left), that are limited to a maximum of 128 megabytes, are becoming obsolete. Photo by Robert Reeves.*

transferred to a computer hard drive or a permanent storage medium like a CD. Once an image has been written to the memory card, it will remain even if the camera is turned off or the batteries die.

With film astrophotography, lengthy exposure times make it difficult to use up an entire roll of film on deep-sky images in one evening. Digital photography approaches deep-sky imaging in a different manner. The inherent noise in long digital exposures makes it necessary to combine large numbers of shorter exposures, and calls for as much memory card capacity as we can get. 128-Mb Smart Media cards for three megapixel fixed-lens cameras are inexpensive, costing little more than several rolls of color slide film. Intermediate and DSLR models use more expensive CompactFlash memory cards, but the six- to eight-megapixel images they produce are memory-hungry. A six-megapixel DSLR outputting images in RAW format will fit less than 40 images on a 256-Mb memory card. Digital astrophotography's inherent need for memory dictates the use of the highest capacity memory card that you can afford.

If small memory card capacity is a problem, another option is that most cameras, when connected to a computer through a USB cable, will by default download to the computer's hard drive. To make this happen, you must start the camera without a memory card installed, and the camera must be started while connected to a computer.

There is no clear standard camera memory card. Confusing the issue is the fact that new memory card styles have been recently adopted by manufacturers for their new cameras. There are three popular types in use now, and some cameras can accept several types of cards. The very thin Smart Media cards are slowly going out of favor and are being replaced by the smaller, higher capacity Secure Digital cards. Most DSLR models use the thicker, high capacity, and more cost-effective CompactFlash cards. Sony also makes the Memory Stick, used only by their products.

Some of the memory options in use are:

- **Built-in memory** Inexpensive cameras have built-in flash memory and no removable memory card.

- **SmartMedia (SM)** SmartMedia are small thin Flash memory modules that can be removed from the camera. SmartMedia memory capacity has topped out at 128 Mb, a capacity generally regarded as too low for to-day's higher-resolution digital cameras. While still supported for older models, this style of memory card is being phased out.

- **CompactFlash (CF)** CompactFlash is another form of Flash memory but the cards are larger, thicker and have much higher capacity, as high as 8 Gb per card. CompactFlash cards measure about 2-inches square and are among the most popular cards in use.

- **Hard disk** Cameras capable of accepting CompactFlash cards can use Microdrive hard disks that are miniature high-capacity hard drives with a small spinning disk inside.

- **Memory Stick (MS)** Memory Stick is a proprietary form of Flash mem-ory only used in Sony cameras. The card's shape is like a small stick of chewing gum, and it comes in capacities of up to 1 Gb. An anti-erasure feature to prevent inadvertent data loss is standard.

- **Secure Digital/MultiMediaCard (SD/MM)** A new memory standard called Secure Digital has evolved, which utilizes a card about half the size and twice as thick as SmartMedia. Secure Digital comes in capaci-ties up to 1 Gb and has an anti-erase tab that can be set to prevent data loss. Confusing the issue is a similar-shaped MultiMediaCard, because in some cases the two are not interchangeable. There is also the MiniSD card that is about the size of a U. S. dime with capacities up to 256 Mb.

- **XD Picture Card (xD)** Is a format adopted by Fujifilm and Olympus that uses a card about the size of a large postage stamp and comes in capaci-ties up to 1 Gb.

- **Floppy disk** Some older, essentially obsolete, cameras write images di-rectly to a 3.5-inch computer floppy disk.

- **Writable CD** Some cameras have the ability to write images directly to a writable CD.

Some early digital cameras used PCMCIA cards for image storage. The acronym stands for Personal Computer Memory Card International Association and is the same type of device widely used in laptop comput-ers as a networking interface, or used for expanding the storage, input/output capability or processing capability of the host device. In the case of digital camera image storage, the card is used in the camera and not connected to a computer.

A PCMCIA card resembles a thick credit card and has a 68-pin connector on one end. The form factor, or design of the card, allows three types: the Type I, Type II and Type III PCMCIA card. Overall, their length and width is the same and they all have a 3.3-mm thick guide rail along their edges to align them in the receiving slot of the host machine, but the Type II and Type III cards are thicker; 5 mm and 10.5 mm, respectively. Type II cards will also fit in Type III slots, and a Type I card will fit any PCMCIA slot.

CompactFlash cards are the memory cards of choice because of their high capacity, small size, very low power consumption and built-in controller. Although their high capacity, reaching into the gigabytes, can make an individual card an expensive investment, CompactFlash cards achieve the highest megabyte per dollar capacity of all memory cards, making them the cheapest to use per megabyte. Like all memory cards, they do not need power to retain their data. CompactFlash cards are about half the length of a laptop PCMCIA card and have a 50-pin interface instead of 68. They are very rugged, being able to function in temperatures between –40 and + 85 C and withstand a 2000 G shock impact. These devices come in two versions: a Type I or Type II CompactFlash. The difference is their thickness; the Type II being twice as thick at 5 mm. Type I cards will fit the interface for the thicker Type II cards. The popular high capacity Microdrive, a miniature rotating hard drive, is a Type II card; but the majority of CompactFlash cards are solid-state memory devices. They were introduced as 32-Mb devices and have steadily evolved to multiple Gigabyte capacities. They allow very fast read times, now theoretically being able to write data at 5.2 Mb/sec and read data at 6.1 Mb/sec if the camera is capable of supporting such speeds. Cameras use either a 3.3- or 5-volt operating system, and the CompactFlash card can be used with either voltage.

It is easy to be confused about the capacities of CompactFlash cards. They are advertised as having a specific megabyte capacity, but often show less capacity when installed in the camera. This is because the manufacturers advertise the capacity by counting 1,000,000 bytes as a megabyte, while the camera operating system will count the megabyte capacity by 1,048,576 byte increments like the operating system of most PCs. Thus, a card such as a Sandisk Type I CompactFlash card with a label of 512 Mb will have an actual in-camera capacity of only 487 Mb. Not all manufacturers use this misleading rating. The IBM Microdrive CompactFlash card states a capacity of 1024 Mb and delivers a true 1024 Mb capacity in the camera.

The data transfer speed for a CompactFlash card is rated similar to that of computer CD drives—the speed is a certain number times "x." Thus, a CompactFlash card rated at 12x will theoretically transfer data at 12 x150 kb/sec, or 1800 kb/sec. In practical application, this transfer speed is not reached. Manufacturers also do not state how their transfer speeds are tested, or whether the given figure is read or write speed, that differ for a given application. Type II cards are generally faster than Type I, but the individual camera's operating system determines the actual achieved read and write speeds for a given card. One of the fastest CompactFlash cards is the 1-Gb IBM Microdrive. However, its performance varies greatly from one camera to another. For instance, the Microdrive will write 510 kb/sec, or 3x, in a Nikon Coolpix 995 using RAW format. The same card installed in a Canon EOS D-30 will achieve 836 kb/sec write speed, or 6x, with RAW format. More advanced professional cameras, such as the Canon EOS 1D using the same 1-Gb Microdrive will achieve a blazing 16x write speed, recording 2,345kb/sec.

Lesser-known or generic cards can be as fast, or even faster, than brand name cards. Confusing the issue is the fact that some card manufacturers will offer a "standard", as well as a "premium", card that is supposed to achieve faster performance. In use, the advertised gain in performance from a premium card is not realized, and it is probably not worth the added cost. Until makers arrive at a common standard for measuring Compact-Flash card capacity and the read/write speed when interfaced with a given camera, it is up to the buyer to research the expected performance of a particular card in their camera. A good place to find comparisons is dpreview.com.

IBM invented the Microdrive CompactFlash card. After the device became very popular, the name "IBM Microdrive" began to be used generically to identify all hard disk memory cards. However, IBM no longer markets the device, having sold its hard drive business to Hitachi. Hitachi also makes one-inch Microdrive Type II CompactFlash cards in up to 4-Mb capacity.

Another CompactFlash card using a small hard drive is made by GS Magicstor, Inc. and is resold under the brand names of Transcend, PNY and U.S. Modular. However, this new mini-drive, particularly the 2.2-Gb versions, has a history of compatibility problems with cameras that operate normally with the IBM and Hitachi Microdrives. However, other options exist. Seagate offers both 2.5- and 5-Gb 3600 RPM one-inch drives. More manufacturers will enter the one-inch drive market, not only for photography, but for devices such as MP3 music players, hand-held video games,

Table 10.1 Memory Card Write Speed

Card Designation	Minimum Sequential Write Speed
4x	600 Kb/sec
8x	1.2 Mb/sec
12x	1.8 Mb/sec
16x	2.4 Mb/sec
32x	4.8 Mb/sec
40x	6.0 Mb/sec
80x	12.0 Mb/sec

and advanced GPS location finding equipment. There should be, in the future, sufficient competition to insure that these high-capacity devices will be an affordable storage medium for the digital astrophotographer.

A word of caution about high-altitude operation with a Microdrive CompactFlash card: the maker cautions that the drive needs air to float the drive heads above the spinning disk. Operation above 10,000-foot altitude is not recommended because the density of air is not sufficient for the design of the drive head, which could strike the media surface and cause loss of data. High altitudes or vacuum do not affect the unit if it is not in operation. This limitation is usually of no consequence to astrophotographers unless they are lucky enough to visit high-altitude observatories such as Mauna Kea.

While for various reasons the actual data writing speed in the camera is usually less than the card's rated write speed, this value can be used as a benchmark for performance comparisons. The faster professional-level cards are, of course, the more expensive. In general, entry-level cameras will operate satisfactorily with a 4x card, while intermediate cameras aimed at photo enthusiasts will give good results with a 12x card. Professional DSLR cameras outputting 6-Mb image files need a 40x or 80x card to prevent a noticeable slowing of performance while recording images. See Table 10.1.

No matter what camera you use, an important accessory is a memory card reader to import images into your computer. The devices attach via a USB connection and cost between $30 and $50. Most models accept all common styles of memory card and are well worth the expense in added convenience and camera safety. Computers with a memory card reader are now in production, and they should soon become a standard feature.

Table 10.2 Approximate Image Capacity for Memory Cards *

Camera Type	File Size	32 Mb	64 Mb	128 Mb	256 Mb	512 Mb	1 Gb	2 Gb	4 Gb
2 Mpix	900K	35	71	142	284	568	1137	2275	4551
3 Mpix	1.2 M	26	53	106	213	426	853	1706	3413
4 Mpix	2 M	16	32	64	128	256	512	1024	2048
5 Mpix	2.5 M	12	25	51	102	204	409	819	1638
6 Mpix	3.2 M	10	20	40	80	160	320	640	1280

* Capacity assumes the highest JPG quality and most complex image possible. Actual capacity will likely be higher.

10.17 In-Camera Image Processing

Most digital cameras have in-camera image enhancement options, such as differing levels of contrast adjustment, sharpening and dark frame subtraction for long exposures. For the most part, it is best to disable contrast and sharpening functions when using the camera for astronomical imaging. These features modify the image before it is saved to the memory card, and in the process you lose some of the original image information. It is best to retain control of these functions and perform them yourself with an image processing program after the exposure. This is especially true if you plan to do image stacking and dark frame subtraction. To be really effective, image processing requires that you have all the image data your camera is capable of capturing. Remember, if you have all the data to begin with, you can always undo and correct any processing mistakes; but if the camera performs the processing before saving the image, the data is permanently lost.

Your camera may have the option to completely disable in-camera image processing functions. If it does not, the contrast controls in some cameras may have settings such as "Low," Normal" and "High," and the sharpening controls may be similarly labeled "Soft," "Normal" and "Hard." Read the camera instructions carefully because these settings can be misleading. Contrast can always be changed in an image processing program, so leaving the contrast control on "Normal" will allow you to change that parameter. With the sharpening function, however, "Normal" would not be the least amount of processing done by the camera. If you want complete control over this function, it should be set to "Soft" so the camera does minimal modification to the image, and leaves you with a "less processed" image containing more original data to work with. The lesson here is to read the instructions for your particular camera as these functions may work differently for various camera brands. If the instructions are not specific enough, run some tests and determine for yourself what is acceptable to you.

In-camera dark frame subtraction can be useful with fixed-lens cameras that have limited time-exposure capability, but care must be taken to insure that the software algorithm does not treat pinpoint stars as "noise." If this happens, the camera will eliminate the signal we are seeking in the image, as well as the noise. These cameras generally limit time exposures to between 16 and 60 seconds, depending on the model. For those who are not fully comfortable with the procedures for dark frame subtraction, having the camera perform this function with fairly short exposures can be a benefit. Cameras that allow multi-minute exposures, however, can lead to excessive periods of time where the camera is not actually exposing the sky. Dark frame subtraction requires that a dark image be taken of equal exposure length to the image the subtraction process is being applied to. If your target calls for, say, 15 successive two-minute exposures of an object for later stacking, each actual exposure would be followed by a two-minute period where the camera is taking a dark exposure. This would push a series equaling a 30-minute total exposure to a process taking over an hour, plus the time it takes to record each image to the camera's memory card or computer hard drive. With more advanced cameras that do allow such lengthy exposures, and if the noise characteristics of the imaging do not change significantly after several exposures, it is best to take dark frames before the exposure sequence and then apply that dark frame to all the subsequent exposures using an image processing program. Details of such procedures are in Chapter 14.

10.18 Calculating Image Field Size

A characteristic of photography through the telescope is a limited field-of-view. On the other hand, piggyback photography can, with various lenses cover large areas of the sky, often encompassing multiple constellations. When planning a photography session, we may want to know what the field-of-view is for either various zoom settings or for a lens of a particular focal length. There is a simple formula that will tell us field size for various focal lengths used with a given image sensor size.

Let's determine the field-of-view for a DSLR camera using a 28-mm lens and the four-thirds standard imaging sensor, that has a 13.5- by 18-mm dimension with a diagonal measurement of 22.3 mm. We use the following formula to determine each dimension of the image frame:

$$F = \frac{57.3}{L} \times X$$

where

$F =$ field (width, height or diagonal) in degrees

$L =$ lens focal length in length units

$X =$ image sensor (width, height or diagonal) dimension in same length units.

Applying this formula to the 28-mm focal length lens, we come up with the following horizontal field width

$$F_{\text{Horizontal}} = \frac{57.3}{28} \times 18 = 36.836°$$

We can round this off to 36.8 degrees and complete the calculations for the other image sensor dimensions. We find the field coverage with this lens and sensor is:

13.5 mm = 27.6 degrees vertically

18.0 mm = 36.8 degrees horizontally

22.3 mm = 45.6 degrees diagonally

The same formula will work for prime focus telescopic photography just as well as it works for camera lenses. Applying the formula to an 8-inch *f*/10 Schmidt/Cassegrain telescope, we find the following horizontal field-of-view with the four-thirds sensor

$$F = \frac{57.3}{2000} \times 18 = 0.516°$$

We can round this figure off to a half-degree horizontal field-of-view.

The value of 57.3 in the above formula is a significant number that will enable us to determine a number of useful facts about the scale of an image when photographed at the prime focus of a telescope:

a **57.3-mm** focal length will project **one degree/mm** on the focal plane,

a **57.3-inch** focal length will project **one degree/inch** on the focal plane,

a **57.3-foot** focal length will project **one degree/foot** on the focal plane.

10.19 Image Storage

The digital files created by your camera are the equivalent of film negatives. Without them, there is no image. The proper handling and storage of these digital files is as important as preserving and storing old negatives and color transparencies. In some ways, the problems of storing them are greater; the digital media used to store your images must be capable of being read by an electronic device in the future, or the image is lost. Today, the storage media of choice are recordable CDs or DVDs. There is no guarantee that CD or DVD readers or players will even be a consumer item in 20 years. Obsolescence of the format is a real possibility. Think of how many 8-track music players you see in stores now . . . none. Everyone with music stored on 8-track tapes that were popular in the 1970s can no longer purchase a means of playing them. The same may be true of the digital CD and DVD format in the near future.

A storage medium must also be long-lived to prevent the loss of the image. While CDs and DVDs are advertised as having a lifetime measured in decades, this level of archival storage is heavily dependent on the environmental conditions in which the medium is stored and the quality of the CD itself. Exposure to sunlight or UV radiation is detrimental to recordable and re-recordable CDs, because it alters the dyes in the plastic base. Under adverse conditions, a recordable CD can become unreadable within six months. People think that because their CD burner successfully burns all the CDs on a spindle that it naturally follows that the quality of those CDs is good. This is not necessarily true. This just shows their CD burner is compatible with that particular brand of CD. In fact, the manufacturers of CD writers have become very good at making their hardware compatible with all kinds of CD media. Successfully creating a CD says nothing about how long the data on that CD will be readable.

Some CD brands have a reputation of marketing more reliable CDs than others, but the situation becomes confusing when a brand switches from one factory to another for a different production run. Comprehensive tests of recordable CDs have revealed some surprising results. Brand names we would normally trust sometimes fare poorly on data longevity tests. For an informative discussion about this issue, perform an Internet Google search using the keywords "CD reliability." The overall conclusion, however, has to be that you get what you pay for. Higher quality CDs cost more than the blank "no name" CD at discount electronics stores. However, even the higher quality CDs need to be stored at moderate temperature and humidity, and protected from UV and bright light.

Chapter 11
Digital Astrophotography Techniques

Before you venture out under the stars to do your first imaging with a digital camera, practice operating it in full daylight until you feel confident that you can easily work all the controls. It is quite easy to get lost dealing with the many features, options, and operating modes found in even the most basic digital camera, especially when you do it in the dark! Once you feel comfortable in daylight, move to a dark room and test your skills there. If you can set up your camera in a dark room, your first trip under the night sky will be far easier and more productive.

11.1 Focusing Through a Telescope

For most people, fast f/ratio telescopes seem to be easy to focus. The image appears to "snap" into focus quickly. But, are you really focused? Fast systems have a narrow range where the image is actually in focus. So, while the process may seem easy, actually finding the precise zone of where exact focus is achieved is not a trivial task. Table 11.1 illustrates why. A fast f/4 system, when compared to an f/10 system, has a 2.5 times narrower in-focus range and that range is measured in the thousandths of an inch!

Table 11.1 Depth of Focus (inches) vs. Focal Ratio

Focal Ratio	Depth of Focus	±Total Range
f/4	0.002	0.004
f/5	0.0025	0.005
f/6	0.003	0.006
f/7	0.0035	0.007
f/10	0.005	0.010

Focus is also dependent on the thermal stability of the telescope. As the temperature varies throughout the night, a telescope tube will cool and shrink. This is especially true with the aluminum ones used in popular

Schmidt-Cassegrain telescopes. The tube radiates heat and becomes slightly cooler than the nighttime air. You know this is happening when dew forms on a surface. Field experience has shown that the focus point for a metal tube 8-inch f/10 Schmidt-Cassegrain shifts nearly 0.035 inches with a 15°F drop in temperature. This is more than three times the acceptable depth-of-focus range for this instrument, and demonstrates that not only do we need to focus precisely, we need to refocus often as the telescope cools throughout the evening. While you cannot see this easily through the telescope while imaging, you can verify this effect through a guiding eyepiece. It has been my experience that after an hour, the guide star image has softened, and I need to slightly twist the focus knob to bring it back into focus. Frequent verification of focus during the evening is much easier with a GOTO telescope mount. You command your telescope to quickly slew to a nearby bright star for refocusing and then easily slew back to the target for more imaging.

Focusing through a telescope requires more precision than simply looking through the viewfinder and bringing the image on the camera's monitor into focus. It is impossible to pick the point of best focus with this method. You are going to have to resort to something more precise like a Hartmann mask or crossed wires in front of the telescope aperture to create diffraction spikes (discussed in Sections 12.10 and 12.11) or specialized focusing software like DSLRFocus. It is necessary to use a bright star when using a Hartmann mask, because the mask blocks most of the incoming light. Whatever method you choose, you will also need to attach a numeric scale, to your telescope focusing knob. Having such a scale will allow you to mark and easily return to that point where you achieve best focus. This scale can be easily made with card stock and double-stick tape.

11.2 Focusing a Digital Camera for Astrophotography

While digital cameras make many of the tasks associated with imaging the heavens easier, focusing on a dim celestial object is not one of them. The optical viewfinder on many fixed-lens digitals is useless for focusing, as is the low resolution LCD preview screen. The situation is little better with SLR-like cameras that present what the imaging sensor sees through a small LCD display in the viewfinder's eyepiece. Similarly, true digital SLR cameras present very dim views on the focusing screen. If we are to accurately focus a digital camera, we are going to have to resort to methods other than what the manufacturer provides.

11.2.1 Focusing a Compact Fixed-Lens Digital Camera

For bright targets (Sun, Moon and bright planets), and an afocally-coupled camera, the autofocus feature found on compact fixed-lens digital cameras will usually achieve the best focus once the operator has achieved a coarse visual focus. However, this will only work with cameras using through-the-lens autofocus—not those that focus with an external rangefinding sensor.

When focusing a fixed-lens digital camera, remember the following points:

1. Even though the camera's lens is set for widest aperture it may default to a smaller aperture until the shutter button is partially depressed. A lens that remains partially stopped down presents a dimmer image for focusing. If this is the case, partially depressing the shutter button will brighten the image during the focusing process.

2. Some cameras present a brighter image for focusing if they are set to their highest ISO speed setting. Experiment to see if this works with yours. Again, the shutter button may have to be partially depressed in order for the image to brighten. Remember to change the ISO setting back to the desired level after focusing.

3. Partially depressing the shutter button while also operating the telescope's focus control can be a challenge. Holding the button will also make it difficult to keep the telescope steady while watching the LCD preview screen. This task can be simplified by building a device as shown in Section 12.7 to allow using a standard mechanical camera cable release to lock and hold a partially-depressed shutter button while focusing.

Some early Nikon Coolpix and Olympus cameras have a software flaw that improperly calibrates their infinity focus. The Coolpix reached infinity at the 30-foot setting. Olympus also focused at a point less than infinity, regardless of the zoom setting. In most cases, this problem can be cured by shipping the camera back to the manufacturer for an internal software upgrade.

When using the afocal method, the optical axes of the camera and telescope must coincide. The front lens of the camera must be square with the eyepiece. If this is not the case, no amount of focus refinement will achieve sharpness across the entire field-of-view. One side or the other of the image, or the center but not the sides, will be in focus while the rest is not.

Beware of field curvature in Schmidt-Cassegrain telescopes. If a significant amount of field curvature is present, no amount of focus refinement will produce sharpness across the entire field. Focusing in the presence of severe field curvature will be a best-fit compromise.

A stellar image is far too small for the autofocus option to lock onto; a different approach is needed. If possible, set your camera to manual focus and adjust it to infinity. Attach it to the telescope and zoom in as far as possible. It may also be useful to apply digital zoom if the camera has that option. Point the telescope at a bright star and adjust the focuser so the star's image is as small as possible on the preview screen. Once you get the image focused as best you can, turn off the digital zoom, since it does not provide any real gain in magnification. Since the resolution of the preview screen is far less than the camera's imaging sensor, this method does not guarantee precise focus.

If your camera has a video outlet jack, the ideal focus method would be to hook up an external monitor or small television. These devices have much more resolution than the camera's own LCD preview screen, and have the added benefit of avoiding heat buildup on the image sensor that can happen when the LCD preview is running.

A more accurate, but time consuming, focusing method uses a Hartmann mask. The Hartmann mask has dual apertures that render an out-of-focus star as a double image. When the two images merge, the telescope is in focus. If your camera will download images directly to a laptop, each focus test image can be easily checked, and the focus improved upon by successive iterations. If no computer connection is available, I have found a workaround using a USB memory card reader on a laptop to shuttle the memory card between the camera and computer after each exposure. With the camera set for manual operation, I simply pull the flash memory card from either my Olympus or Canon 10D. Both of these cameras shut down when the card door is opened and restart when it is closed. Both also retain their setting when the memory card is re-installed. Check your manual to determine if your camera has this capability. Of course, I do this when the camera is not recording data onto the card.

Once camera and telescope are focused, you should normally expect to find no change in focus when pointing to various celestial targets. However, if you are using a Schmidt-Cassegrain telescope, expect mirror movement and focus shift—a common problem with this class of instrument—and verify focus after each significant movement from one object to another.

11.2.2 Focusing Digital SLR Cameras

The optical viewfinder in a digital SLR, although dim, provides a far better focus than the LCD preview screen on any digital camera. Still, the view leaves much to be desired. Fortunately, there are two superior ways to fo-

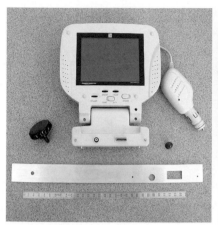

Fig. 11.1 *The Interact mobile monitor (left) is a popular 12-volt powered video monitor for checking the focus of a video outlet equipped digital camera. The lightweight monitor can be mounted directly on the telescope (right) for use with the camera. Photos by Paul Hyndman.*

cus a digital SLR; one is a software program and the other is a knife-edge focuser.

11.2.3 Software Based Focusing

A laptop computer interfaced with a digital SLR camera will allow you to use a powerful but inexpensive program called DSLRFocus (see Appendix B). One of the many features of this software allows critical analysis of the focus for any image that you can download from your camera. This program allows one to scroll around an image to find a suitable star for focus analysis. Once a star is found, you click on it and a close-up window showing that star opens up allowing detailed examination of the image. The window displays the star image, plus graphical and statistical information on the area around it, within a 100 by 100 pixel zone. By comparing that image to succeeding images, each taken after a focus adjustment, the true point of best focus can be found. The statistical data about the star image is contained in parameters like full width half maximum, radial sum value, and peak value. The full width half maximum parameter should be as narrow as possible, the radial sum value should be as small as possible and the peak value should be as high as possible, but below 255. If a peak value of 255 is reached, the star is too bright for focus analysis, and either a shorter exposure or a dimmer star needs to chosen. With the statistical information, the exact focus point can be determined with far greater accuracy than by visual examination of the star's image on the computer display.

DSLRFocus provides a wide selection of control functions for a steadily increasing number of cameras as they become available. Control

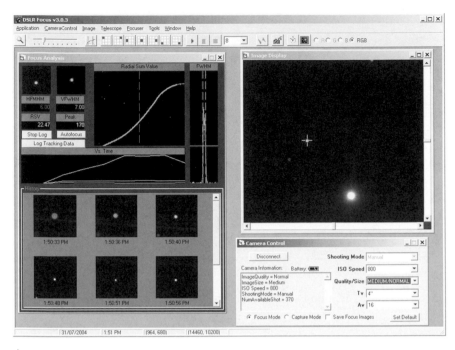

Fig. 11.2 *DSLRFocus software assists in achieving exact focus by analyzing a selected star image and providing a graphical analysis of its parameters. Successive images are analyzed until the best focus is reached. Image by Chris Venter.*

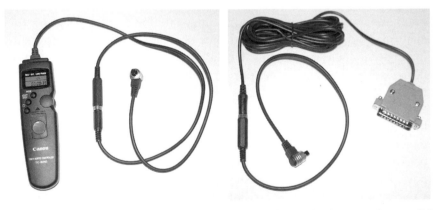

Fig. 11.3 *Camera control with DSLRFocus must be performed through either a serial or parallel cable interface instead of the traditional USB interface. Hap's Astrocables will adapt the owner's remote shutter switch (left) with a quick disconnect so the proprietary camera connector can be attached to the appropriate interface connector (right) for use with DSLRFocus. Photos by Hap Griffin.*

of extended exposure on "bulb" and programmed sequences of lengthy exposures over 30 seconds in length is possible. This important astronomical feature is not available with Canon's own Remote Capture software, that is limited to 30-second exposures. For instance, with DSLRFocus it is possible to enter parameters that allow your Canon camera to take 10 exposures of 60 seconds each at a particular ISO setting, set the time between exposures at 20 seconds to allow the images to fully record or download into the laptop, and then activate the mirror lockup when the exposure sequence begins with a click of the computer mouse (or Enter key on a laptop). If the telescope is being controlled with an autoguider, this allows totally hands-off automated control of a lengthy digital imaging sequence.

A special serial or parallel interface cable must be purchased in order to use all of the DSLRFocus features (on supported cameras). The standard USB interface cable that comes with the camera will not work. These cables can be purchased from Hap Griffin (see Appendix A).

Macintosh computer users also have a camera focus program, written by Steve Bryson, called iAstroPhoto (See Appendix B). This program was inspired by the DSLRFocus Windows program and allows control of Canon cameras through a USB interface with computers using the Mac OS X 10.2 or higher operating system.

11.2.4 Knife-Edge Focusers

A DSLR camera can be focused using a knife-edge focuser. This is a device that is attached in place of the camera body on the telescope, and uses a Ronchi screen to verify focus. The Ronchi screen is located at the same distance from the camera's lens coupling ring as the focal plane of the camera being focused. Once the telescope is focused with this device, the camera body is reinstalled, and the camera will be properly focused. Knife-edge focusers for almost all DSLR cameras are available from Stellar Technologies International (see Appendix A).

The idea behind knife-edge focusing is simple. The telescope is centered on a bright star, and the Ronchi screen is placed into the light path near the approximate focus point. Each individual line on the Ronchi screen acts like a knife-edge, and when the proper focus is achieved, the star will quickly wink out as the knife-edge occults it. If the star is not in focus, either multiple lines of the Ronchi screen are visible, or the star will slowly wink out as the knife-edge seems to sweep across the image from one side to the other.

It is best to align the Ronchi screen so the target star will cross the lines at a right angle. An example of the focus procedure is shown in Figure

Fig. 11.4 *Knife-edge focusers, such as the Stilleto IV made by Stellar Technologies International, are substituted for the camera body and use a Ronchi grating to achieve exact telescope focus. The telescope focuser is then locked (lower left), and the camera body installed in place of the knife-edge focuser (lower right). Photos by Robert Reeves.*

11.5. Image A shows the star when it is not in focus—multiple lines will be visible. As the focus improves, fewer lines will be visible (see Image B). At very near focus, only a single line will be visible. At this point the telescope is slewed slightly to make the star cross the line. The star's image will fade as the line "cuts" the star's image as shown in Image C. When perfect focus is achieved, the star will evenly gray out and disappear as shown in Image D. If focus is adjusted too far and overshoots the point of perfect focus, the star will fade from one side to the other, but this time the Ronchi line will seem to sweep across the star's image from the opposite direction, as shown in Image E.

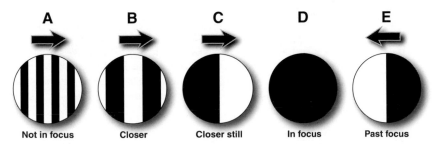

Fig. 11.5 *As the focus improves when viewing a star image through the Ronchi grating in a knife-edge focuser, the number of grating lines visible decreases until the star disappears instantly when occulted by a grating line.*

As the point of best focus is approached, a knife-edge focuser will reveal many things about your optical system. Imperfect optics and poor seeing may present a confusing image. When very near either side of focus, as in Image C or E, poor seeing may cause the line crossing the star to waver. Poor optics may cause the Ronchi line not to be straight or the image to gray out in an uneven manner leading to uncertainty about just where the exact focus point is. Also, short focal length optics will snap into focus quickly, but those of long focal length have a greater depth of focus, or zone, within which the image will appear to be in focus, leading to uncertainty as to just where the exact point is. With practice, you will learn how to go back and forth through the focus point and pick the mid-range area where the star seems to gray out evenly, indicating best focus. If telescope shake caused by manually turning the focus knob makes this procedure difficult, a motorized focuser will simplify the process by helping keep the telescope steady.

While the theory behind the knife-edge focuser is simple, a high degree of precision is required in the instrument's construction. The Ronchi screen has to be held within 0.001 inch of where the DSLR sensor will be positioned. The easiest way to get a knife-edge focuser is to buy one from Stellar Technologies (see Appendix A). The units are pre-calibrated for your camera brand and attach with a T-ring adapter that is an exact match for your model.

For those who like to fabricate their own gadgets, an old 35-mm camera body can be converted into a knife-edge focuser by placing a Ronchi screen at the film plane and attaching a magnifier to view the star image. The camera needs to be the same make as the one to be used on the telescope; that is, if a Nikon D70 is to be used, you must use an old Nikon SLR so that the T-ring adapter-to-focal plane distance remains the same between the knife edge focuser and the DSLR.

11.3 Solar Photography

WARNING! Observing the Sun is one of the few things in amateur astronomy that is dangerous. If a solar filter falls or blows off your telescope, your eye can suffer irreparable damage; or, serious damage may render optical and photographic equipment useless. Always insure that a solar filter is attached to the telescope in a fail-safe manner. The finder scope should be filtered or removed from the telescope. Take extra care if children are present.

Today, photography of the Sun through a telescope is little different than photographing the Moon, except for the critical need to filter out most of the Sun's energy before it enters the telescope. Decades ago when I started doing astrophotography, good solar filters were a rarity, and it was standard practice to project the solar image onto a white screen, then photograph the screen. This procedure involved aiming an unfiltered, unprotected telescope directly at the Sun. This was not a good practice because the telescope could be damaged; and there was serious risk should somebody inadvertently view the Sun directly when setting up the instrument. There are now two general types of solar filters: white-light models that simply darken the Sun enough for us to see and photograph its surface features, and hydrogen-alpha filters that isolate a single wavelength of light allowing us to examine solar prominences and features of the chromosphere that are invisible in ordinary white light.

White light solar filters come in two general varieties: the full aperture type that significantly reduces the Sun's energy before it enters the telescope; and the filter, such as a Herschel wedge, that rejects a majority of the Sun's energy at the focus of the telescope. Although not everyone will agree, my opinion is that filters of this latter type should not only be avoided, but they should be disposed of so novice observers do not injure themselves. It is an uncontested fact that telescopes are giant magnifying glasses, and we all know that concentrating solar rays with a magnifying glass can quickly burn most materials, including people and expensive camera mechanisms.

Safety need not be costly. Good, reliable filters for white-light solar photography are inexpensive, often costing less than a good eyepiece. Specialized solar filters enable photography of solar prominences, calcium floculae, and other narrow-band phenomena are very expensive. For this reason, the majority of amateur solar photography is done in white light. This allows imaging sunspots, solar granulation and bright faculae.

The Sun and Moon are nearly identical in apparent size; thus, what works for imaging the full Moon will also work to image the entire prop-

Fig. 11.6 *The Sun is a bright enough target that excellent solar images can be taken by manually holding the camera to the eyepiece of a properly filtered telescope. Photo by Becky Ramotowski.*

erly-filtered solar disk. Daytime solar heating of the air and surrounding terrain can be a problem if the observing site does not take seeing into consideration. Locations next to buildings, hot roofs, asphalt parking lots and other structures that heat up during the day should be avoided, since they all radiate and seriously degrade seeing. An ideal site is one that is elevated about 10 feet in the middle of a large grassy field. A majority of the heat-induced turbulence is located close to the ground, a phenomenon sometimes called "ground clutter."

Traditionally, small telescopes up to about eight inches use a full aperture solar filter. Beyond that size, a stop-down or sub-aperture filter is usually employed. This is a smaller solar filter installed in an opaque full-aperture mask that utilizes only a portion of the telescope's aperture. An example would be a 3-inch filter mounted on a 12-inch telescope. The loss of aperture should be of no concern, as the smaller filter will allow plenty of image illumination, and daytime seeing seldom permits large apertures to resolve detail to their theoretical maximum anyway.

Since the 1970s, the long-time standard solar filter was Solar-Skreen, a silvery flexible aluminized Mylar marketed by Roger Tuthill, Inc. A few

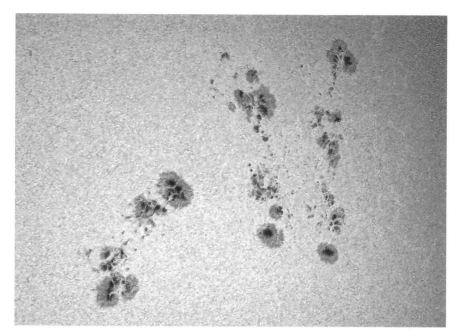

Fig. 11.7 *By applying the same eyepiece projection or afocal techniques used for high magnification images of the Moon and planets, research-grade images of sunspots can be achieved with a solar filter-equipped telescope. An 8-inch Maksutov-Newtonian with 17-mm eyepiece was used with a Nikon Coolpix 995 to capture this sunspot group. Photo by Paul Hyndman.*

years back, the Baader Planetarium developed a new but similar-appearing solar filter material called Baader AstroSolar Safety Film. The Baader filter material is internally darkened to reduce light scattered within the plastic separating the top and bottom of the reflective coatings. The older Solar-Skreen, which does not have this technology, produces a bright sky around the Sun. Though this effect does not greatly hinder seeing solar detail, some observers claim that the Baader film provides higher contrast.

There are many similarities between the Baader solar filter and Tuthill's Solar-Skreen. Each is coated on both sides so pinholes or scratches on one side will be blocked by the coating on the other side. Neither filter should be stretched tightly over the optics, or they will be damaged; they should be mounted loosely and wrinkly. While this runs counter to the idea that optical surfaces should be as smooth as possible, this is the preferred way to mount the filter material. Indeed, the instructions that come with the Tuthill model advise the user to ignore the wrinkled appearance and consider that they are supposed to look through the filter, not at it. Even though both the Baader and Solar-Skreen material appear very flimsy, they are quite durable in use.

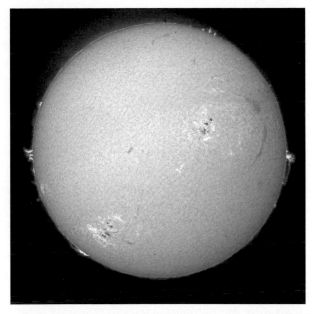

Fig. 11.8 *The image-recording characteristics of digital cameras and the availability of specialized hydrogen-alpha solar filters have allowed amateurs to capture exotic details on the Sun that once were the realm of professional astronomers. Photo by Paul Hyndman.*

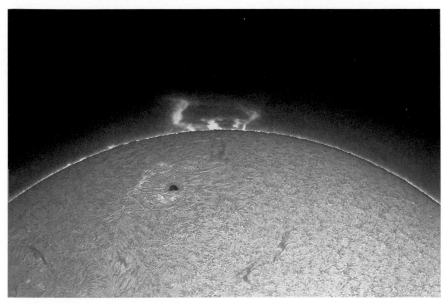

Fig. 11.9 *Hydrogen-alpha solar filters allow capturing images of solar prominences that are normally invisible except during a total solar eclipse. Photo by Paul Hyndman.*

The Baader filter produces a more natural white image instead of the bluish tinge seen through the Solar-Skreen. Solar-Skreen scatters more light, giving the sky around the Sun a bluish tint while the Baader filter makes it look a dark gray. The Baader filter produces a brighter image, and for photography, an even brighter version is available with a photographic density of 3.8 to allow shorter exposures in turbulent heated daytime air.

As an alternative to plastic filters, several manufacturers market glass solar filters. While the overall image quality is good with them, the plastic Baader solar filter is generally considered to produce better images.

Two representative glass filters are the Thousand Oaks Type 2+ and the Orion solar filter. The Thousand Oaks unit is similar to their standard Type 2, but the "+" designation refers to the addition of a steel alloy that improves the filter's durability. Orion's filter is made by J. M. B. Inc., that markets the same filter under their trade name of Identi-View Class A. The Class A has the added steel alloy for durability that the same company's Class B does not have. The Thousand Oaks filter produces an image of the Sun that is orange, while the latter renders it a more natural pale yellow.

Almost all solar filter products claim to pass only about $\frac{1}{100,000}$ of visible light. This works out to a photographic density of 5.0. Often, a glass solar filter arrives from the manufacturer with tiny pinholes. Tiny defects can also develop on the filter after extended use. A filter can be checked for pinholes by backlighting it with a bright illumination. These tiny defects should not be cause for alarm, and the accepted repair procedure is to dab a spot of flat black paint on the pinhole with the tip of a toothpick. The small spots do not affect the overall performance of the filter. Be sure to mount your filter to the telescope in a secure manner, so it cannot come off under any accidental circumstances.

11.4 Hydrogen-alpha Solar Photography

Photographing the Sun in the single wavelength of hydrogen-alpha (H-a) light is a relatively new field for amateur imaging. The new affordable sub-Ångstrom filters that can isolate this spectral line are a natural combination with today's DSLR cameras. A good starting point in the discussion of this topic would be to look at just what the solar H-a signal is, both frequency- and color-wise, and assessing the camera's abilities to handle it. This will determine how one goes about gathering and processing the data, and minimize the amount of trial-and-error before satisfying results are consistently produced.

Singly-ionized hydrogen radiates at a wavelength of 656.28 nanometers, a point that lies in the visible red band of the spectrum. There is no blue or green emission from the Sun at this wavelength—nothing but red. Once this facet of solar photography is understood, all else will be clear. Blue and green filters only reduce the red signal level, as there are no blue or green components to be filtered. Similarly, an additional infrared blocking filter will not enhance the image because the hydrogen-alpha radiation occurs only in the red, not the infrared, band.

An important question in solar H-a imaging is how wide is the frequency range of the feature that we want to record? Even though the "laboratory" wavelength of H-a radiation is always exactly 656.28 nm, the complexities of the Sun's energy production cause the wavelength of its H-a light to vary slightly from that benchmark. (For instance, the light of a prominence may be either slightly blue- or red-shifted depending on whether the eruption that caused it was directed toward, or away, from Earth. Furthermore, they tend to be billowy, diffuse features.)

So to photograph wispy prominences, which often have rapid velocities both toward and away from an observer, a filter with a relatively wide bandpass centered on 656.28 nm—say about 0.7 Ångstrom—is advantageous. When the bandpass is above 1.0 Ångstrom prominences tend to dissipate into the sky background, and at 0.5 Ångstrom they often appear as thin, wiry whips.

Details on the solar disc itself, however, are usually better seen with a narrower filter. The tighter the bandwidth, the more contrast features—such as solar filaments—will show. Surface detail appears extremely washed out at 0.9 Ångstroms, looks good at 0.5 Ångstroms, looks great at 0.3 Ångstroms, and is spectacular at 0.2 Ångstroms.

Filaments and coronal mass ejections on the disc of the Sun will often look best when they are slightly blue-shifted. This does not mean there is any blue in the image, but merely that the frequency of the light passed by the filter has been shifted a miniscule amount toward shorter wavelengths to pick up signal that would otherwise spill outside the 656.28 Ångstrom frequency. Several solar filter manufacturers include a feature to provide this adjustment, and when tweaked it can enhance the contrast of the filaments dramatically. The following discussion assumes the use of a Coronado brand filter.

The first step in imaging the hydrogen-alpha Sun is to set up the telescope and filter assembly without the camera, as it is simpler to determine key filter adjustments while viewing through an eyepiece. Etalon-based systems use an adjustable blocking filter. An etalon works on an interference pattern principle, with multiple harmonic frequencies arriving at the blocking filter. This allows the user to set the frequency tuner ("T-max" when using the Coronado filter) to pull in the filaments with the highest amount of contrast, but not so far as to shift away from the surface features. If the etalon filter is exactly perpendicular to the optical path of the telescope, the surface features will usually be at their sharpest. Slowly turning the T-max while at the eyepiece will often yield a considerable increase in contrast of the eyebrow-like filaments that dance across the surface of the Sun. You can often find a very small area of adjustment where surface features are still sharp and filaments jump out at you. Newer filter models are extremely stable over a wide temperature range, and you can often tune out a little more red background "haze" by tweaking them.

Once the filter is tuned, what remains is to determine exactly how the camera will behave with the hydrogen-alpha signal. It is time to take a series of images and examine them in an image processing program. Use camera coupling methods and whatever optical devices (Powermates, Barlows, eyepieces) are needed to provide the desired image scale. If the camera supports it, use RAW format or one that allows storing the data without camera processing. JPGs may irretrievably alter the data making accurate analysis of the image difficult.

Most camera imaging sensors utilize a Bayer pattern color filter array to produce a color image. This results in a matrix that places either a red, green, or blue filter over each photodiode in a pattern that looks like this (R = red, G = green, B = blue):

```
G B G B G B G B G B G B G B
R G R G R G R G R G R G R G
G B G B G B G B G B G B G B
R G R G R G R G R G R G R G
G B G B G B G B G B G B G B
R G R G R G R G R G R G R G
G B G B G B G B G B G B G B
R G R G R G R G R G R G R G
```

Half of the photodiodes are covered by green filters to mimic the center wavelengths of our own visual acuity (usually considered to be 532 nanometers, or greenish yellow); while red and blue filters each cover 25 percent of the remaining photodiodes. In solar H-a imaging, however, ONLY the red channel should contain image information; hence, 75 percent of the pixels in a Bayer pattern sensor are not used, with this sparsely populated image matrix as the result:

```
...................................................
R...R...R...R...R...R...R...R.
...................................................
R...R...R...R...R...R...R...R.
...................................................
R...R...R...R...R...R...R...R.
...................................................
R...R...R...R...R...R...R...R.
```

An interesting facet of many digital cameras is signal leakage, an effect where image data is recorded in other than the correct color channel. For example, some models of the Nikon CoolPix have such high signal leakage at 656 nanometers (the desired H-a wavelength), that one is often better off overexposing the image and saturating the red-sensitive photodiodes, and then working with the green channel in lieu of the correct red channel. This technique will often allow using a single image to capture both the brighter surface detail in the green channel and the dimmer prominences from the now overexposed red channel.

Now, take images at different exposure times that show up on the camera's LCD preview screen as ranging from bright red down to almost dark black. Pay attention to what Powermate, Barlow, eyepiece, and camera zoom settings were used because the EXIF exposure data recorded by the camera will not store this. Usually the lowest ISO setting will provide the most noise-free images to work with, and the exposures are still fast enough to reduce atmospheric smearing.

11.5 Lunar and Planetary Photography

An advantage of lunar and planetary imaging is that local light pollution will have little effect on the results. Urban backyard imaging of solar system objects can be quite rewarding as long as the atmospheric seeing does not blur the image. You should, however, insure that a bright light is not shining directly on an SCT corrector plate or into an afocal camera setup.

You can personally select the technical details of the telescope, camera, and image processing; but the most important factor, atmospheric seeing, is generally beyond your control. Nothing can be done about high altitude conditions, but local environmental items, such as observing location are somewhat controllable. As with any serious observing, locations near a hot roof, asphalt parking lot, or other sources of localized heating should be avoided. Without steady air, your best hope for high resolution imaging cannot be achieved.

The factors that you have the most control over are as follows:

1. A telescope with clean, well-aligned, high quality optics.
2. A solid, polar-aligned mount with an accurate drive.
3. A camera that is accurately aligned with the optical axis of the telescope.
4. Optics that are at ambient temperature
5. Optics that are precisely focused.
6. A high signal-to-noise ratio image must be produced. Underexposure or overexposure robs the image of optimum data.

The best way to get sharp lunar and planetary images is to use the

Fig. 11.10 *These Jupiter and Saturn images are typical of what can be achieved by amateurs with modest telescopes and digital cameras. Saturn was imaged through a 17-mm Nagler eyepiece on an 8-inch Intes-Micro Maksutov-Newtonian telescope using a Nikon Coolpix 995 camera. The planet is a stack of 8 seperate ¼-second exposures, while the moons are the sum of four 2- to 8-second exposures. Jupiter was imaged through the same telescope using a Canon D60 viewing through two stacked TeleVue Powermates (2x and 4x). 200 separate ⁷⁄₁₀-second exposures at ISO 100 were taken and the 30 best images were stacked. Lengthier series of exposures are difficult because of Jupiter's rapid rotation. Photos by Paul Hyndman.*

shortest exposure possible to minimize the effects of atmospheric turbulence. The exposure must be long enough to maintain the image density needed to preserve details. A popular method for achieving a high signal-to-noise ratio is to stack a series of short-exposure images in an image processing program. Ideally, the images are taken during periods of steady seeing. The more good images you collect, the more likely your final processed image will be good. Accumulating images rapidly for stacking is important for quick rotating planets like Jupiter, Saturn and Mars. The rotation of these first two planets can be detected in images taken just ten minutes apart.

Use the entire memory card during a lunar or planetary imaging session. There is strength in numbers. Even on turbulent nights, at least one of a string of 100 Moon shots will be steady. The drawback is that you have to manually scan dozens of images to find the best one, and then delete the sub-par images. If individual lunar images are being processed, a good way to sort through and select the best image from a sequence is to use the freeware program Irfanview (see Appendix B). This program enables you to view and judge a series of images by pressing the space bar. If an image is not a "keeper," you can delete it on the spot. You also have the option to back up one image should you decide the next one is better. The process is repeated through the entire image sequence until only the best one is left.

11.6 Deep-sky Targets

If you are imaging deep-sky objects from an urban location, try to shoot after midnight. The local sky at zenith can become a magnitude darker when most business activity has diminished and road traffic is reduced. The sky is also darker after midnight on weekdays because of increased after-hours human activity on a weekend night; so if work schedules allow, try to do astrophotography during these darker periods. Using a broadband light pollution filter will also help prevent those unwanted wavelengths from reaching the camera's imaging sensor.

Man-made skyglow is greatest nearer to the horizon, so the higher the target is in the sky, the farther it will be away from light pollution sources. If the air is clean and not carrying a lot of dust or humidity, there is noticeably less reflection from stray lighting at the zenith. Unfortunately, not all celestial targets will pass within 15 degrees of the zenith; but by shooting them when they are passing the meridian, you will at least have them placed in the area of the least light pollution.

Celestial targets outside the solar system can be categorized by their size. Digital cameras as a category have different fields-of-view for the

Fig. 11.11 *Nearly half the Milky Way fills this image taken with an 18-mm lens on a Canon EOS 300D. Three 5-minute exposures at ISO 400 were combined to create this image. Photo by Terry Lovejoy.*

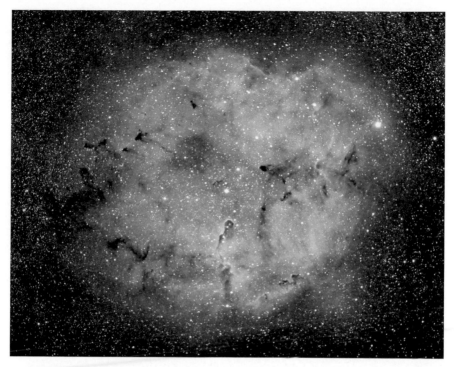

Fig. 11.12 *The large emission nebula IC1396 was imaged with a Canon 10D through a 300-mm f/2.8 lens. The low surface-brightness of the nebula, coupled with the low red sensitivity of digital cameras, required combining 15 five-minute exposures taken at ISO 800 through a Lumicon UHC filter with 15 more eight-minute exposures taken at ISO 1600 through a hydrogen-alpha filter. Photo by Johannes Schedler.*

same focal length lens as do 35-mm film cameras, so these categories have rather fuzzy boundaries. Depending upon the brand and model of a digital SLR, a particular lens can provide the same field-of-view as a 35-mm film camera, or the imaging sensor may be smaller, introducing a lens crop factor. With this in mind, the following general categories are established with the assumption that the sensor being used is the size of a full 35-mm film-frame. If various crop factors are applied to your particular camera, the resulting field size has to be scaled accordingly. These categories include:

- **Objects for wide-angle and normal lenses (15 degrees or wider)** include the entire Milky Way, constellations, the zodiacal light and the aurora. The Milky Way itself is full of possibilities for all focal lengths from fisheye to long-focus telescopic views of individual nebulae within it. Constellations similarly cover a span of focal lengths with large constellations like Ursa Major requiring an extreme wide-angle lens while a small one like Lyra can be imaged with a long telephoto.

Fig. 11.13 *The star clouds of the Milky Way are a natural target for wide-angle camera lenses. This image of the central part of the Milky Way is a combination of 5 two-minute exposures through a 35-mm lens on a Canon 10D. Photo by Antonio Fernandez.*

Fig. 11.14 *M45, the Pleiades, is a beautiful sight when imaged through any kind of telephoto lens. Photo by Hap Griffin.*

Fig. 11.15 *This image of M64, the Black Eye Galaxy, was taken through a 10-inch LX-200 at f/ 6.3 with a Canon EOS 300D using 78 separate ninety-second and two-minute exposures. Photo by Michael Howell.*

- **Objects for short telephoto lenses up to 135 mm (10 – 15 degrees)** include the galactic bulge of the Milky Way in Sagittarius, extended nebular complexes such as Barnard's Loop in Orion, or groups of nebulae like the North America Nebula and the Gamma Cygni Nebula, IC 1848 (the Baby Nebula) and IC 1805 (the Valentine nebula) in Cassiopeia and the very large star clusters such as the Hyades.

- **Objects for medium telephoto lenses up to 200 mm (5 – 10 degrees)** include nearby galaxies such as the Large and Small Magellanic Clouds and M31. Objects like IC1396 in Cepheus and the Lambda Orionis nebula in northern Orion are also popular, but these are dim and shine mainly in the H-a portion of the spectrum where digital cameras lack sensitivity. To effectively image these red objects you will have to digitally stack many individual exposures to build up an image with sufficient contrast.

- **Objects for long telephoto lenses 300 mm and above (2 – 5 degrees)** include the huge number of objects that populate the Milky Way. These range from large open star clusters like the Pleiades in Taurus and M6 and M7 in Scorpius to bright nebulae like the Lagoon and Trifid Nebulas in Sagittarius. The larger globular clusters such as Omega Centauri and M13 will begin to be resolved into stars with a long telephoto. The Andromeda Galaxy and the Pinwheel Galaxy in Triangulum will also stand out. Other large emission nebulae, again requiring multiple image stacking because of digital photography's inherent insensitivity to red light, include the California, Veil, and the Helix.

- **Telescopic targets (less than 2 degrees)** include virtually all galaxies, star clusters and nebulae that are visible through the telescope you are using. The numbers are so vast that they literally provide a lifetime's worth of targets. Prominent examples include the Andromeda Galaxy, the Orion Nebula, the nebulosity surrounding the Pleiades, groups of galaxies like M81 and M82 in Ursa Major, and M65 and M66 in Leo.

Wide-field imaging of dim deep-sky objects using wide-aperture camera lenses is sensitive to urban light pollution. These lenses have a field-of-view wide enough so that varying amounts of light pollution at varying heights above the horizon will cause a shading gradient across the image. This will result in one side of the image having a brighter sky background than the other. Such artifacts can often be reduced or eliminated through image processing, but the best line of defense is to avoid light pollution altogether. If possible, shoot deep-sky objects when they are highest in the sky where the glow of light pollution is usually at its least. Also, the use of light pollution rejection filters may improve the contrast of images taken in urban areas. Not all forms of light pollution are the same, and your results with a particular filter may differ from that achieved at another location. For best results, you must experiment under the conditions through which you will be imaging.

11.7 Filters and the Wave Nature of Light

One important specification to consider when choosing a filter for astronomical purposes is it light acceptance angle. Some only pass light waves that strike the filter nearly perpendicular to its surface. Such filters will vignette images taken with wide-angle camera lenses, but will work well in telescope optical systems where the light rays are more nearly parallel at the optic's focus.

To understand how filters work with astrophotography, we have to understand the wave nature of light. Each color of the visible spectrum has its own unique wavelength expressed in units called Ångstroms, each of which measures 0.0000000001 meter, or a ten-billionth of a meter. In practical terms, however, spectral wavelengths are expressed in nanometers (nm)—one nanometer being equivalent to one billionth of a meter.

The visible spectrum ranges from red light, that has a wavelength of 700 nm, through orange, yellow, green, blue, and indigo to violet, that has a wavelength of 400 nm (see Figure 11.17). Wavelengths just beyond the longest and shortest ends of the visible spectrum are called infrared and ultraviolet. Although the eye cannot see them, digital cameras are very sensitive to these unseen wavelengths and must use special filters to prevent them from reaching the imaging sensor. While ultraviolet and infrared radiation will not harm the sensor, they will alter image coloration and interfere with proper autofocus operation during normal terrestrial photography. Table 11.1 shows the special characteristics of popular photographic filters.

The term "cut-off" refers to a wavelength above or below which one wishes to filter light, transmission at the cut-off wavelength itself being 50 percent and transmission dropping sharply above or below the cut-off point, depending on which end of the spectrum is being filtered out. The last column refers to the percent of H-a radiation transmitted by the filter.

There are several different types. A cutting filter will not pass a wavelength shorter than the band it was designed to transmit—red #29 or red #92 are examples. A band-pass filter transmits a limited band of wavelengths. Lumicon's Deep-Sky filter is of this type. Monochromatic or interference filters transmit a single wavelength of light, such as that of hydrogen-alpha and hydrogen-beta radiation.

Filters affect the focus of the optical system by an amount dependent on the refractive properties of the filter itself and the wavelength of light passed by it. If one is placed in front of a lens, the plane glass portion supporting the filter coatings has no effect on the lens's focus. The color of the filter may, however, shift the focus. If the filter position lies inside the focused beam of the lens, the amount of focus shift caused by the filter glass itself is equal to the thickness of the filter multiplied by one divided by the refractive index of the filter glass; that is, shift = filter thickness x (1/refractive index).

In terrestrial photography, the lens is usually stopped down, masking minor focus changes caused by the filter. In astrophotography, however, the lens is almost always wide open, and the effects of a filter upon the focus will degrade star images if they are not compensated for. A narrow-wavelength H-a filter requires very precise focus. The extra effort to achieve a precise focus is well spent because most lenses will be twice as sharp in red light as they are when admitting the entire visible spectrum.

11.8 Color Imaging Limits Filter Selection

One of the advantages of digital astrophotography is the fact that the final product is a color image. We, therefore, want to take steps to control light pollution and still maintain a good color balance in the final image. Filters that reject light pollution wavelengths may also create color tints that render the sky or the celestial object in a color that is not natural. This is not a

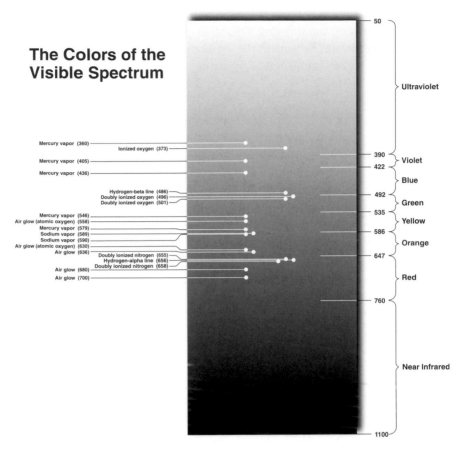

Fig. 11.16 *The colors of the visible spectrum.*

problem with black and white film imaging, but it limits our selection of filters that will work with color images.

One type that can help digital imaging is the common ultraviolet (UV) filter. Most stars emit significant amounts of UV radiation, and because most camera lenses are unable to focus wavelengths in this region of the spectrum properly, the results are fuzzy halos around otherwise well-focused stars.

A skylight 1B filter does not degrade star images and will cut most UV radiation, including the 360-nm UV emission from mercury vapor street lighting. By eliminating the ultraviolet, star images are slightly sharper. However, a 1B filter does pass about 10 nm of the near-UV spectrum. Therefore, a more effective UV filter is Lumicon's Minus-violet, that cuts wavelengths shorter than 420 nm in the blue region of the spec-

Fig. 11.17 *A problem with the use of filters with a camera is reflections between the filter and lens elements. The pointers in this image of the area around the stars Altair and Tarazed, and dark nebula B143, show reflections opposite the brightest stars in the field-of-view. See insert. Photos by Robert Reeves.*

trum. The Minus-violet will help control the bluish halos around stars caused by chromatic aberration and by unfocused UV wavelengths. It thus sharpens stars without adversely affecting their colors. However, since only the blue end of the spectrum is filtered with the Minus-violet, some chromatic aberration will be evident from longer wavelengths, and the stars will only be about half as sharp as they would be if a more aggressive monochromatic filter were used, such as one for H-a.

A Minus-violet offers no protection from light pollution, except for 360-nm emission from mercury vapor lighting. The photographic results and image contrast are generally the same as if no filter were used, although the stars will be sharper. The Minus-violet does, however, slightly increase the deep-red color of emission nebulae by cutting the oxygen emission line at 373 nm. This line turns reddish emission nebulae pink when they are imaged. Unless a digital camera has been modified by removing the infrared blocking filter, the color difference in nebulae with the filter will be minimal because of the reduced deep-red sensitivity of digital cameras.

With traditional black and white film imaging, various yellow and blue filters measurably improve the focus of star images. While these fil-

Table 11.2 Filter Spectral Characteristics

Filter	Color	Lower Cut-off	Upper Cut-off	Passes % of H-alpha
Wratten #1A	UV	380 nm		91.0
Lumicon M-V	UV	420 nm		96.0
Wratten #2B	Yellow	390 nm		91.5
Wratten #2A	Yellow	405 nm		91.0
Wratten #2E	Yellow	415 nm		91.0
Wratten #8	Yellow	495 nm		91.0
Wratten #12	Yellow	520 nm		91.0
Wratten #15	Yellow	530 nm		91.0
Wratten #21E	Orange	505 nm		90.3
Wratten #47A	Blue	415 nm	475 nm	0.0
Wratten #47B	Blue	430 nm	430 nm	0.0
Wratten #47	Blue	440 nm	440 nm	0.0
Wratten #80A	Blue	495 nm		18.0
Wratten #44	Blue	476 nm	510 nm	0.0
Wratten #25	Red	600 nm		87.9
Wratten #29	Red	620 nm		88.5
Wratten #92	Red	640 nm		76.4
Lumicon H-alpha	Red	640 nm		90.0
Schott RG 645	Red	650 nm		60.0

ters can be also used for digital imaging, they will alter the colorcast of the image. If the final image is to be reduced to a grayscale black and white one, the color shift is of no consequence, and the filters can be used to good effect.

A pale yellow filter such as a Wratten #2B or #2E will narrow the incoming band of wavelengths and reduce the halos around stars in color images without seriously affecting their color balance. These filters are weak and may not produce much improvement in image quality. Deeper #8, #12, or #15 yellow filters will eliminate the blue end of the spectrum, and while they will sharpen star images, they will noticeably alter the image color. Again, if the final image is being converted to grayscale black and white, the color shift is not important.

Yellow filters can be used to accentuate star clusters embedded in emission nebulae. By suppressing the H-a emission, the yellow filter will give the cluster a more "visual" appearance.

Blue filters can also be used if the image is destined for black and white output. Star clusters are typically bluish and thus benefit from blue filters, which also suppress H-a radiation from any nearby emission nebulosity around them. The open cluster at the center of the Rossette Nebula

Fig. 11.18 *Most lenses render ultraviolet and shorter blue wavelengths slightly out of focus while wavelengths in the middle of the visible spectrum focus well. This causes blue halos around brighter stars (left). A Minus-violet filter will eliminate the far end of the blue spectrum and reduce the blue halos around bright stars (right). Photos by Robert Reeves.*

in Monoceros, for example, will be drowned out by the reddish glow of the nebula unless the latter is filtered out by a blue or some other filter that blocks H-a emission.

Some useful blue filters for astrophotography are a Wratten #44, #47, #47B, and #80A. If the objective is a black and white image of galaxies, a #44 blue filter will reject enough light pollution to allow a three-fold increase in exposure time on such objects.

Kodak publishes a data book, titled *Kodak Wratten Filters for Scientific and Technical Use*, which gives the transmission characteristics of all filters that use the Wratten numbering system. Investigate the astronomical usefulness of various ones by matching their transmission characteristics to the celestial target. For instance, a common #48 is perhaps the worst yellow filter to use for astrophotography because it does not transmit more than 33 percent of incoming light in any wavelength. A much better filter would be a #15 yellow; it blocks blue and green while transmitting 91 percent of the important H-a wavelength.

11.9 Hydrogen-alpha Filters Enhance Digital Images

WARNING: hydrogen-alpha filters designed for long exposure deep-sky astrophotography are not designed for solar use. Do not look at the Sun through a hydrogen-alpha filter designed for light pollution control. Permanent eye damage can occur by viewing the Sun through a nighttime light pollution control hydrogen-alpha filter.

The red H-a wavelength is one of the most important in astrophotography. Stars, galaxies and emission nebulae shine brightly in this spectral region. While reflection and planetary nebulae radiate most of their energy in shorter wavelengths, they still emit enough H-a to be recorded. Therefore, by using a filter tailored to pass only light in the H-a region of the spectrum, we can eliminate most of the light pollution sources in the blue, green and yellow wavelengths without drastically reducing the desired H-a light. This gives us a powerful means of penetrating light pollution and sky glow.

Deep red filters are a good tool to combat light pollution, but unfortunately two things work against their use in digital imaging. First, all digital cameras are color cameras. Isolating the H-a wavelength may penetrate light pollution, but it turns a digital camera into a monochrome imager, and the resulting image must be converted to grayscale and viewed in black and white. The second problem is that the imaging sensor in these cameras is especially sensitive to infrared, and a rejection filter has to be built into the system to prevent such radiation from causing autofocus and image coloration problems. These blocking filters are so aggressive that they significantly reduce the camera's sensitivity to red light. For example, the Canon 10D and 300D block all but 15 percent of the H-a wavelength. This means many exposures must be stacked together to build up a sufficient red signal to match the red light performance of popular astronomical films such as Ektachrome E200 and gas-hypered Technical Pan.

To digitally build up the H-a signal requires a longer exposure, often several times longer with the red filter than without, but the results are a more pleasing image showing colors closer to the rich reds captured by Ektachrome E200 film.

Two representative sources for hydrogen-alpha light pollution control filters are Lumicon and Hutech. Hutech markets the IDAS Light Pollution Suppression (LPS) filter for color imaging and the Kenko R64 deep red filter; both of which are closely matched to the H-a wavelength of emission nebulae, for about the same cost.

Lumicon's filters are opaque to all wavelengths below 640 nm but pass between 90 and 98 percent of the wavelengths between 657 nm and the infrared region. They also pass at least 90 percent of the deep red 656 and 658 nm hydrogen-alpha and doubly ionized nitrogen lines prominent in emission nebulae. They only pass three percent of the oxygen airglow lines while rejecting virtually all of the light pollution spectral lines. The monochromatic filter also virtually eliminates the effects of chromatic aberration within a lens.

While Hutech does provide specific wavelength numbers for their

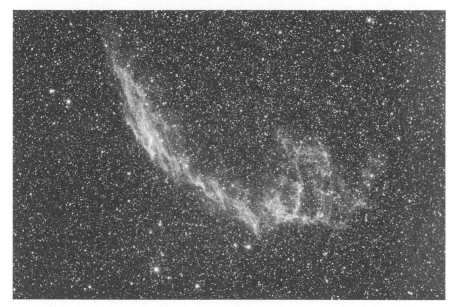

Fig. 11.19 *The use of filters during part of the exposure will enhance the appearance of red emission nebulae. This image of the Veil Nebula was taken with a Canon EOS 10D and a Takahashi FCT-150. Fifteen 5-minute exposures at ISO 800 using no filter were combined with ten 5-minute exposures at ISO 1600 through a hydrogen-alpha filter and ten more 5-minute exposures at ISO 1600 through an O-III filter. Photo by Mark Hanson*

Kenko R64 filter, my analysis of their spectral response chart indicates that this filter is fractionally less effective than the Lumicon's at the H-a wavelength; but since digital exposures must be short compared to lengthy film exposures, the practical difference between the filters is negligible. Either is adequate for the job.

No available filter blocks the airglow wavelengths at 680 and 700 nm, but the sky fogging capability of these wavelengths is so slight that it will not affect digital imaging.

Kodak Wratten gel filters come as 3-inch or 4-inch squares. They used to be very popular for celestial imaging. Today, their availability is limited because some are out of production, and fewer camera stores stock what remains. The red #25, #29, and #92 are all useful for red light imaging as described above. The #25 is usually available as a threaded glass filter from most manufacturers. The deeper red #29 is available as the B + H 091 glass filter. The Lumicon and Kenko H-a filters are a close match for the #92 and statistically outperform it for celestial imaging. Glass filters cost more, but they are more durable than the soft, easily scratched, gel filters. Digital astrophotographers will find the glass filter alternatives a better choice for their work.

11.10 Broadband Light Pollution Rejection Filters

Broadband light pollution filters are widely available. They reduce the effect that light pollution has on color astrophotography. Because these filters must pass many important wavelengths besides just the hydrogen-alpha, they invariably also pass some of the light pollution wavelengths we are trying to avoid. Therefore, none of these are as aggressive as the H-a filter in combating light pollution.

All brands of interference filters used for light pollution control work best with telephoto lenses and, to a lesser extent, prime focus photography. This is because the wavelengths blocked, or those transmitted, by the filter are dependent not only on their placement in the spectrum but their incidence angle to the filter. Light rays that strike these filters at large angles do not pass into the system, resulting in a highly vignetted image. Using a full-frame 35-mm sized sensor as a standard, the shortest focal length lens that will give unvignetted views is 85 mm. Because of this sensitivity to incidence angle, the filter must be square to the incoming light beam.

The most widely used broadband filters are Lumicon's Deep-Sky filter, Orion's SkyGlow filter, and Hutech's IDAS Light Pollution Suppression filters. Lumicon offers their filter in thread sizes up to 72 mm, while Hutech offers an 82-mm size for large telephoto lenses. The largest Orion version, although considerably less expensive, is available only in 48 mm.

The Lumicon Deep-Sky filter produces a somewhat reddish sky background, but does allow longer exposures in slightly light-polluted areas. The filter renders nebulae in truer colors because it passes both red and blue wavelengths that are prominent in deep-sky objects while blocking yellow and green where most man-made light pollution exists. Since the Deep-Sky filter is selective in the wavelengths it does pass, the nebulae look normal, but star colors may be changed. It also passes infrared radiation in the 850 to 900 nm range, but the infrared blocking filter built into digital cameras will eliminate this. This filter typically allows exposures about three times longer than possible without it. Because they are not monochromatic, broadband filters have no effect on chromatic aberration and will not help in sharpening star images. Typically, they pass about 90 percent of the light between 400 and 480 nm.

The IDAS filter is most effective against mercury vapor light pollution. In recent years, many places in the United States have switched outdoor lighting to sodium vapor rendering the IDAS filter less effective. However, some of the wavelengths associated with the orange sodium vapor lighting do fall within this filter's cutoff. Since the mix of mercury vapor lighting is unpredictable, it is not possible to state with conviction that the filter will achieve a certain level of performance in your area. However, a typical image can be exposed for twice as long when using this filter to filter out mercury vapor light pollution.

Digital imaging rarely uses 10-minute exposures, and until better image sensor technology is perfected, exposures beyond 10 minutes are not likely. The effectiveness of the IDAS filter is, therefore, not as dramatic with digital imaging as it is with film because each exposure is inherently shorter than with film imaging under similar sky conditions. The elimination of light pollution wavelengths by the filter does create a better signal-to-noise ratio, allowing exposures to be stacked without the sky background overwhelming the final image. The general consensus among users is that the IDAS LPS filter renders a more natural looking sky, but it is not quite as effective as the Lumicon filter in controlling light pollution in color images.

11.11 Soft-focus Filters to Enhance Star Sizes

As much effort as we put into achieving the sharpest possible focus, it seems strange to use a filter to deliberately defocus the stars. However, this technique can produce some extremely pleasing and artistic images of the sky. The proper grade of diffusion filter will blur the brighter stars into colored round blobs whose size is roughly proportional to their magnitude. The effect can be used to produce images where the stars appear similar to the way they are represented in printed star charts with the dot size scaled to show apparent brightness. These filters encourage experimentation. Typically, the filter is used for a portion of the overall exposure to blur the brightest stars, then removed for the remainder of the exposure so the dimmer stars will register normally.

11.12 Star Trails

Star-trail images are created by holding the shutter open on a fixed camera for extended periods of time, sometimes for hours. The darker the sky, the better will be the results. This imaging method relies upon the rotation of the night sky. Over time, the stars appear to move across the sky, leaving deliberately trailed images on the film. In their own right, star-trail images have little scientific value, except to illustrate sidereal movement. However, when thoughtfully combined with foreground objects, they can be quite striking.

As we have seen earlier, because of thermal noise buildup, long digital exposures do not produce a good single hour long exposure. Therefore,

Fig. 11.20 *This 20-minute digital star trail image from the author's backyard under bright urban sky could not have been taken with a film camera. Confining each exposure to the sky fog limit of 15 seconds prevented overexposure while digitally combining 80 separate images resulted in an attractive star-trail image taken from an urban location. Photo by Robert Reeves.*

the digital imager may feel left out when it comes to creating images with long star trails. However, this need not be the case. A digital camera can create star trails and produce stunningly beautiful exposures of a quality that film cannot match. In fact, lengthy, colorful, high-contrast star-trail images can be taken in bright urban locations or under bright moonlight that are impossible with film.

The key to producing digital star trails in bright locations is to have a camera capable of taking a series of short images with very little time between each individual exposure so that the motion of the stars does not produce gaps in the trail. If the camera takes too long to initiate the next image, the star trails will resemble a dashed line. Each of the individual short exposures is limited by the amount of ambient lighting or inherent camera noise. The long star-trail image is produced by combining each exposure into a composite image with image processing software.

If your camera does not have an intervalometer to program a sequence of automated exposures, you will have to trigger the shutter manually. Each exposure in the series should begin as soon as the camera allows it. Some cameras have an image buffer that allows the next image to be taken without having to wait for the last image to be completely recorded on the memory card. Fast downloads and a return to imaging will create a seamless sequence of images with no gaps in the star trails. If your camera has too small of a buffer, or the image sequence planned is longer than the buffer can hold, then each exposure will have to wait until the previous one has been recorded. The exact time to download an image to a memory card is highly variable between cameras and storage devices, so you must determine this interval yourself.

In regions close to the celestial poles, the star appears to rotate slowly and gaps between succeeding images will be insignificant; but in equatorial regions, the stars appear to move across the sky at a faster rate, making it important to minimize the interval between images. To minimize this interval, it is best to disable any in-camera dark-frame subtraction for long exposures.

To capture star trails, disable the camera's autofocus and manually focus on infinity. Experiment with different ISO settings and exposure lengths to see which works best with your sky background and any foreground scene. Also, if possible, use an external power supply since the camera will be imaging continuously over an extended period of time, which may run down the camera's internal batteries before you have completed the sequence. Dark-frame exposures should be taken at the beginning of the star-trail sequence and then again at the end. You do this by leaving the camera lens cap in place and making an exposure that is equal to the exposure time for each image segment. So, if you are doing 60 one-minute exposures, you would have two additional dark frame exposures; the first is taken at the beginning of the segment, and the last at the end. These two dark frames are combined and averaged to produce a master dark frame that is used to remove electronic noise from each of the 60 one-minute exposures in our example. The second dark frame helps average out any change in noise over the exposure sequence.

From bright urban locations, light pollution will limit each exposure segment to 20 or 30 seconds, while image noise will set the exposure limit from a dark-sky location. If possible, record the images with a lossless image format like RAW or TIF. The steps and procedures to assemble a dig-

Fig. 11.21 *Comet Neat was imaged through an Astro-Physics AP130EDF refractor using a Canon 10D at ISO 800. Seven images, ranging from two- to four-minutes exposures, were combined to create this view that preserves the delicate structure in the comet's coma while still displaying the tail. Photo by Jeff Ball.*

ital star-trail image from numerous shorter exposures is explained in Section 14.12.

11.13 Comet Photography

Few celestial events stir up interest like the spectacular arrival of a grand comet. Comets offer a photographic challenge like no other celestial target because they can range from being very small telescopic objects to a huge naked eye spectacles. In fact, the same object can be both in a period of a few months. When accompanied by a fully-developed tail, a comet is a natural target for wide-field astrophotography using telephoto lenses. Close-up telescopic views of its nucleus may be scientifically interesting, but nothing is more thrilling in the sky than the sweeping panorama of a majestic comet tail.

If a comet can be seen with the naked eye, it can be photographed with virtually any camera capable of short time exposures. Here are some points to consider when photographing a bright comet:

- Set the camera's ISO speed to 400, or if possible 800. Avoid higher ISO speeds than these, due to the possible excess noise that will degrade the delicate structure in the comet's tail.

Fig. 11.22 *These two images show the telescopic imaging options available for fast moving comets; guide on the stars and have a trailed comet (top), or guide on the comet and have trailed stars (bottom). Both images were taken with a Canon 10D using a 200-mm f/4 lens and 23 separate two-minute exposures at ISO 800, calibrated with the average of eight dark frames. The comet-tracked image was composed of 14 frames selected from the 23 used with the star-tracked image, but were manually aligned on the comet. Images by Phil Hart.*

- If the camera is mounted on a fixed tripod, set the controls to manual and the lens aperture to its widest opening. With short focal length lenses, limit exposures to 30 seconds. Telephoto focal lengths will show star movement in as little as 10 seconds. Do not be afraid to extend the exposure in order to capture the comet's tail, but expect to see trailed star images in the field around the comet.

- If the camera is mounted on a telescope or tracking platform, the lens can be stopped down one *f*-stop from wide open to reduce bloated star images arising from common lens aberrations.

- Comets often display color in their structure. The nucleus may appear greenish, the straight ion tail bluish, while the curved dust tail has a yellowish color. A series of exposures may be necessary to establish where color might be visible.

- Comet tails are extended objects; therefore, the aperture of the lens used governs their limiting magnitude for photography.

- Comets are brightest when near the Sun and thus are usually close to the western horizon after sunset or above the eastern horizon before sunrise.

- The piggyback camera bracket used on most commercial Schmidt-Cassegrain telescopes will orient a camera at an angle near the horizon, resulting in a tilted picture. Installing a swivel ball-head adapter will allow the camera to be leveled with the horizon for a more natural view.

- Match your lens to the comet. The longer the comet's tail, the shorter the focal length. Conversely, short-tail comets need a telephoto lens to reveal details.

- Since comets orbit the Sun, their apparent motion in the sky is dependent on their orbit relative to the Earth. Comets displaying rapid apparent motion across the sky can noticeably shift position when imaged even briefly through a long-focus lens or telescope. In such cases, it may be necessary to guide on the comet's nucleus instead of a background star.

- When you frame a comet, allow some extra space at the end of the tail for any unseen, but photographically visible, extensions. Offsetting the comet nucleus toward the Sun will center the entire comet in the field-of-view.

Comets often exhibit bizarre behavior, and it is interesting to document these phenomena with photography. Their tails sometimes seem to break off and new ones grow in their place. This phenomenon is called a disconnect event and occurs when the comet passes through the different polarity boundaries in the Sun's magnetic field. A comet may display an anti-tail, a spike or brush-like appendage that extends toward the Sun. This is an effect of perspective that may be seen when the comet passes through the plane of the Earth's orbit, and a part of the dust tail appears to fan toward the Sun.

11.14 Meteor Photography

Meteors are actually bits of asteroidal material or particles cast off from comets. The vast majority of meteoroids, as they are known while they travel through space, are little larger than grains of sand. What makes them appear as bright streaks across the night sky is their tremendous speed as they collide with Earth's upper atmosphere. Depending on their original orbit around the Sun, meteoroids can enter the atmosphere at speeds between 20 and 70 miles per second. At these velocities, the particles quickly burn up in the atmosphere between 50 and 70 miles altitude. What creates the bright flare we see in the sky is not the glowing meteor particle itself, but the ionization of air molecules in the atmospheric shock wave that builds up ahead of the decelerating particle. Occasionally, a larger meteor will survive its incandescent encounter with the ionosphere and fall to Earth where we pick it up and call it a meteorite.

Normally, there are six to ten random meteors per hour. Meteors are more frequent after an observer's local midnight, appearing in highest numbers just before dawn. This is because before local midnight, a given location is in the trailing hemisphere of Earth as it orbits the Sun and any meteors must "catch up" with our planet's night side. After midnight our location is in the leading hemisphere as Earth orbits the Sun, plus Earth's diurnal rotation is now turning our location straight into its orbital direction—thus we have a greater chance of seeing meteors being swept up by Earth. Meteor showers occur at certain times of the year as Earth's orbit intersects the debris stream from dead comets that still circle the Sun along the comet's original orbit. Increased numbers of meteors are visible during the peak of these showers. Table 11.3 shows the dates and approximate meteor frequency during the major meteor showers.

Except at maxima of major showers, on average it takes about 50 hours of open shutter time to capture a meteor. For this reason, photographically seeking meteors is simply a matter of setting up a camera as a meteor trap and hoping for the best. Since a meteor can appear randomly at any time in any part of the sky, there is no one part of the sky that is better than another to aim a meteor camera. Geometrically, however, a camera aimed lower to the horizon views a wider portion of the upper atmosphere where meteors are seen than a camera that is aimed at zenith. This slant view through greater portions of the upper atmosphere increases the chances of seeing a meteor, but increased air mass, air and light pollution, and greater distance to the meteor may make it too dim to be recorded by the camera. I routinely set up several unattended cameras at my observing site

Table 11.3 Annual Meteor Showers

Name	Active	Peak	v	ZHR	Radiant
Quadrantids	Jan 01–Jan 05	Jan 03	41	120	15^h20^m, +49°
Lyrids	Apr 16–Apr 25	Apr 22	49	15	18^h04^m, +34°
Eta Aquarids	Apr 19–May 28	May 05	66	60	22^h32^m, –01°
Delta Aquarids	Jul 12–Aug 19	Jul 28	41	20	22^h36^m, –16°
Perseids	Jul 17–Aug 24	Aug 12	59	200	03^h04^m, +57°
Alpha Aurigids	Aug 25–Sep 05	Aug 31	66	10	05^h36^m, +42°
Orionids	Oct 02–Nov 07	Oct 21	66	20	06^h20^m, +16°
Leonids	Nov 02–Nov 21	Nov 17	71	40+	10^h12^m, +22°
Puppid-Velids	Dec 01–Dec 12	Dec 07	40	10	08^h12^m, –45°
Geminids	Dec 07–Dec 17	Dec 13	35	110	07^h28^m, +33°
Ursids	Dec 17–Dec 26	Dec 22	33	10	14^h28^m, +76°

Velocity (v) expressed in km/sec
ZHR = estimated peak number of meteors per hour if the radiant were at zenith

solely for the purpose of catching any random meteors. In order to make the image as aesthetically pleasing as possible, frame a well-known bright constellation or circumpolar star trails.

11.15 Meteor Photography Techniques

If there are dozens of meteors each evening, why are they so hard to photograph? There are a number of reasons. First, a normal camera lens will image only about $\frac{1}{20}$ of the sky. This is a small "keyhole view" when compared to the entire visible sky. Because of their high altitude, meteors essentially appear as points of light. Photographically, they can be regarded the same as stars; that is, it is the usable light-gathering area of the lens, not the focal ratio, that is important. Another factor in photographing meteors is the focal length of the imaging lens. The longer the focal length of the lens, the greater the image scale, and thus the faster the meteor will cross the field-of-view. If the meteor crosses the field-of-view too quickly, it will not stay on any one spot on the image sensor long enough to be recorded.

If a normal camera lens only covers about $\frac{1}{20}$ of the sky, it would seem to be logical that using a wide-angle lens would increase the chance of imaging a meteor because of increased sky coverage. This turns out not to be true for very wide angles because of another factor when photographing point sources like meteors and stars; wide-angle lenses usually have smaller apertures and thus gather less light. This factor significantly reduces the number of meteors that can be recorded because it eliminates the dimmer meteors. Both short and long focal lengths quickly enter a zone of diminishing returns with meteor photography. If the focal length is too long, the image scale is too big and swift meteors may not be recorded; if the focal length is too short, the resulting smaller aperture limits the detection of dimmer ones. The correct answer is to arrive at a balance point where there is sufficient aperture yet a short enough focal length so meteors will still be imaged. This usually works out to be the "normal" lens used with a particular camera's image format (50 mm for 35 mm, 80 mm for 2¼ by 2¼, etc.).

For the sake of simplicity, we will assume that a full 35-mm film size image sensor is being used in a camera dedicated to meteor photography. We can determine which lens would work best by looking at the mathematics of each lens: a 28-mm $f/2.8$ and a 50-mm $f/1.4$. A 28-mm focal length lens will cover 3614 square degrees of the sky, or more than three times the 1136 square degrees covered by the 50-mm lens. Three times the sky coverage might lead one to expect that this lens is three times more likely to capture a meteor. But this is not the case. The 28-mm lens has an aperture of just 10 mm, giving it a light-collecting area of 78.5 square mm. On the other hand, the 50-mm lens has an aperture of 35.7 mm and a light-collecting area of 1000 square mm, or 13 times that of the smaller lens. So for a given exposure length, the 50-mm lens will record objects 13 times dimmer than the 28-mm one, a gain of nearly three magnitudes.

Our numbers in the above example are not absolute because of another factor: plate scale. The 28-mm lens will have a plate scale of 2.04 degrees per mm, while that of the 50-mm lens will be 1.15 degrees per mm. The plate scale of the 50-mm lens is thus just over twice that of the 28-mm one, meaning the image of a given meteor will travel twice as fast across the sensor when using the 50-mm lens, and thus have less chance to be recorded. Dividing the doubled plate scale of the 50-mm lens into its 13 times greater light grasp tells us that it is still over six times, or two magnitudes, more sensitive to faint meteors than the 28-mm lens.

Yet another factor that will determine whether a particular meteor will be recorded by the camera is its angular velocity across the sky. The angular velocity is the meteor's apparent motion across the sky in angular degrees per second, not its actual velocity in miles per second. For example, two identical meteors, both traveling at the same speed through the atmosphere and appearing at the same brightness as they flare in the sky, will record differently if one is traveling on a path close to "head-on" toward the camera while the other appears to travel across our line of sight. The "head-on" meteor will apparently travel across the sky more slowly, and thus remain on any one spot of the camera's sensor for a longer period of time and make a deeper exposure. This factor is important when photographing meteor showers. By aiming the camera at or near the radiant, or the point in the sky from which all the meteors appear to radiate, we can take advantage of the meteor's foreshortened trail to record brighter images with the camera. When aiming near a shower radiant, be sure to align the long axis of the camera's field-of-view parallel with a line leading from the radiant so any meteors will have the longest possible path to travel within the image.

The meteor photographer has a choice to make: seek only the brightest meteors with an aesthetically more pleasing wide field-of-view, or seek the maximum number of meteors with a narrower field of the more sensitive normal lens. The rate of success depends greatly on just plain luck. During a meteor shower when there are 50 visible meteors per hour, it statistically takes five hours of exposure time before a bright enough meteor enters one camera's field-of-view. During non-shower times, this may increase to 100 hours of exposure between hits. Multiple cameras aimed at different areas of the sky will improve the odds of recording these elusive targets.

11.16 Digital Advantage for Automated Meteor Photography

Some digital cameras, and virtually all DSLR cameras, can be automated to image sequences with remote control software and a laptop computer. If your camera can be automated, you will have a significant advantage. Capturing meteors on film cameras is a hit-or-miss proposition; you expose the sky frame after frame until eventually, you catch one. This process uses a lot of film. However, the only "expendable" for a digital camera is the electrical power needed to run it and a laptop computer. Most cameras, if connected to a computer while the memory card is removed, will default to downloading the image directly to the computer's hard drive. This frees you of the camera's limited memory card capacity and allows automated sequential imaging all night long. If the resulting sequence of images shows no meteors, simply deleting the captured images from the hard drive prepares it for the next attempt at no additional cost. Unlike imaging meteors on film, your digital camera will work in an urban backyard, even with bright moonlight. If the sky is bright, you will use shorter exposures.

A 30-second exposure is sufficient for patrol work from bright urban locations. A meteor flash is so brief that the length of each exposure is not

Fig. 11.23 *This image of a bright Perseid meteor recreates the naked eye view of the sky while watching for meteors. This 30-second exposure was taken with a Nikon D70 at ISO 1600 through a Russian Peleng 8-mm f/3.5 lens. Photo by Dale Ireland.*

important. Meteors seldom last more than a second, at best. The flash itself will record the same no matter how long the exposure, and as long as there is sufficient overall exposure to reveal some stars in the background, the image will be aesthetically pleasing. Indeed, one will find that once a meteor is imaged, the main limitation is that all too often the meteor will exit the edge of the field and be framed incorrectly.

There is very little difference between an urban and a dark-sky location when shooting meteors. From the dark site, exposures can be longer, with the maximum length dictated by the appearance of objectionable image noise. As is the case with any long digital exposure, take dark frames from time to time because the amount of noise in an image will vary as temperature changes. Be sure to log which dark frame is to be used with which series of images.

Multi-minute time exposures can be manually triggered with a remote shutter switch, which is basically an electronic cable release; but the operator must start each exposure separately. Additional items needed for automated exposures longer than 30 seconds, at least with Canon EOS digital cameras, are a camera control program like DSLRFocus and either a serial or parallel computer interface cable like that used with DSLRFocus. Canon's Remote Capture software normally interfaces with the camera

Fig. 11.24 *The 2004 Perseid meteor shower was captured in this composite of 57 separate thirty-second exposures taken with a Canon EOS 1D Mark II camera at ISO 3200 and a 17-mm f/4 lens piggybacked on an 8-inch Schmidt-Cassegrain telescope. A total of 680 30-second images were taken. Of these, nine were stacked to create the background star image, and 48 images containing meteors were added to show the meteor shower. A total of 51 meteors are visible on the original print. Photo by Fred Bruenjes.*

through a USB cable and will not allow exposures longer than 30 seconds. Canon makes a programmable TC-80N3 remote switch that can be set up to take unattended sequences of images, but this switch costs more than DSLRFocus, including the special serial or parallel interface cable this software requires.

11.17 Asteroid Photography

A succession of digital images can easily be "blinked," animated, or streaked; all three will reveal the motion of a minor planet. Digital SLR cameras capable of 30-second exposures can image a dozen of the brightest asteroids with moderate telephoto lenses. The photographic system must be capable of recording objects down to 12 magnitude without its background noise overwhelming the image. Longer exposures will, of course, capture fainter asteroids, but the majority of minor planets are quite dim and beyond the reach of a simple telephoto lens. Telescope-mounted digital cameras will allow you to go deeper to bring more asteroids within reach.

For binocular users, the popular astronomy magazines provide location predictions for the brightest asteroids. Advanced star charting programs, such as Megastar, have a feature that allows you to load a database containing all known asteroids from Lowell Observatory or Harvard University's Minor Planet Center. Megastar will then filter visible asteroids by magnitude so as to not overwhelm the observer with objects beyond their equipment's capability. It will also plot the location (and path) on a printable star chart.

Since asteroid hunting has been around for some time, those looking for new objects should plan on working 18 magnitude objects and dimmer. Discovery searches at magnitudes this faint require specialized CCD imaging and large telescopes. However, the digital camera user can still image the brighter asteroids and chart their motion against the star background. Indeed, it is a thrill to actually see these objects move within a period of an hour or two.

As mentioned earlier, there are three ways to photographically show asteroid motion: create an animation sequence that references the asteroid's motion against a field of stars; blink two images, thus causing the object to appear to "hop" back and forth against a star field, or create a single image that shows the asteroid's motion as a streak superimposed on a background of stars. For blinking and streaking, each image is exposed for as long as sky conditions and electronic noise allow.

To make animations, the sequence of images showing the asteroid slowly moving across a star field is assembled using one of the many popular GIF animated programs. The sequence can be exposed with the telescope tracking the stars so the asteroid slowly moves across the stars, or the telescope can track to show the star field drifting behind the stationary asteroid. To show the asteroid's motion as a streak against a background of stars, the images are guided to follow the stars and assembled using the same image processing technique used in creating star trails from a series of short images (see Section 11.12). Both methods can be done using either a telephoto lens or a telescope-mounted camera.

11.18 Lessons Learned at the Telescope

Here are a few tips that help make a celestial imaging session go smoothly:

- **First, and most important, set up early while it is still daylight.** You will have a power source, computer, telescope and guiding equipment all of which will be interconnected with cables. It never ceases to amaze me how inanimate power and control cables tangle themselves in the dark. Misplaced items become more than a simple inconvenience in the dark;

Fig. 11.25 *A flexible USB-powered plug-in light can provide low-level continuous keyboard illumination. Photo courtesy of Hoodman.*

they become trip hazards and roadblocks to enjoying the night sky.

- **Give thought as to how you will lay out your cables.** For example, individual cables for camera power, shutter release, guiding eyepiece illumination or autoguider add up to four cables on the back of one scope. I have had my eyeglasses pulled off in the dark by camera cables. Route them so they remain snugly against the telescope instead of hanging across open space where they can be snagged in the dark. Use finder scope brackets, camera and counterweight mounts to help keep cables snugly out of harm's way.

- **Keep a record of what file name each image is stored under.** A good imaging session will create a lot of images. Repeated dark frames taken as the night cools will also be interspersed with astrophotos. After dozens of images, it becomes hard to keep track of which is which.

- **Use a Laptop screen filter.** A laptop without a screen filter can be incredibly bright at night. I do not need a flashlight because of the screen glow. A bright laptop screen is also a bug attractor; all sorts of insects will soon be crawling on your screen; there is little you can do about this.

- **Use a small keyboard illumination lamp.** The laptop display screen glow alone is not enough to properly illuminate the keyboard. If there is no filter on the screen, it will dazzle; and if it is filtered, the glow is too feeble. A small keyboard illumination lamp makes the evening a lot less frustrating. An LED model mounted on a USB-connected cable that is flexible yet rigid enough to remain where it is aimed is available from

Hoodman. Continuous, low-level illumination is much more convenient than having to pull out a flashlight each time you need to use the keyboard.

- **Plan ahead.** Astrophotography is a complex operation performed late at night when you are tired. To make the work go easier when you do late night astrophotography, plan ahead. The fewer decisions you have to make to get the job done, the better. Know beforehand what you are going to shoot, with what lens and how much exposure. A star-charting program like MegaStar can create and print custom camera fields-of-view and finder charts where the camera field size parameters can be set to any size in arcminutes. See Section 10.18 for calculating your camera's field-of-view. Using the imaging field overlay in a star charting program will show exactly how you camera's field-of-view will overlay a star pattern and reveal the celestial coordinates of the aim point. An interesting cloudy night exercise is to explore familiar constellations with MegaStar display options set to show all deep-sky objects and your imaging field-of-view. This will allow you to photograph previously unattainable objects by taking advantage of the increased capabilities available with digital image stacking techniques. Once you have chosen your targets, have a pre-listed set of bright alignment stars for setting circle verification if the targets cannot be seen through the finder.

- **Remember that DSLR cameras are very sensitive.** The image builds up linearly—that is, the camera keeps on recording at the same rate throughout the exposure—there is no reciprocity law failure as with film. As a film astrophotographer, I am used to the results expected from Ektachrome 200. Thus, on my first deep-sky exposure with a Canon 10D, I took eight one-minute exposures of M13 at ISO 400 only to discover that a single exposure produced a good image and only two were needed to produce a pleasing image where the cluster was not overexposed. On summer evenings the 10D produced no noticeable noise on three-minute exposures even after the camera had been on for four hours.

- **It will take longer than you expect.** A stack of 10 two-minute exposures will not take 20 minutes to obtain; it will be closer to a half hour. There is a period between exposures when the image is recording to the memory card or downloading into a laptop. With my older Pentium II laptop and a USB cable camera connection, it takes my Canon 10D 55 seconds from shutter close to the point where the shutter is ready to open again for the next exposure in the sequence. This adds one-third more time to what would have been a straight 20-minute exposure on film.

- **Take a break.** There are several advantages to the download delay between images if you are guiding the old-fashioned way using a cross-hair eyepiece. Guiding 10 separate two-minute exposures is far less fatiguing than a continuous one of twenty minutes. This helps you stay more alert and makes the evening a lot more pleasant. Also, if there is a minor error in guiding or a passing cloud interferes during any segment of the exposure, it can be discarded from the image stack and another segment exposed to take its place. If clouds interrupt an imaging session, additional exposures of the same object can be added on another night. With film astrophotography, a guiding error, even in the last minute of a long exposure, ruins the entire shot. But with digital imaging, you lose only a small segment of the total image that can be easily replaced. Indeed, I have inadvertently bumped the tripod midway into a series of 45 three-minute exposures of a deep-sky object. If this had happened at the two-hour mark of a continuous film exposure, I would have been very upset. But with digital imaging, I simply discarded the ruined three-minute exposure, recentered the guide star and add another exposure to make up for the lost one.

- **Convert your files later, not out in the field.** Clear Moonless skies are few and far between for most of us. Set up your computer to do the RAW conversion process while you are driving home. I have more than an hour's drive on a smooth highway before I am home from my dark-sky location. I simply cradle the laptop on the car floor with a cushion and blanket to help isolate it from road vibration and let it perform the RAW to TIF conversions. By the time I arrive home, the images are in a format that can be used by an image processing program the next morning.

It takes far longer to process a digital astronomical image than it takes to acquire one. The actual exposure is one small link in the chain leading to the final image. Beyond taking a quick peek at the images to see what you accomplished, don't try to perform extensive processing late at night. Processing is a complex task requiring a clear mind, and a good night's sleep will make it go more smoothly and efficiently.

Chapter 12
Homemade Accessory Projects

12.1 Multiple Outlet 12-Volt Power Supply

As more telescope accessories are designed for use at remote locations, the need for multiple sources of 12-volt power is growing. The telescope drive, the autoguider, a small inverter to run a digital camera battery eliminator and a laptop computer; and perhaps creature comforts such as a music player or radio all require their own sources of portable 12-volt power. A simple solution is to modify a plastic marine battery box to house a number of fused 12-volt cigarette lighter-style sockets in its cover. Plastic and fiberglass battery boxes are widely available from outlets such as West Marine. Radio Shack sells 12-volt power outlet receptacles that are equipped with an in-line fuse and a convenient rubber cap as PN# 270-1556. As many as four outlets will conveniently fit in the cover of a standard group 24 automotive battery case. To monitor the charge of the battery, a Radio Shack "Vehicle Battery Tester" PN# 22-112 can be installed in the battery box cover.

This accessory should use the safer, spill-proof gel cel instead of regular car batteries that can spill highly corrosive acid if tipped over for any reason.

12.2 Portable 120-Volt Power in the Field

Digital imaging places a premium on the use of electrical power at the telescope. Sometimes electrical power is not available at dark locations, but cameras, telescope drives, and laptop computers still ask for their usual voltage before they will operate. I solved the remote location power problem by building a combination battery and inverter box with two built-in ground-fault circuit interrupt (GFCI) protected 120-volt outlets as well as twin 12-volt cigarette lighter outlets. The box was built from ½-inch plywood, and all the electrical and cabinet hardware was found at the local

Fig. 12.1 *A multiple-outlet 12-volt power source can be made from a standard group 24 automobile battery-size gel cel and a plastic boat battery case (left). Four fused cigarette lighter outlets are installed in the lid of the battery case. An inexpensive battery-charge monitor such as a Radio Shack vehicle battery tester, is installed on the battery box cover (right). Photos by Paul Hyndman.*

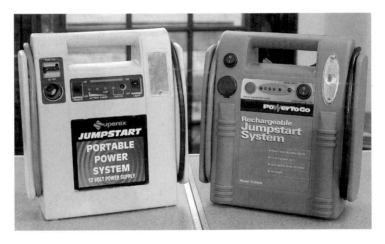

Fig. 12.2 *Automotive jumpstart battery packs can supply 12-volt power for telescope accessories. Photo by Paul Hyndman.*

Lowe's home improvement store. The only items purchased elsewhere were the two cigarette lighter outlets, that are available from an auto parts store. The power box was scaled around the size of the inverter used, in this instance a 1000-watt X-Power inverter by Xantrex available at Costco and other major chain stores. Xantrex makes other inverter models with built-in GFCI protection designed for use in RVs and marine applications, and these models are available at RV and boating stores. If the user is not intimately familiar with electrical 120-volt wiring, I recommend using an in-

Fig. 12.3 *The author powers all 12-volt and 120-volt telescope and camera accessories with this battery box and inverter package made from ½-inch plywood. Two 33-amp gel cells power a 1000-watt Xantrex inverter and two 12-volt cigarette lighter outlets. Twin GFCI-protected outlets distribute power from the inverter. Photos by Robert Reeves.*

verter model that contains the built-in GFCI instead of wiring GFCI outlets to a non-GFCI inverter.

When in use, the inverter needs to be exposed to open air for cooling purposes. For this reason, it is mounted under the lid of the box on a removable platform above two 33-amp/hour 12-volt Power Sonic gel cell batteries. The lid is opened to expose the unit for cooling while it is in operation.

The box dimensions were scaled to allow the 12-volt input wiring and the 120-volt outlet plug leading to the two GFCI receptacles to clear the lid of the box when it is closed. To access the two batteries, the outlet plug is disconnected from the inverter, and the inverter platform lifted out and swung aside. The 12-volt input wires are sufficiently long to allow the inverter to be removed without difficulty. The inverter only needs to be removed to occasionally check the battery connections for corrosion. The batteries are held in place by a short plywood framework that cradles each one and prevents its movement during transportation. When in use, the batteries need ventilation so the inverter platform has a large cutout under the inverter to allow air circulation.

It is not the purpose of this book to give instructions about cabinet-making, but a point to remember about building a wooden case with a hinged lid is that it is easier to build an enclosed box, then cut it in half, than it is to build two open boxes that fit together at the hinge.

The box shown in Figure 12.3 is a basic design that can be enhanced with additional features or scaled down if the user desires. The unit shown allows the inverter to power twin 120-volt GFCI wall receptacles so more than one person can use it. Twin GFCIs are used so that if one user trips their GFCI for some reason, it will not shut down the other person's equipment. A six-outlet computer power strip extension cord allows the connection of multiple items to the GFCI outlet.

With the two batteries installed, the power box weighs in excess of 60 pounds. User-friendly enhancements like recessed carrying handles can be added to the sides to aid in transporting the unit. Wooden skids can be attached to the bottom so the box can be placed on the bare ground, and brass corner protectors can be attached to prevent damage during transport. An on/off switch can be added to control the 12-volt outlets. If heavy amperage use is expected on the 12-volt circuit, the on/off switch can control two standard 30-amp automotive relays, which in turn supply power to the outlets. When the inverter is not under an electrical load, the inverter fan cuts off making it difficult to tell if the inverter is operating. An LED status light can be added to indicate the 120-volt inverter circuit is turned on.

12.3 Laptop Case With Built-in Power Supply

The good news is that used, inexpensive, but very capable Pentium II and III laptops with CD drives and wireless communication capability are readily available from many sources, and adding one to your equipment may now cost less than a premium eyepiece. The bad news is operating a laptop in cool, often moist, nighttime conditions can be challenging. Laptops do not like cold. In low temperatures hard drives can fail, LCD screens become unreadable, and keyboards can become unreliable.

The solution to cold weather computer problems is to keep it running all the time and to insulate it with a foam-lined carrying case. If external 115-volt power is not available to keep the computer continuously running, you can use a battery-powered inverter. An inexpensive Xantrex X-power Mobile Plug 175w inverter (Figure 8.13) can be connected to a 10-amp/hour sealed lead-acid battery such as the Power-Sonic PSH-12100. This will supply enough power to operate the laptop for several hours beyond its internal battery capacity. The inverter can be placed close to the rear of the laptop where it will add more warmth for winter operation.

In really cold climates, simply leaving the laptop on will not provide enough heat. In this case, 12-volt heating mats designed specifically for laptops can be purchased from Kendrick, the supplier of telescope anti-dew heaters (see Appendix B). If 115-volt power is available, electric heating pads from a drug store will keep a laptop warm and operating. (On the same note, where power is available these heating pads go a long way toward keeping astronomers warm and operating when the temperature is sub-freezing.)

On cool nights with high humidity, parts of a laptop may start to gath-

Fig. 12.4 *Carrying a laptop computer in a padded self-powered carrying case not only protects the computer at all times; it insulates it in very cold weather and protects vital components from freezing. Photo by Robert Reeves.*

er dew. Leaving the laptop on, but keeping the screen folded down when not in use can minimize dewing. If dew becomes a problem even with the lid down, you can throw in the towel, literally. Drape a small terrycloth towel over the computer when it is not in use to protect it from dew.

The laptop carrying case with internal power supply shown in Figure 12.4 was based on a unit originally built by Roger Groom using an aluminum photographic/electronics case available from many sources. He added layers of foam to fill all the voids between the power supply and laptop, and even added an outlet board so other small accessories can run off the inverter. With everything housed in a common case, he can power up his computer and operate it in the field for hours without the need for a 115-volt wall outlet.

12.4 Cooling Compact Cameras to Reduce Noise

Early model compact digital cameras such as the Olympus Camedia and the Nikon Coolpix series are limited to maximum exposures between 16 and 60 seconds. They also produce a considerable amount of noise after several long exposures because of heat buildup within the camera during its operation. This would seem to limit their application for deep-sky photography; but if the noise can be minimized, they can produce surprisingly good images of the brighter Messier objects.

For those who like to tinker, there are several modifications that can be performed on inexpensive older digital cameras to reduce noise in long exposures. Used, outdated two- and three-megapixel models are readily available from sources such as eBay for pennies on the dollar of their original price. These can be dismantled for such modifications as adding air-cooling of the imaging sensor, or attaching Peltier coolers and associated

Fig. 12.5 *Small digicams can be effectively cooled for short periods with reusable ice packs. The ice packs are cut to fit (left) the back of the camera and are secured by rubber bands (right). Photos by Paul Hyndman.*

insulation, or cooling them directly with frozen ice packs.

If an inexpensive camera fails to function after such modifications, the financial loss is minimal. For example, astrophotographer Gary Honis modified his Olympus C-2000Z by removing the threaded insert in the tripod socket hole and attaching a small 12-volt CPU cooling fan over the opening with tabs of Velcro. This forces air into the camera and significantly reduces the noise during maximum-length exposures. Power to run the 12-volt fan is supplied by the electrical system on his Starmaster telescope. Although the C-2000Z is just a two-megapixel model and considered outdated by today's standards, Honis is able to use his modified camera to take many sequential noise-free 16-second exposures through his equatorial platform-mounted 20-inch Dobsonian. By digitally stacking them, he produces good images of deep-sky objects with a camera now available on eBay for less than $20. A two-megapixel camera can also take excellent lunar and planetary images, so the capabilities of these older models should not be underestimated. The disassembly of early model compact Olympus cameras is not a complicated task.

In warmer climates, modifications such as a fan may not offer enough cooling capacity to significantly reduce an older camera's long-exposure noise levels. Paul Hyndman took the idea of cooling the imaging sensor of his Nikon Coolpix 995 a step further by applying ice packs directly to the camera back to cool the sensor down to winter night temperature levels. Strapping ice to the back of a camera constitutes operating the camera in a manner the manufacturer did not intend and *may* cause later problems with it. The procedure shown here is for information purposes only and Paul Hyndman, I and Willmann-Bell, Inc., the publisher of this book, take no responsibility for any problems encountered because of this procedure.

The main problem with using cooling packs is condensation. The outside of the camera can be protected from condensation and dewing by separating the camera body and ice pack with a plastic bag with a hole cut in it for the lens. Moisture inside the camera is harder to control. Condensation on the optics, image sensor, or electronics will cause later problems with the camera's operation. For the climate near Paul Hyndman's home, the camera's CCD sensor seems to produce enough heat to prevent dewing. A solution for more humid areas may be to keep the camera in an air-conditioned area of low humidity for a lengthy period to "dry it out" before use, and then store it in a sealed plastic bag until it is ready for use.

Paul's cooling procedure is simple; he straps strips of flexible Coleman ice packs to the back of the camera with large rubber bands. These can be purchased in 4 x 11-inch strips for about $2. The strips are segmented into individual squares and can be cut up as needed to conform to the camera back. He calls his cooling modification "Internal Cooling Enhancement" or ICE for short.

Each frozen package allows about 20 minutes of on-camera cooling and enables the Nikon Coolpix 995 to take relatively noise-free 60-second exposures even on a summer night. Cooling Paul's CCD sensor in this manner virtually eliminates noise in the image, even after 10 minutes of operation. With the camera's own noise-reduction routine also being applied, noise is absent from the image. With the cooling packs applied, his camera produces stunning deep-sky images reminiscent of those taken with more advanced models. After the ice packs are removed, the camera should be placed in a sealed plastic bag while it warms up in order to protect it from condensation. Figure 12.6 shows the Coolpix 995 noise levels with and without the ice cooling.

12.5 Cooling DSLR Cameras to Reduce Noise

DSLR cameras generally have much lower noise levels than compact digital models, but their extended exposure capability increases the likelihood that noise will become a problem. The cooling solutions mentioned above either produce minimal cooling or are short-term solutions. To reduce the noise in large cameras such as the Canon 10D, a liquid cooling system can be devised such as the one shown in Figure 12.7 which was engineered by R. A. Greiner. This system allows the camera to remain cooled at the telescope for extended periods of time.

Greiner's system uses a Peltier cooler to chill the camera 24° C below ambient temperature. Noise is reduced by a factor of two for each 6° C reduction in temperature; thus, this system will reduce the noise in Greiner's camera by a factor of 16. On relatively noise-free cameras like the Canon

Fig. 12.6 *This composite of 60-second exposure dark frames shows how the ice packs reduced noise in Paul Hyndman's Nikon Coolpix 995. The lower row of images is the same exposure as the upper row, but with the camera's internal noise reduction turned on. The left vertical column shows the camera's traditional noise after it had been on for 10 minutes. The middle vertical column shows the reduction in noise with the ice packs applied. The right vertical column shows that even after 20 minutes, the ice packs still significantly reduced noise. Photos by Paul Hyndman.*

and Nikon DSLRs, cooling systems can reduce long exposure noise to the vanishing point.

The Peltier mechanism shown here uses ethanol alcohol mixed with water to prevent the coolant from freezing. The liquid is pumped by an Eheim aquarium pump that circulates it through the heat exchanger, to the camera, then through a reservoir to eliminate bubbles. Using automotive antifreeze as coolant is not recommended for several reasons. First, it is very toxic to small animals such as cats and dogs. Additionally, pure anti-freeze will not circulate through a small centrifugal pump. Its molecular structure is such that it must be mixed with water before it will flow through the pump.

An insulated jacket that houses the coolant tubes is wrapped around the camera body (Figure 12.7 left). This prevents condensation on the camera. In extremely humid climates, it is a good idea to place some desiccant packages inside the insulating jacket to control any condensed moisture. Camera power and control cables, as well as coolant lines, protrude from the insulated jacket. Coolant hoses are attached between the camera box and the cooling unit by quick-disconnect couplers. The cooler unit is housed in a plastic storage container. The Peltier cooler and fan assembly is divided from the pump and reservoir side by insulating foam. Hot air from the heat exchanger is pushed out of the cooler by a fan through open-

Fig. 12.7 *This unit built by R. A. Greiner uses a large Peltier cooler and an aquarium pump (left) to circulate an alcohol solution into an insulated camera cover (right). Automotive antifreeze should not be used because it is toxic. Photos by R. A. Greiner.*

ings cut in the side of the plastic container.

The Peltier cooler is a model LC200 from TE Technologies, the pump is an Eheim model 1250 available from many aquarium supply outlets, and the hose quick-disconnects are from Cole Parmer. Be aware that Greiner's cooler design is a heavy-duty unit, and the "new parts" price for the components is high, with the Peltier cooler alone costing nearly $500. Used components may be found on eBay at lower prices. But the fact is, this design does reliably cool the camera while others using cheaper, smaller coolers have failed to perform.

12.6 Easy to Make Stable Telescope Mount

A heavy, stable, vibration-free mount is just as important as good optics in achieving sharp astrophotos. After admiring the excellent commercial mounts available today, I decided to build something similar. I have had very good results with two heavy mounts fabricated from common steel supply items. Not wishing to transport a heavy mount to my remote observing site, I decided to build two separate but similar mounts, one for home and one permanently left at my remote site. The fact that the latter would be left out in the open dictated that it had to be essentially weatherproof.

My solution was to build two mounts that are a cross between a pier and a tripod. The resulting design, shown in Figure 12.8, displays exceptional stability and is built entirely from items available inexpensively from a steel salvage yard and hardware stores. Some welding is required to attach the end plates to the vertical center pier. The twin mounts are based on a 30-inch long, 6-inch square pipe for the "at home" version,

 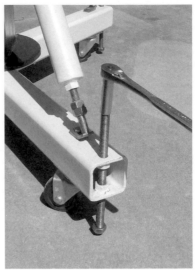

Fig. 12.8 *A combination of characteristics from both a pier and a tripod (left) make this telescope mount design exceptionally strong, stable, and vibration free. Details of the extendable tip on the stabilizer legs are visible (right) along with a leveling jackscrew. Photos by Robert Reeves.*

while the field version uses a 36-inch pipe. The home version is shorter because it is mounted on swiveling casters to allow it to be rolled out of the garage for use.

The piers have a round 12-inch diameter ½-inch steel plate welded to the base to adapt the three legs. The top of each pier is covered with an 8- x12 ½-inch steel plate that is drilled to accept the telescope mount adapter. Both piers are filled with dry sand to damp out any vibrations, so the bottom plate inside each is covered with a thin sheet metal plate to seal the open bolt holes that attach the three legs. Without the sheet metal shield, the sand would trickle out of the pier when it is transported with the legs removed. A square hole is cut into the side of each pier just below the top to allow sand to be poured into it. The hole is covered with a plate and sealed with silicone caulk to prevent water from penetrating the pier.

Three legs, made from 3-inch diameter square pipe, are bolted to the round bottom plate with ⅜-inch bolts. ⅝-inch jack-bolts thread through two ⅝-inch nuts welded into the ends of the legs to level the mount and raise the casters off the ground. Initially, the legs were stabilized with large turnbuckles that tensioned them against the pier. This system proved to actually significantly enhance vibration. To cure the vibration, the turnbuckle attachment brackets were turned 180 degrees, and a system of large jackscrews was devised to push between the top of the pier and end of the

Fig. 12.9 *A drawback of heavy steel telescope mounts is the loss of portability. To solve that problem, I made two mounts: one without casters (left, on furniture moving dollies) to permanently leave at a remote observing site, and the other with casters (right) which can be rolled onto the home driveway for use. Mary Reeves, the wife of the author, shows the scale of the mounts, which in spite of being just waist-high, each can carry several hundred pounds. Photo by Robert Reeves.*

legs. The jackscrews were made by welding a cap onto each end of a 2-inch diameter round pipe. A "pencil point," made from a tapered ⅝-inch bolt, was fixed to the top of the jackscrew. The bottom features a tapered ⅝-inch bolt threaded into the jackscrew assembly. In practice, the tips of the jackscrew assemblies are engaged into brackets on the top of the pier and the ends of the legs. When the threaded bolt is extended to tighten the jackscrew into place, the pier and leg assembly becomes exceptionally stable and vibration-free.

High quality exterior enamel paint has protected the field mount from the elements for the past three years and the mounts provide a highly stable platform for any telescope I have placed on them. These mounts can be fabricated by anyone with a half-inch drill and a saber saw. The welding required is a simple task for any garage or small welding shop.

12.7 Cable Release Adapter Frame

Many early model compact digital cameras lacked some of the features we have become accustomed to with 35-mm film cameras. The lack of a cable release accessory to operate the shutter without shaking the camera or telescope is one of the first items that will be missed if not available. Later model compact cameras use infrared remotes to actuate the shutter, but my Olympus 3020Z has no built-in remote operation capability. My solution to this problem is shown in Figure 12.10. A simple frame that attaches to the camera with the tripod socket hole and wraps around it is used to secure a standard mechanical 35-mm film camera cable release above the shutter

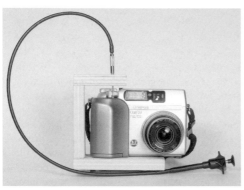

Fig. 12.10 *If your digital camera has no provision for remote shutter release, a lightweight frame can be fabricated from wood to adapt a mechanical cable release to it. When any camera is attached to the eyepiece with an adapter, loop the camera strap over a telescope part as a safety in case the adapter accidentally breaks loose. Photos by Robert Reeves.*

button on my Olympus 3020Z. The base plate is attached to the camera tripod socket with a ¼x20 thread bolt. If the camera and frame assembly both have to be attached to a tripod, drill a ⅜-inch hole in the wood base and install a ¼x20 T-nut which will accept the tripod bolt.

12.8 Small Tracking Platforms

Wide-angle, wide-field imaging can be accomplished without the need for piggybacking the camera on a telescope by using a small tracking platform. There are two ways to do this: use a Poncet platform designed for a small Dobsonian telescope, or use a Scotch mount, commonly known as a "barn door" tracker. Figure 12.11 shows my Poncet platform, a unit constructed by Kurt Maurer of Houston, Texas.

If a Poncet mount, or equatorial tracking platform, is properly aligned, a camera mounted atop it and aimed anywhere in the sky will track the stars. All that is needed is a method for attaching the camera to the platform while it is tracking. The simplest method is to attach a swivel ballhead tripod adapter to a piece of plywood. With the camera attached to the ball adapter, the plywood base is laid on the tracking platform. Compose the picture by aiming the camera with the ball head adapter, turn on the platform to begin tracking, then open the camera's shutter. A well-aligned platform can easily track a camera equipped with a normal or wide-angle lens for several minutes without star trailing.

A Poncet platform can be polar aligned using the star-drift method explained in Section 9.5. Place a small tabletop telescope mounted on the platform to will enable you to view star drift.

Fig. 12.11 *A Poncet platform from a small Dobsonian telescope makes an excellent star-tracking device to carry a cluster of cameras for meteor shower photography. The author's wife, Mary Reeves, sets up a camera cluster on such a device. (In this pre-DSLR era photo from 1999, they are all film cameras). Photo by Robert Reeves.*

An equatorial tracking platform such as this is an ideal way for a group of people, such as an astronomy club, to pool their photographic equipment for wide sky coverage of meteor showers. Figure 12.11 shows eight separate cameras attached to my platform. In this case, film cameras are shown, but the same principle can be used with digital cameras. Each is attached to a pre-aimed bracket to allow it to image its own section of sky.

The Scotch mount, so named because G. Y. Haig of Scotland popularized it, is a simple hand-operated tracking platform that is attractive for its portability and ease of construction. Often called the "barn door mount," this device also allows piggyback-like star tracking without the need for an equatorially-mounted telescope. In its most basic form, the barn door tracker consists of two horizontal wooden boards hinged on one end. The hinge is the polar axis. The bottom board is the fixed base of the tracker; the top board is driven upward around the hinge by a jackscrew, or pushbolt, to follow the motion of the stars.

A barn door tracker will allow your camera to follow the motion of the stars either manually, or automatically with a small electric motor, depending on how sophisticated you wish to make it. Because a tracking platform only needs to support the camera, there is no need for a massive equatorial mount, counterweights, large tripod, and other paraphernalia accompanying a large telescope. Such a mount is compact enough to fit on a large photo tripod.

The basic single-hinge mount does suffer from tracking error after about five minutes of operation because of the changing geometry between the two hinged boards and the push bolt. Wide-angle images should be able to track for nearly ten minutes without noticeable star trailing. This should not be a significant problem for the digital astrophotographer because most of today's cameras are limited to exposures of five minutes or less by electronic image noise. If greater accuracy is needed to allow stacking of successive digital exposures of the same object, there are advanced variations of the barn door tracker that use multiple hinges to maintain extreme tracking accuracy for up to an hour. Details about more advanced designs can be found in the *Handbook for Star Trackers* by Jim Ballard. The book is out of print but is available from used book dealers on eBay.

A ¼x20 T-nut on the bottom board will allow it to be attached directly to a tripod tilt head. Latitude adjustment is made by tilting the tripod head upward until the hinge angle equals the local latitude. If needed, making a simple gun sight can refine aiming at the pole. This can be constructed by placing a 0.335-inch hole 12 inches from a peep sight. This gives a 1.6-degree field-of-view. Align the peep sight axis parallel with the polar hinge. To polar align the platform in the northern hemisphere, place Polaris on the side of the sight hole opposite Alkaid, the end star in the handle of the Big Dipper.

To use a barn door star tracker, place the unit on a stable tripod and attach the camera to the swivel ball-head camera adapter. Polar align the tracker by aiming the hinge at Polaris (or the south celestial pole if shooting from the southern hemisphere). Aim the camera at your desired target and make sure it is focused on infinity. Set the *f*/stop to the desired setting and open the shutter. Begin operating the tracker's jackscrew mechanism for the duration of the exposure, and then close the shutter after the desired interval.

12.9 Barn Door Mount Construction Details

The best hinge for barn door trackers is a classic piano hinge. Use a hinge that is as long as the mount's two boards are wide. The longer the hinge, the less play in the hinge, and the easier it will be to get the two boards parallel. Piano hinge is found in most good hardware stores and must be trimmed to fit the application.

The key to the barn door mount is the drive mechanism that propels the top board in synchronous motion with the stars. An ordinary ¼x20 thread bolt passing through the bottom board 11.42 inches from the centerline of the hinge and turned at one RPM will raise the top board at the sidereal rate. These parameters are related by the formula

$$D = R \times 228.56 \times P$$

where

$D =$ is the distance from the centerline of the pushbolt to the centerline of the hinge in units of length

$R =$ is jackscrew revolutions per minute (RPM)

$P =$ length of one jackscrew thread in same units of length D.

The formula will be helpful in computing the proper distance D of the pushbolt from the hinge when the barn door is being propelled by a motor that either delivers more or less than one RPM, or if a different thread is used for the pushbolt. If a motorized version is built, a counterweight may be needed to keep the weight on the pushbolt from being too much for the motor to handle.

The precise 11.42-inch dimension from the pushbolt to the hinge necessary for a ¼ x 20 bolt to accurately track at one RPM is almost impossible to achieve using common hand woodworking tools. The best thing is to make the pushbolt's thrust location adjustable. A slotted attachment point will allow the screw's thrust point to be moved to the ideal location, which must be found by trial and error. A simple way to test the accuracy of the thrust location is to secure a laser pointer at a right angle to the polar hinge and aim it horizontally at a perpendicular wall exactly 50 feet away from the polar hinge. To simplify the calculation of how far the laser point should travel up the wall while the mount is tracking, the figure is approximately 2.64 inches per minute. A more easily measured distance would be 26.4 inches after 10 minutes of tracking. If the spot moves too slowly, adjust the thrust point closer to the hinge. If the spot moves too quickly, move the thrust point further away from the hinge.

Timing the turns of a manually-driven right ascension pushbolt when using a wide-angle lens can be as simple as giving a quarter turn every 15 seconds as gauged by a wristwatch. On the other hand, short telephoto lenses will require that corrections be made every 5 seconds to prevent star trailing. In this case, tape-recorded tick marks every five seconds can time the advancement of the pushbolt in $\frac{1}{12}$ revolution increments. To aid in determining the proper amount of rotation, a round card with 12 equally-spaced clock marks can be attached to the pushbolt and illuminated in a way as to not interfere with the camera exposure.

12.9.1 Suitcase Star Tracker

For my own hand-driven star tracker, I made a self-contained device that I can easily carry while traveling. It combines the features of a barn door tracking platform with a mount that allows control of both right ascension

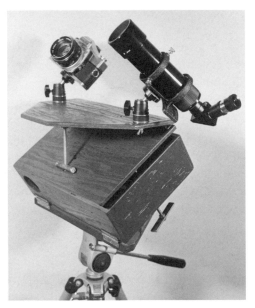

Fig. 12.12 *Small hand-powered star trackers can be as simple or as complex as the builder desires. The author's tracker (shown with a film camera) has both right ascension and declination controls (left), and the components all store within the box-like body of the tracker (right). Photos by Robert Reeves.*

and declination. The entire apparatus folds up into its own 5x9x14-inch carrying case. The camera mounting, a small low-power guide scope, polar alignment scope, and tracking hardware are all stored within the box (Figure 12.12). The platform can be mounted on a tripod, or attached to a tabletop or other flat surface with C-clamps, also stored within the unit.

The dimensions of the platform are dictated by the size of the components stored within it. The basic platform is a plywood box with an extra bottom panel hinged to it so the box can be tipped upward with the short dimension (width) aimed at Polaris. This flap is the latitude adjustment and can be eliminated if the platform is to be used exclusively on a tripod that itself can tip the unit toward Polaris.

The lid of the box is hinged along its "western" edge; this hinge acts as the polar axis and the entire lid as the right ascension control. In order to provide declination control, a second flap is hinged along the "north" side of this right ascension flap. This double-hinged box top thus allows control of both celestial axes.

All three flaps—the latitude adjuster, the right ascension control, and the declination control—are adjusted using ¼x20 all-thread jackscrews.

Because the geometry of the right ascension and declination flaps continuously changes while following celestial rotation, a means must be provided to allow the jackscrews to tip as the flaps move; otherwise, the screw ends will scrape against the underside of the flaps as they are slowly pulled off to one side. My solution was to T-drill and thread short pieces of ½-inch shaft to use as rocking pivots for the ¼-inch jackscrews. The ½-inch shafts nestle into grooves cut into the plywood.

The larger the hand controls on the jackscrews, the easier they are to use in the dark. Consequently, for the latitude and declination controls, I cut two-inch aluminum disks with a hole saw. For the right ascension control, I made a large T-handle to provide more leverage and reduce vibrations.

For the polar finder scope, I used a discarded 6x30 finder scope. It is permanently mounted inside the box, which has a small lift-out window at each end to allow viewing through it. The guide scope is an 8x50 right angle finder purchased at a swap meet. The camera and guide scope mounts are inexpensive swivel ball-head camera mounts purchased at a camera store.

This platform was originally built to allow long exposure wide-field imaging with film cameras. In use, the 8x50 finder scope is aimed at a bright star near the middle of the photographic field and defocused so the blurry star will show the junction of the crosshairs in the eyepiece. Hand tracking is performed at a rate that keeps the star centered on the crosshairs. For use with digital cameras, the platform is used in a similar way, but the exposures are shorter, allowing for less rigorous tracking.

12.10 Hartmann Mask Focusing Aid

A Hartmann mask focusing aid can be made out of any opaque material that can be placed in front of the camera lens or telescope aperture. This device is also known as a Scheiner disk because it was perfected in 1619 by Christoph Scheiner. Technically, a Scheiner disk has only two holes while a Hartmann mask can have three or more, but today both devices are commonly called a Hartmann mask. In its basic form, the mask has two holes, each about one-third the aperture of the optic it is being used on, and separated by a distance equal to the hole's diameter. The smaller and closer to the edge the holes are, the better defined the image is; however, there is a practical limit on how small the holes can be and still be useful. For visual focusing, the holes have to be large enough to allow an image bright enough to see. For photographic focusing using one of the popular software aids, the holes can be smaller because the camera can use longer exposures and is more sensitive than the eye.

Fig. 12.13 *A two- or three-hole Hartmann mask can be cut from thin cardboard and laid on the corrector plate of a Schmidt-Cassegrain telescope. A mask on a Newtonian telescope will require a mounting frame to attach to the open tube. Photo by Robert Reeves.*

If the seeing is not good, some users prefer using only a two-hole Hartmann mask because when focus is nearly achieved, the resulting image is still elongated and the need for further focus refinement is indicated. A three- or four-hole mask can produce some confusion as the point of good focus is approached because the three or four independent star images begin to merge into a larger triangular or square pattern that appears like a focused round star image. If the seeing is exceptional, the three-hole mask will produce better focusing performance.

Foam core artboard can be used to make an inexpensive mask for wide aperture telescopes, while any stiff cardboard or heavy construction paper can be used to make masks for camera lenses. For durability with repeated use, small cardboard masks for camera lenses can be mounted in filter step-up adapter rings. Check hobby stores for embroidery or needlepoint hoops that may fit over the end of your telescope tube to hold a mask in place. Hobby stores may also have artwork templates useful for cutting out various size circles and triangles in the mask. If a more durable mask is needed for heavy use on a large telescope, thin sheet metal can be cut with a jigsaw and fabricated into a mask that will survive repeated handling.

In operation, the Hartmann mask produces two out-of-focus images that merge together when the target is focused, much like a rangefinder camera viewfinder. The advantage of the Hartmann mask is that it can be used for extended objects like the Moon, while knife-edge focusers can only be used on point sources like stars. The disadvantage of the Hartmann mask is that it darkens the viewfinder, and in turn, makes it harder to see

Fig. 12.14 *A diffraction focusing aid can be quickly installed on a telescope by stretching two crossed bungee cords over the aperture. Photo by Robert Reeves.*

the target for visual focusing. This will not be a problem for those using an image capture program to display the image on a computer screen.

Ron Wadaski, a CCD imaging authority, has suggested an excellent enhancement to the Hartmann mask by using triangles at an angle to each other instead of the traditional circle-shaped holes. Each triangle creates six diffraction spikes that help determine the center of the star image; when the diffraction spikes overlap, the image is in focus.

Paul Gitto has perfected another improvement to the Hartmann mask by adding a third triangular hole to the standard two round holes. Paul's technique places the triangular hole at the north end of the telescope tube and the two round holes on the east and west side. After some experimentation to see how the modified mask works with your telescope, the triangle will allow determining which way to move the focuser. For instance, if the triangle is seen on the northern side of the two round images, determine which way the focuser must be turned in order to merge the triangle toward the two round holes. If true focus is overshot, the triangle will appear on the southern side of the two round images. The direction the focuser needs to be turned, either in or out, can thus be calibrated by analyzing how the north-mounted triangle reacts through your optical system. Focusing targets in the future will be easier because the north-south position of the triangle will indicate whether to turn the focuser in or out to improve the focus.

12.11 Diffraction Focusing Aid

An alternative to the Hartmann mask is diffraction focusing. With this method, an artificial spider is placed over the aperture of the telescope.

This can be two rods that are attached to a needlepoint embroidery hoop that attaches to the front of the telescope, or it can be something as simple as two small bungee cords stretched at right angles to each other over the telescope aperture. Either system will produce the same result; an out-of-focus bright star image will display a dual diffraction-spike pattern. When the diffraction pattern merges as the telescope is focused, the image will be in focus. After focusing, the spider assembly can be removed prior to taking the picture.

Most users feel the diffraction-pattern method produces a crisper focus than that achieved with a Hartmann mask. To see if this is the case with your system, focus as best you can with a Hartmann mask, then without disturbing the aim of the telescope, substitute two cords strung across the aperture and see if the focus can be further improved.

12.12 Schmidt-Cassegrain Mirror Lock

Maintaining good focus with a Schmidt-Cassegrain telescope will be almost impossible without a means of locking the primary mirror into a fixed position. An SCT mirror is supported by a collar that slides up and down the central Cassegrain light baffle. The mirror is designed to move on this device in order to focus the telescope, and thus some degree of mirror flop is a normal characteristic of an SCT. The natural "mirror flop" in the SCT design will alter a focus point when slewing from one part of the sky to another to focus on a bright star, then slewing back to the target. Experienced Schmidt-Cassegrain users lock the mirror at a point of good initial focus, then refine it afterward with an auxiliary rack and pinion focuser attached to the visual back at the rear of the telescope.

To secure the mirror after initial focus, some people install nylon screws through threaded holes drilled in the back of the telescope. These screws gently push on the mirror in three locations to prevent it from rocking as the scope is moved. Be advised that such modifications requiring drilling into the tube or body of the telescope are done entirely at the owner's risk. The concepts provided here are for information only, and the telescope owner who performs these modifications does so without liability from the author or publisher of this book.

I use a different approach for preventing SCT mirror flop; holding the mirror by its edge with small plastic tabs that grip the rim of the mirror. This technique, however, may pinch the mirror and create undesirable image artifacts. My SCT is used solely as a visual guide scope, so perfect star image quality is not an issue; but if the mirror lock tabs are excessively tightened, the telescope produces triangle-shaped stars. Fortunately, when

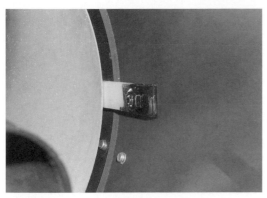

Fig. 12.15 *One of three mirror lock tabs on the author's 8-inch Schmidt-Cassegrain guide scope. The mirror has been retracted to show the plastic tab that is tightened against the edge of the mirror with a setscrew from the outside of the tube. The forward portion of the tab has been painted black to reduce reflections. Photo by Robert Reeves.*

using an SCT as a guide scope for digital imaging, mirror flop will not be the disaster that it is with film imaging. If the mirror shifts 50 minutes into an hour-long exposure, the entire exposure is wasted. However, if the mirror shifts during the 10th of a series of 12 five-minute digital exposures destined for stacking, only that one five-minute image is lost.

To perform the mirror lock modification on my own 1970s vintage Celestron-8, I devised a system of three plastic tabs that gently contact the rim of the primary mirror at 120-degree intervals. Because the in-focus location of the primary can change with different eyepiece and camera combinations, a single location for mirror lock setscrews may not be wise. Thus, the setscrews do not contact the mirror directly, but instead, push on three plastic tabs, which in turn flex to contact the mirror anywhere along one inch of mirror focus travel.

To find the correct placement for the lock tabs, first focus to a typical infinity position. Next, remove the corrector plate to access the inside of the tube. On a classic SCT, this is accomplished by removing a retaining ring secured with eight screws. Be sure to mark the corrector plate so it can be aligned in the same position when reinstalled. Then measure the distance from the front of the tube to the mirror and mark the outside of the tube at the same distance. Locate the desired position for the three setscrews at 120-degree intervals at the marked location from the front of the tube. Now move the mirror to its lowest position with the focus control.

Stick masking tape as drilling chip traps to the inside of the tube where the holes for the screws and plastic tab anchors will be drilled. The setscrew holes should be tapped with 1/4 x 28 thread to accept a fine-thread 1/4-inch bolt. Each stiff plastic tab should be 2 inches long, 1/2-inch wide, and anchored with a machine screw and nut one inch above each setscrew (see Figure 12.15).

Before an exposure begins, the telescope is focused, and the mirror

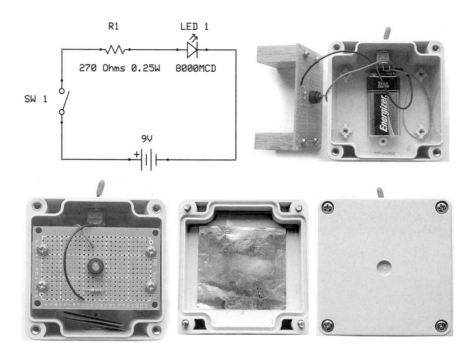

Fig. 12.16 *(Top Left) The basic circuit of the artificial star shows the layout of the components. Note that any DC voltage can be used to power the circuit as long as the resistor is changed to match the required current limitation to the LED. (Top Right) The interior of the project box shows how the LED circuit board is raised up to the artificial star pinhole using small pieces of wood. (Bottom Left) The mounted circuit board shows the current-limiting resistor and the LED held in place with a plastic mounting bezel. (Bottom Center) The project case lid contains a piece of aluminum foil with a pinhole centered over an aperture drilled into the cover. The pinhole has to be as small as possible to recreate a point-source of light. A single strand of fine-gauge wound copper wire will make a small enough puncture in the foil. (Bottom Right) The completed artificial star shows the light being emitted from the pinhole. Photos by Peter Langsford.*

locks set in place. They should be tightened in sequence and just sufficiently that there is a slight movement of the star images; any tighter and the plastic tabs will distort the rim of the mirror and present the viewer with triangular stars. If the focus is altered because of temperature change, the setscrews must be released before the refocusing. Be sure to release the mirror locks when the telescope is not in use.

12.13 Artificial Star for Collimation and Focus Tests

A good cloudy night project is an artificial star like the one fabricated by Peter Langsford to allow daytime collimation of telescope optics or focus tests of camera optics (see Figure 12.16). The parts required are all available from electronics supply stores:

LED (bright white, 3.6-volts, 20 mA, 8000mcd)
LED mounting bezel
9-volt battery clip
270 Ohm ¼ watt resistor
Toggle switch
Small plastic project case
Small piece of circuit board cut to fit
9-volt battery
Single strand copper wire.

Since the LED requires only 3.6 volts to operate, and it is more convenient to use a small 9-volt battery instead of the physically larger AA batteries, a 270 Ohm ¼-watt resistor limits the current reaching the LED (R = (9V – 3.6V) / 0.020, R = 270 ohms).

Chapter 13
Real World Astrophoto Problems

13.1 Atmospheric Turbulence Limits High Resolution

In the age of space-borne telescopes providing us with unprecedented high-resolution images from their lofty perch above the Earth's turbulent atmosphere, we are constantly reminded of how much the air above us limits our astrophotography. It's a fact that we can't live without this blanket of atmosphere, but astrophotographers often can't live with it either! The great French optician and astronomer, André Couder, has candidly stated, "The worst part of a telescope is ... the atmosphere."

It creates trouble for the high-resolution skyshooter because the density of air and thus its refractive index, changes with temperature. The atmosphere is far from uniform in temperature, especially near sea level where most people live, because of winds, local air currents, and the interaction of the atmosphere with local surroundings and the telescope itself. As the refractive index of air constantly changes, it effects incoming light rays from space and causes them to deviate from their original path. This results in light arriving at the telescope or camera not as a uniform wavefront like a plane, but more like an irregular surface. The temperature change needed to create such refractive changes is amazingly small. A variation of only one degree Centigrade in a layer of air only 15 centimeters (six inches) thick is enough to cause stars to oscillate 0.2- to 0.3-arcseconds in an 8-inch telescope, and will render the star as a complete blur in larger instruments.

The quivering and oscillation of a star image due to atmospheric turbulence is called scintillation. The relative amount of scintillation is commonly known as "seeing." The better the seeing, the more stable the star image is. Numerical scales exist to grade seeing, with 10 being perfect movement-free star images while 1 represents a star rendered into a total blur by turbulence. The atmosphere effects shorter wavelengths to a greater extent. Also, the closer an object is to the horizon, the greater the amount

Fig. 13.1 *All of the advantages of digital imaging for planetary photography go for naught if the atmosphere won't cooperate. These five images of Venus, all taken in the space of 30 seconds, demonstrate how turbulent air can spoil attempts at high-resolution imaging. The only option available when the air is very turbulent is to try again on another night. Photos by Robert Reeves.*

of air and turbulence through which the object is viewed. Thus the seeing deteriorates proportionally to the viewing angle from the zenith.

In his 1995 book, *High Resolution Astrophotography*, Jean Dragesco provides an excellent discussion of the technical aspects of the effects of atmospheric turbulence on the imager. But such scientific treatment of the subject is beyond the scope of this book, so we will only look at the causes of turbulence that disturb astronomical seeing and some of the things we do to try to limit its effects on our celestial imaging.

There are three main types:

1. **High-level turbulence.** This can be caused by the recent passage of a strong weather front or the jet stream passing over your local area. This produces turbulence at extremely high altitudes, and in general, there is nothing an astrophotographer can do to limit its effects except wait for it to go away. Weather front movement can be followed through a web weather forecasting service, and the present and predicted future location of the jet stream can be found on Internet weather forecasting sites such as weatherimages.org. High-level turbulence can be viewed directly in the telescope by placing the edge of the Moon or a bright star in the field and defocusing the telescope to a large degree. When the focus is racked sufficiently away from the telescope's primary optic, we can directly observe the movement of the atmosphere across the image of the celestial object. It is often possible to have more stable conditions just before dawn when the atmosphere has cooled overnight thus minimizing the temperature fluctuations.

2. **Local turbulence.** This is caused by heating of the local environment by the Sun during the day or by objects and devices that create heat, such as furnace or air conditioning vents on the roof of a nearby building. Objects near the telescope that absorb large amounts of heat, such as black asphalt roadways or dark-shingled roofs and buildings, will radiate this heat after the Sun goes down. The radiated heat will create local atmo-

spheric heating, sometimes called "heat waves," that have a very detrimental effect on telescopic images. Local turbulence is even worse during the day when constant solar heating stirs the atmosphere. This makes solar photography difficult. Local turbulence can be mitigated in several ways. One technique is to elevate the telescope above the ground. A 10-foot elevation will significantly decrease the amount of "ground clutter" or turbulence created from warm local objects. However, mounting a telescope on a stable platform may be impractical in many cases. Another useful technique is to set up in areas of uniform local terrain, such as a wide-open grassy field with no changes in local topography to create unequal local heating. Astronomers often build their own observatory to keep the telescope ready for use at all times. But the observatory itself can create heating problems and turbulence across the open slit of a dome, or the heavy walls of a roll-off roof observatory can radiate heat that will affect seeing. Painting the building with white titanium dioxide paint can help minimize the amount of heat absorbed by the observatory's structure. Domed observatories will always have more problems with local turbulence than roll-off roof structures because of heat flowing out the slit. Roll-off structures that completely remove the building from around the telescope work best for reducing local turbulence created by heat. Some well-known astronomers live in locations that are prone to better seeing than most, such as Don Parker, the high-resolution planetary imager from Coral Gables, Florida. His location near the ocean provides him with stable temperatures and laminar airflow, resulting in steady seeing nearly all year.

3. **Instrument turbulence or "tube currents."** Daytime heating of the telescope creates local heating of optical components. This effects the figure of a mirror as well as creating heat waves inside the tube of larger instruments that can refract light rays in the optical path. Covering a telescope with a plastic tarp by day may actually make the heating worse because the tarp absorbs heat. Many imagers instead cover their telescopes with aluminized Mylar "survival blankets" that reflect much incoming solar radiation. If a telescope is protected from the Sun during the day and is in thermal equilibrium with outside air temperature, instrument turbulence will not be a problem. Traditionally, older telescopes were painted white to help control heat absorption. More recently, manufacturers have produced telescopes that are painted black. This may seem to aggravate the instrument heating, but in fact black telescopes cool off quicker after dark, as long as they were not exposed to the Sun during the day. However, a black telescope is difficult to keep thermally stable during the day when observing the Sun. If the temperature inside the telescope remains three to four degrees Centigrade higher than outside, the images produced will be inferior. Closed tube assemblies like Schmidt-Cassegrain designs do not allow ready circulation of air within the telescope, and optics take a long

time to cool off. Telescopes with internal fans to speed cooling will reach thermal equilibrium more quickly. Schmidt-Cassegrain telescopes are prone to dewing on their corrector plates, but imagers should resist the temptation to clear dew with an electric hairdryer. A long plastic dew cap is best to maintain thermal equilibrium.

13.2 Predicting Seeing and Transparency

Anyone with Internet access has the same weather data available to them that professional forecasters have. With some study of how the atmosphere in your region reacts to certain influences, you can learn how to interpret and improve on forecasts for your area.

I have found the web weather data and real-time satellite images presented by the Space Science and Engineering Center at the University of Wisconsin-Madison to be quite useful (see Appendix B).

For astrophotographers, one of the most useful weather sites on the Internet is Atila Danko's Clear Sky Clock. Danko is a Canadian amateur who became tired of being frustrated by clouds just as he wanted to do his astronomy. He thus set out to create an accurate method of forecasting the astronomically relevant weather in his area. This project evolved into the Clear Sky Clock that has become an indispensable tool for amateurs throughout North America. In conjunction with the Canadian Meteorological Center, Danko created a script that interprets weather data for a particular location in the U. S. or Canada and displays it in a graphical format representing a 48-hour forecast in three-hour blocks. My observatory site is one of over 2000 locations with its own Clear Sky Clock. To check if there is an existing location near your area, see cleardarksky.com and check the state or province you live in. If there is no other Clear Sky Clock already in existence for a location within 15 miles of yours, contact Danko and ask him to create one.

The Clear Sky Clock provides cloud cover, transparency and seeing forecasts that attempt to predict turbulence and temperature differences that affect seeing for all altitudes. The seeing scale is calibrated for instruments in the 11- to 14-inch range. There are gaps in the forecast because the prediction does not consider daytime heating and thus only forecasts nighttime seeing. Additional seeing forecasts for extended areas of the U. S. and Canada are also available at the Canadian Meteorological Centre (CMC) (See Appendix B).

Wide-field imaging only needs a clear sky. However, high-resolution work requires steady seeing. By monitoring a number of weather factors, such as the jet stream, the passage of cold fronts, and the proximity of thick

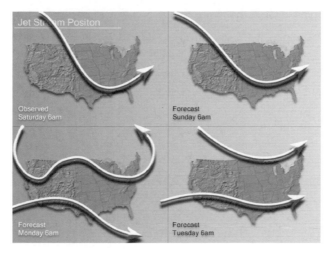

Fig. 13.2 *The World Wide Web has many weather sites that are of value to the astronomer. The jet stream plays a critical role in high resolution astrophotography. The location of the jet stream can be tracked at Weather Images. See **Appendix B** for this and other addresses. Courtesy Weather Images*

clouds, a reasonably good assumption can be made about local seeing conditions. If any of these conditions are present, there is a high probability of poor seeing. A major contributing factor to poor seeing is the jet stream, the band of high-speed winds that arcs across the continental United States at high altitudes. The jet stream can cross at varying locations, depending on regional weather conditions, and create high-altitude turbulence that degrades astronomical seeing. Any time the jet stream is within 225 miles of your location in the winter or within 300 miles in the summer, the local high-resolution seeing conditions are likely to be poor. The current location of the jet stream can be tracked at the California Regional Weather Server. The passage of cold fronts will also create increased atmospheric turbulence. Anytime a cold front is within 250 miles in the winter or 300 miles in the summer, there is a probability of poor seeing. The movement of frontal systems across the United States can be tracked at wunderground.com (see Appendix B). Additionally, most local television stations have a web site featuring the display from their Doppler weather radar that will show local storm systems. Thick clouds form when ascending air reaches its dew point at cooler altitudes. Thick clouds have colder cloud tops. By monitoring the infrared cloud images from ssec.wisc.edu, you can detect the nearby presence of thick clouds even at night. Anytime there are thick clouds with upper temperatures lower than −50 degrees Fahrenheit within 150 miles of your location, the atmospheric turbulence will likely create poor seeing.

13.3 Light Pollution

It is a sad fact that astrophotography is affected by light pollution more than any other aspect of amateur astronomy. This is because photos accumulate light pollution, blotting out the target just as surely as a bright light blinds a visual observer. As urban areas grow larger and brighter, the imager is faced with the choice of traveling further to escape light pollution, or to resort to the use of specialized filters to block it from entering the camera.

There are two kinds of light pollution—natural and man-made. The two main forms of natural light pollution are scatter and airglow. The daytime sky is blue because air particles scatter sunlight. At night, the air scatters starlight in a similar way to make a photographically detectable sky fog. Fortunately, the digital astrophotographer does not need to worry about the effects of scatter or low level airglow because the nature of current digital imaging sensors prevents lengthy continuous exposures that would encounter such phenomena. An extreme form of airglow is the aurora. This event occurs when intense solar activity channels energetic solar particles into the Earth's upper atmosphere. Even a low-level aurora that is not visible to the eye can be bright enough to interfere with digital astrophotography. When such an event is happening, there is nothing that can be done except await another clear night.

Man-made light pollution is primarily from nighttime outdoor lighting. High-pressure sodium lights radiate strongly in the yellow but appear pink to the eye, because they also weakly emit from the ultraviolet to the green portion of the spectrum. Mercury vapor lights are blue to green and photograph as green.

Just as filters can be used to enhance terrestrial photographs, there is an arsenal of filters that can be used to combat light pollution. Our goal is to block the man-made and offending natural wavelengths with the proper kind of filter before they enter the camera, but still pass the wavelengths unique to celestial sources. Fortunately, the wavelengths from light pollution sources are not spread uniformly over the spectrum, but are concentrated in a number of peaks that can be filtered out while allowing the desired astronomical wavelengths to get through.

Light pollution restricts the length of exposures, but the use of proper filters allows us to increase the exposure and contrast of a celestial photograph. Indeed, a skilled astrophotographer working with filters can often achieve results equal to those obtained by traveling an hour's drive away from the city and not using a filter. A map charting the sky brightness from nighttime light pollution for your local area can be found at the web site of

Istituto di Scienza e Tecnologia dell'Inquinamento Luminoso (see Appendix B). With the data presented there, you can compare your observing site with other nearby areas, and perhaps find a darker location.

13.4 Dewing

The formation of dew on a camera or telescope lens will slowly obscure starlight, gradually dim a progressing star trail, and eventually extinguish a star's image during a guided exposure. In severe dewing conditions, a camera lens can be completely covered in five minutes. Dew will end an astrophotography session as surely as if the shutter was closed. Thus, the morning dew, the poetic fairy's breath that makes a sunrise landscape beautiful, plays havoc with astrophotography.

The warmer the air, the more water vapor it can hold, because with increased temperature water vapor molecules move faster and remain suspended more easily. By night, when the air cools, these molecules slow down. When they have slowed sufficiently, they begin to stick together to form liquid water. The temperature at which this happens is called the "dew point."

The dew point temperature is strictly a measure of the amount of water vapor in the air and is independent of the air temperature—except that the former can never be greater than the latter. When the dew point and the air temperature are the same, the air is as saturated as it can get and water will begin to precipitate in the form of fog or dew. If the dew point is below 32 degrees Fahrenheit, frost forms. (A dew point below 32 degrees Fahrenheit can be called a "frost point.") "Relative humidity" is a measure of how much water vapor is in the air compared to how much it can potentially hold. When the air temperature has fallen to the dew point, the relative humidity is 100 percent. If, for example, the air temperature and the dew point are both 45 degrees Fahrenheit, the air is completely saturated, feels damp, and the relative humidity is 100 percent. But if the air temperature is 100 degrees Fahrenheit and the dew point is still 45 degrees Fahrenheit, the relative humidity is very low and the air feels quite dry.

A telescope or camera will cool down to the dew point and begin to become moisture-covered before the surrounding night air itself has cooled to the dew point. The reason is that metal and glass are efficient radiators of heat—more efficient than trees or grass, the cooling of which is what makes the air itself cool. (Air is transparent to heat—which is infrared radiation—and allows heat from solid objects like trees and telescopes to radiate into space. The air itself cools by contact with solid objects that are already cool.) A gentle night breeze is beneficial because it will keep the telescope and camera from cooling and collecting dew.

13.4.1 Dealing with Dew

When Schmidt-Cassegrain telescopes with their thin, exposed, corrector plates became popular in the 1970s, they changed dew from an inconvenience to a major headache. An aggressive approach to the problem was clearly necessary; namely, to prevent dew from forming on the optics in the first place rather than to remove it after it had formed. For convenience the usual anti-dew tool has been the hand-held electric hair dryer. However, a hair dryer will keep dew off a corrector plate for only about ten minutes until it has to be used again. Moreover, care must be exercised to prevent the dryer from unduly heating the optics—this can degrade optical performance by causing the corrector to slightly warp.

The first real line of defense against dew is to shield the optics from the cold of the night sky, keeping the glass warmer than the surrounding night air. But this is difficult to accomplish with a camera lens or corrector plate that must be aimed upwards. In practice, we trap a column of warm air over the optics with a cylindrical extension of the scope's tube called a dew cap or dew shield. At best, this will only delay the onset of dewing; but in moderate humidity, this is usually long enough to complete the observing session.

All brands of SCT telescopes have commercial dew caps available. Or, a very effective dew cap can be made out of the hard ⅝-inch foam-rubber camping mats that are available from sporting goods stores. It should extend beyond the corrector plate or objective lens a distance 1.5 times the telescope's aperture. Roll the foam into a cylinder and glue the overlapping ends together. Make it fit snugly so it will stay put when slipped over the end of the telescope tube. To prevent possible vignetting of the field-of-view, flare the far end of the shield out slightly.

If your camera does not have a commercial lens hood, a lens shade made from stiff cardboard lined with absorbent black construction paper can hold dew off for an hour or so. In times of heavy dew, or dusty conditions, a simple trick to protect a camera is to cut a hole in a zip-lock bag to place it in a plastic bag which has had a hole cut in it for the lens to protrude through.

The most effective long-term defense against dew is continuous low heat to keep the optics just above the temperature of the night air. An excellent commercial device is the Kendrick Dew Removal System that has many options for heating different types of optical equipment. Its single 12-volt DC controller powers as many heaters as needed as long as their total power drain does not exceed three amps. Because it uses a relatively benign 12-volt DC instead of a potentially lethal 120-volt AC, the Kendrick is a safe system. For permanent installations, an inexpensive 120-volt AC to 12-volt DC transformer can be obtained.

Portable 12-volt DC hair dryer-style dew removers can be obtained from several sources. Auto supply stores sell them as portable rear window defoggers. A 12-volt unit can also be obtained from Orion Telescopes and Binoculars (item #05601 in their catalog) that plugs directly into a cigarette lighter receptacle. The dew blower can be powered directly from a 12-volt battery by using an "alligator clip to cigarette receptacle" adapter (Orion's item #05607). For those who also run their telescope directly off a cigarette lighter receptacle, adapter #05605 permits two cigarette lighter-style accessories to run off the same power supply receptacle.

Twelve-volt dew blowers do not work as well as their 120-volt cousins, but in the field they can make the difference between saving the observing session and going home.

13.5 Mosquito Control

Anyone who spends a significant amount of time outdoors in the evening knows that humans are low on the food chain when mosquitoes are present. Astronomers spend an inordinate amount of time outdoors at sunset and early evening when mosquitoes are most active, making the airborne pests a problem. The outbreak of new mosquito-borne diseases in the U. S. like St. Louis encephalitis, dengue fever, and the West-Nile virus, raises mosquitoes from a mere nuisance to a genuine health threat.

There are more than 150 different species of mosquito in the United States alone and 2500 species worldwide. It is only the female that bites humans, doing so in her search for protein for her developing eggs. Male mosquitoes feed on nectar and other sugar sources and are not the cause of the welts and itching that result when a female inserts her blood-sucking probe into the skin.

Mosquitoes live in tall grass where they can keep cool by day. Since they are most active around sundown, just when astronomers are setting up their telescopes, it is best to set up the scope earlier in the day, if possible, then return after sunset.

13.6 Mosquito Repellants and Traps

Much discussion in astronomy centers on the non-celestial topic of which mosquito repellent works best. Indeed, a spray can of OFF repellent is a standard item in most telescope accessory kits. Urban folklore has suggested various items such as garlic or certain brands of fabric softener have value as repellents, but their effectiveness is variable depending on personal chemistry and local environmental circumstances. Sonic devices and smoldering mosquito coils also have varying degrees of effectiveness depending on environmental conditions. Natural repellants such as essential oil of citronella, are available from health food stores. Just rub a few drops on exposed skin. This should not be confused with the citronella oil sold for use in yard torches. Essential oil of geranium also works well. Mix a few drops with water and spray it onto your skin. Some people also have good results using Vick's Vapo-Rub on exposed skin areas. A decidedly low-tech alternative is a simple 15-inch oscillating electric fan to keep the air moving, causing you to appear to be a "moving target" to a mosquito.

Jonathan Day, a medical entomologist at the University of Florida and Mark Fradin, associate professor of dermatology at the University of North Carolina at Chapel Hill, studied commercial mosquito repellents and concluded that those containing the chemical DEET, listed on a product label as "N, N-diethylmeta-toluamide," are best for maximum protection. Introduced commercially in 1957, DEET is the "gold standard" of mosquito repellent products. Those containing 24 percent DEET have been shown to prevent bites for up to five hours. Repellants with lower concentrations are effective for shorter periods, sometimes less than a half-hour with products containing 6.6 percent DEET. DEET is a solvent that will react with certain plastics and synthetic fibers, so if you have to use products containing this chemical, wear old clothes and be careful not to have any on your hands when handling telescope or camera gear.

However, the safety of DEET has been questioned, leading the Environmental Protection Agency to review the product. Their conclusion was that "normal use of DEET does not present a health concern to the general U. S. population." DEET has been classified as less toxic than many people believe and adverse effects are traced to gross overuse of the product. But still, the wisdom of slathering yourself with a chemical that is toxic enough to be offensive to other living beings is debated with some vigor. Logic suggests that if the chemical is that potent, it can't be good for the person wearing it either.

Chemical safety concerns have led to development of a wide range of products that are designed to lure mosquitoes away from humans and then trap them. New products on the market include the Mosquito Deleto that captures the bugs with a sticky cartridge, the Dragonfly that zaps them with an electrical grid, and the propane-powered Mosquito Magnet that baits the mosquito with carbon dioxide emissions then vacuums them into a net where they dehydrate and die. These products are available at most large hardware and home improvement stores. Successful results with mosquito attracting traps seem to be a 50-50 affair. For instance, the Asian Tiger Mosquito, an aggressive biting blood-sucker prevalent in much of the United States, is not attracted by the octenol used in the Mosquito Magnet.

I have personal experience with the Mosquito Magnet and can vouch for its effectiveness in reclaiming a backyard formerly teeming with swarms of vicious mosquitoes. Within a week of using the device, my yard went from "forbidden territory" to a mosquito-free area my family can enjoy, and I can set up my telescope without suffering constant bites and itching. The Mosquito Magnet requires a 110V power supply and operates for three weeks on a standard 20-pound propane cylinder. The device would be perfect for mosquito control at an observing site that is visited weekly, so the propane tank can serviced as needed.

A word of caution to those who do purchase a Mosquito Magnet: closely follow the manufacturer's advice about refilling the propane tank. I found from personal experience that when they say do not use pre-filled exchange propane tanks, it is for an important reason. Exchange tanks have a certain percentage of air in them to allow for expansion while stored outdoors in racks near the retailer selling them. This air will not affect their use with Bar-B-Q grills, but will cause the Mosquito Magnet to shut down. Have the tank refilled at a propane charging station that will purge the air during refilling.

13.7 Other Nighttime Pests and Critters

Wasps and yellow jackets are other flying pests to watch out for in the evening. They can nest inside the roof of an observatory or other structures, and their sting can quickly end an observing session. Be careful not to set up your telescope near an ant nest.

Nocturnal mammals are another nighttime hazard. Raccoons are especially nosey. They will rummage through your telescope goodies while looking for food and cause all sorts of mischief. Skunks will forage within several feet of your telescope if you are perfectly still. Once, one waddled under my tripod as I guided on a celestial target. As long as you don't scare them, skunks are oblivious to people at close range. But we are all aware of what can happen if they are disturbed. In wilderness areas, bobcats, mountain lions, and packs of coyotes can be a hazard. Music from a radio or CD player can be an effective large-animal repellent. Keep your car unlocked in case it is needed as a temporary shelter from an aggressive animal.

13.8 Keeping Warm at the Telescope

Astronomy can be a chilly activity even if it is not winter. For one thing, when we are at the telescope, we are nearly stationary so our muscles might be generating as little as 100 watts of heat. By contrast, walking or some other continuous activity creates up to 2000 watts of heat. Moreover, during the day our body absorbs heat from the Sun, whereas at the eyepiece, it radiates what little heat it produces into space. Consequently, unless we are observing on a warm night, we need to prevent loss of body heat.

Most people have misconceptions about how to keep warm at the eyepiece. A simple coat that suffices during the day will not do under the stars. Two pairs of socks often make shoes too tight and therefore restrict blood flow, making cold feet feel worse. Ski apparel might keep you warm on the slopes, but it will not work at the eyepiece; it is not designed for insulation, but to look good, fit snugly, and allow freedom of movement. A waterproof jacket may actually make you feel colder because it will not "breathe" and let perspiration escape, so the moisture trapped next to the skin will make you feel colder.

The most effective way to stay warm at the telescope is to dress in layers of clothing. This insulates us with the "dead air space" trapped between the multiple layers: the trapped air is isolated from the outside cold and makes us feel warm after our body heats it. In very cold weather, use a foundation of long underwear. A turtleneck shirt is necessary to stop heat loss around the neck. Corduroy pants are warmer than denim jeans. "Warm-up" pants over regular trousers will provide the dead air space vital to comfort. Dress as if it were going to be 20 degrees cooler than expected.

Extremities like the head, hands and feet need special attention. Because one-fourth of the body's heat can be lost from the head, a warm hat is essential. A wool stretch cap will keep your ears warm. Mittens are best for your hands, but gloves will be more practical for operating telescope and camera controls. Feet need loose shoes or boots to allow for the insulating air space between the layers of heavy socks (cotton socks under wool socks). A double boot designed for mountaineering can be obtained from sporting goods stores. A simple blanket draped over your chair and then wrapped over your legs will work wonders when you are sitting at the telescope for long periods. Also, stand on a foam camping pad to insulate your feet from the cold ground. In very cold weather, a down sleeping bag that unzips at both ends is useful. Zip it up as far as will permit work at the telescope, but leave it open at your feet so you can move around.

Chapter 14
Image Processing

14.1 Introduction

Of all the subjects discussed in this book so far, image processing is probably the most intimidating to the beginner. With film photography, all processing could be sidestepped completely by using color slide film, where it was "what you see is what you get." The more adventurous might scan and then process the image in a program like Photoshop or Picture Window to bring out "hidden detail." However, with digital astrophotography, taking the exposure is usually just the first step. To arrive at a satisfactory image, you will have to deal with things like noise reduction, contrast and brightness enhancements, and color correction.

To see why noise is important, put the lens cap on your camera (and cover the viewfinder eyepiece if it is a true SLR to prevent backscatter through the viewfinder), adjust the ISO to its highest setting and take either the longest exposure possible with your camera, or if the camera allows unlimited time exposures, expose for five minutes. Examine the resulting image. It will not be black, but instead is full of mostly green speckles and bright spots. This is the imaging sensor noise that overlays and degrades the image. Fortunately, the subtraction of this noise is one of the easiest processing steps you can do.

One of the things you will discover as you delve into this fascinating subject is that nobody does image processing exactly the same. While a good program will have tutorials that explain its operation, experienced users soon find a combination that works best for them, particularly if the objective is to produce "pretty pictures." While this book is directed to digital cameras, I would be remiss if I did not point out that to be of value, scientific grade images must adhere to a rigorous step-by-step process that can be documented and does not jeopardize the data. We will confine our discussion here to non-scientific grade imaging.

Most experienced imagers use more than one program to supplement the way an image will be processed for brightness, contrast, color correction, etc. For this reason, it is difficult to lay down universal rules or steps for a novice to follow. My approach, therefore, will be to explain what I do. As you gain more confidence with your interaction with the program, you will gradually become more comfortable with the process and eventually develop a style that is uniquely your own.

14.2 The Laws of Astronomical Image Processing

In their excellent book, *The Handbook of Astronomical Image Processing* (HAIP), of which *Astronomical Image Processing for Windows* (AIP4Win), a full featured image processing program, is an integral part, Richard Berry and James Burnell relate the "laws of astronomical image processing." These laws are not based on its mathematics or methodology, but on the recognition of the realities of image processing. These were learned through long hours at the telescope acquiring images and even longer hours at a computer processing them. They are simple but they illuminate the profound possibilities, as well as the limitations, of digital image processing. Although they were derived while performing rigorous CCD imaging, they are equally applicable to the art of digital astrophotography. Before discussing them we need to establish some foundational principles:

- Image processing is not a miraculous process; it is the application of rules of mathematical numeric manipulation.
- Image processing discards some forms of image information in order to enhance other forms of image information.
- Image calibration is needed to eliminate the noise signature of the sensor from the final image.
- Brightness scaling is performed to eliminate unwanted sky background and excess star brightness in order to more effectively display galaxies and nebulae.
- Unsharp masking and deconvolution are applied to eliminate unwanted blurry portions of the image that hide the desired detail.
- Image information is always lost during processing and enhancement. The final result may be more pleasing to look at, but will have less raw data.

And now the Berry and Burnell Laws:

1. **Astronomy is photon limited.** In a nutshell, if there are insufficient photons arriving at the imaging sensor, the target will not be discriminated from the sky background or sensor noise. The nature of the astronomical target, atmospheric conditions, the size and optical configuration of the instrument used for imaging, and the response characteristics of the imaging sensor all come into play, and at some point, will limit the ability

to acquire the desired image.

2. **Artifacts happen**. Artifacts are imperfections in the image caused by outside factors such as dust on the sensor, vignetting of the optical system, optical imperfections and focusing errors, reflections within the optical system, electron noise in the sensor . . . the list goes on and on. We strive to suppress or eliminate artifacts during image processing by calibrating our images with dark frames. But the fact remains that any image will have many artifacts in spite of our best efforts to eliminate them. One should learn to recognize artifacts and not mistake them for reality.

3. **Never trust one image**. This law does not hold if the object is to simply make pretty pictures. In fact, liberties and licenses are often taken in the processing of images destined *only* for visual enjoyment. However, if we are doing science or deliberately seeking something new, the accuracy of what is recorded on the image becomes very important. Digital images are very prone to many kinds of artifacts that mimic astronomical objects. A cosmic ray strike, sensor noise, or even misapplied processing can create realistic apparitions on an image that do not show on popular star charts. There is always the possibility of something new appearing in the sky, but the watchwords are corroboration and verification. If it appears on two consecutive images, the probability of reality increases but is not concrete proof of existence. If something appears in two images taken by separate instruments, the probability of reality increases significantly to the point of seeking verification from other sources.

4. **Image processing always discards information,** no matter if the processing is to gather science by position or magnitude measurement or to simply provide pictorial information about the subject. The very nature of calibrating a digital image by subtracting a dark frame changes it, hopefully for the better. Further processing, or enhancement, throws away more information in order to increase the visibility of other image information. Contrast stretching and brightness scaling alter image information while deconvolution, unsharp masking, and image scale resampling throw away or distort other information. If done properly, the resulting images appear more pleasing to the eye, but they no longer have the same amount of data as the originals.

    Proof that considerable amounts of data have been discarded during the enhancement process can be seen by comparing the histogram of "before" and "after" images. The smooth full scale of the before histogram is often replaced by a spiky histogram with gaps and protruding fingers showing that some of the original data are missing even though the image actually looks better to the eye.

5. **A well-processed image cannot be improved**. Sky backgrounds set to black have thrown away image information about the dark portions of the image while stars that have been brightened to saturate pure white no

longer retain true brightness information. Increasing the contrast of mid-range tones to improve visibility of wide-scale faint detail has similarly reduced the range of available data contained in the raw image. After processing, you can better see what there is, but it is in a more restricted range of shades and tones than contained in the original image. Further attempts to process the image eliminate more of the limited data, resulting in posterized colors, washed out highlights, and garish over-enhancement of minute detail as pixels are stretched more toward black or white.

Keep these laws in mind as you approach your image processing. Seemingly miraculous transformations can occur in your images, but you have to balance the results within the framework of what is in the raw data and what is mathematically possible to achieve with that data. Over-processing an image cannot substitute for the lack of raw data. If there are simply not enough data present, one of the advantages of digital astrophotography is that more data can be gathered later and added to the image by stacking more exposures of the same object.

It is my recommendation that you purchase *The Handbook of Astronomical Image Processing* with *AIP4Win*, which for around $100 is one of the best bargains you will ever find in astronomy. I consider the handbook to be the image processing companion to this book and any other image processing program you are likely to end up with. Read its 500+ pages completely. Just pushing the control sliders on a processing program will eventually get you something you probably will be pleased with, but it will come without an understanding of what the program is accomplishing with your image. HAIP presents both the concepts of image processing and the mathematics of how it is accomplished. I can sympathize with those who are apprehensive about trying to digest complex mathematical formulae or blocks of programming code. That is also not my forte. But take it from me, if the math is beyond your interest, just skip over it and concentrate on the text. This book gave me a much better understanding of, and much greater control over, all the software that I use.

14.3 Don't Overdo Processing

Two things are assured in this endeavor: the first look at a new image processing program will be intimidating, and once the basics are mastered, it will take far longer to process your data files than it took to actually acquire them under the stars. Therefore, I advise that you not tackle a new program with the intent of immediately performing complex operations. The extended menu and selection of possible operations can be overwhelming. There are countless possibilities! Which one to choose?

For starters, practice the basics of image manipulation on a single image. Start with levels, color correction, contrast and brightness, and cropping. Once you are comfortable with navigating these, move on to more complex operations one at a time and become familiar with each before progressing to the next. Just as it took time to learn the art of actually acquiring the images, it will take some time to learn and become comfortable with the procedures for processing them.

Today's desktop computers are fast but still require significant time to perform many processing operations. Full-frame images from six- and eight-megapixel cameras are large files. Operations requiring repeated cycles such as 10 or more iterations of smoothing and deconvolution take several minutes on my admittedly antiquated 700 Mhz P-III desktop. While a state-of-the-art computer would be several times faster, I feel no need to replace it anytime soon, so I perform the time-consuming automated processing while I am doing other things.

If you are already familiar with Photoshop, you may be tempted to use it exclusively for your image processing. But, as powerful as Photoshop, Paint Shop Pro, or similar programs are, they were designed primarily for pictorial and advertising graphics and have significant limitations when applied to the special needs of celestial photography. It need not be expensive to expand your options. For about one-fifth the cost of Photoshop, you can purchase HAIP/AIP4Win and have the best of both worlds, and for modest additional cost, include ImagesPlus to complete your processing suite.

No matter which program you use, there are basic operations that have to be performed on the raw data to separate the desired image from background noise, correct colors and brightness, and enhance faint detail. But it is easy to get carried away during this process. Over-sharpening an image of the Moon can result in less detail in the final image because the sharpening process will drive pixel brightness too far toward black or white. Subtle crater walls bloat into donuts and artifacts begin to appear that overwhelm finer lunar detail. With deep-sky objects, processing steps can be applied too vigorously with the resulting image being unnaturally colorful or overly bright.

Celestial objects are naturally colorful, but because of the limitations of human vision, their faint telescopic appearance renders them essentially colorless. One of the objectives in our image processing is to render these objects as they would appear if they were sufficiently bright to be seen in full color with human vision. So how do we know when an image is properly processed to show this? We don't. Most of our perception of what a celestial object is supposed to look like is guided by a lifetime of seeing photographs taken by others with films that may not respond correctly to long exposures of dim objects, or have been enhanced in the darkroom to show colors, and then reproduced in a publication which introduces its own set of color variations during the printing process. The fundamental fact is that most celestial objects, other than the Sun and Moon, have relatively muted, pastel colors. It is entirely possible that your image of a galaxy or planetary nebula taken with a high-quality DSLR camera may be closer to "reality" than previously published images of the same object, so to represent reality you do not have create an overly contrasty and excessively colorful scene just for the sake of duplicating what you have seen elsewhere.

14.4 Many Processing Choices, Similar Results

All full-featured image processing programs are built around the same basic operations. However, beyond the basics, there is a wide range of variation so it is common to have a suite of programs and to move from one to another throughout the process. For example, I begin with ImagesPlus to control the DSLR during exposures and for grading my images to reject those that are poorly focused or guided. Next, I move to AIP4Win to perform both the basic and advanced operations such as dark frame subtraction, stacking, noise reduction, and smoothing using 16-bit processing. Finally, I import the image into Photoshop for final "artistic" touch ups like color and contrast corrections. Most full-featured programs will also perform these touch up operations, and it is not mandatory that the user purchase Photoshop as well as a "dedicated" astronomical image processing program. However, I find Photoshop easier and more intuitive to use for these operations.

It is a fact that no two images (or resulting stacks of images) will respond exactly the same to processing, regardless of what program is used. Variations in how the image was taken, its noise content, the nature of the celestial target, and atmospheric conditions are just a few of the variables that conspire to complicate the issue. So what works beautifully for one image *may* not give the same results when applied to another. Knowing what to do requires experience and a certain amount of experimentation to achieve consistently good images.

So as you experiment to find out what works best, I recommend that you never overwrite the original raw image data, or overwrite any version of a processed image. Always rename new versions of any image so that older versions are not lost. Different corrections and enhancements can be applied later as you gain greater understanding of how processing works.

You can easily do this by adding alphabetical suffixes to file names ("a," "b," "c," etc.) as you save progressive versions of an image.

14.5 Basic Processing Operations

The processing operations performed by any program can be broken down into categories. The basic operation groups are geometric transformations, point operations, linear operators, conditional operators, image operations, and deconvolution. We call upon various functions within these operation groups to manipulate the digital data in raw image files in order to eliminate unwanted image data that are obscuring the desired image data, or change the range of digital data to render some feature more visible to the eye.

- **Geometric transformations** are functions that move the entire image but the content and appearance of the image itself are not changed. An example is a translation where the image is shifted a specified distance and direction from its original location. The image can be moved up or down, or from side to side. Flipping the image backwards, inverting, or a combination of the two is also a geometric transformation. Rotation is a transformation where the image is turned in either direction around a specified axis that can be located anywhere within the image. Scaling transforms the image to a different size measured in pixels. Enlargement of small portions of an image will enlarge the original pixels and produce a blocky image with jagged square stars. The solution to blocky stars is to resample the image to a larger size and allow the program to smooth the jagged blocky borders using a feature known as interpolated pixels. This technique will not produce smaller stars or refine the focus, but will mitigate blocky artifacts. Any combination of these geometric transformations can be applied. The most useful is the process of aligning a series of images so they can be stacked and combined into a single image.

- **Point functions** find the desired range of pixel values needed to display image information and alter the values of those pixels to make the pictorial information they contain more visible. Point operations are therefore the most powerful computations that can be applied to an astronomical image. They involve working with histograms to remap the pixel values, stretch brightness values, and set the dark point.

- **Linear operators** generate pixel values based on an individual pixel's neighboring pixel values. This is useful in smoothing an image to reduce the effect of noise or sharpening one to give it a more focused appearance by increasing the contrast along a border. These operations are often presented as a series of filters in a processing program. Smoothing filters, called low-pass filters, have their uses in astronomical imaging, but their side-effect is an overall reduction in image detail; so they are rarely used.

Sharpening filters are often called high-pass filters, although there are technical differences between sharpening and high-pass filters. For the most part, the everyday user ignores these differences because their effects are so similar. Significant improvements can be achieved using high-pass filters to increase the visibility of details in an image. However, the parameters of the filter must match the image data characteristics or noise will be emphasized, artifacts created and high-contrast details will be overemphasized to the point of distraction.

- **Conditional operators** are a series of complex calculations that are not found in normal graphics programs. They are extremely powerful and sometimes difficult to control but are very useful in bringing out fine hidden details. While they are useful in astronomical image processing, they throw away data that do not meet the control parameters of the operator being invoked. Digital development and sharpening filters are conditional operators. The "digital development process" is a procedure that displays the linear data acquired by a digital imaging sensor in a manner that mimics the non-linear response of standard chemical photography. This procedure squeezes the large dynamic range of an electronic imaging sensor into a more restricted range by simultaneously sharpening and compressing the brightness scale to simulate the characteristic curve of film instead of the stark, linear, straight-line response of a digital sensor. Most astronomy-based processing software implements digital development, but Photoshop does not have an option that directly compares. In Photoshop, the Apply Image add option under the Image menu produces results that are similar.

- **Image operations** include those functions that make modern digital astrophotography a viable alternative to film astrophotography. Among these are dark frame subtraction, aligning, and stacking. Another image operation is the alternate blinking of two different images; a tool commonly used for asteroid and comet searches. Without the mathematical magic performed by image operations, digital astrophotography, as we know it today, would not be possible.

- **Deconvolution** is a tool used to deal with the effects of turbulent atmosphere. Unfortunately, while it would be nice if such a program were powerful enough to act as "software adaptive optics," too many variables conspire to prevent that from working. But the good news is deconvolution operators can do a remarkably good job of guessing what is needed and the result, if properly applied, is a greatly improved image. Of course, it is possible to overdo things and create artifacts that detract from the result, so care must be taken to avoid overly aggressive deconvolution.

14.6 Getting Organized

The first thing you should do is organize the night's images into folders according to the target. Within a target's folder, the RAW images are placed in a sub-folder labeled RAW. All the RAW images should then be converted to your working format (TIFF, FITS, JPG, etc.) along with their associated dark frames. These should then be placed in a sub-folder labeled ORIGINAL + DARK. After dark frame subtraction and any other calibration steps, these new files are placed in another sub-folder labeled CALIBRATED. Any files that you are working on should be kept in a WORK sub-folder. Completed images should be moved to sub-folders titled FINISHED.

Logically arranging your files to follow your workflow will consume a lot of file space if you are shooting six- to eight-megapixel DSLR images. For instance, a ten-image stack will result in ten 5-megapixel RAW images plus several dark frames, ten 32-megapixel converted TIF images, ten more calibrated 32-megapixel TIF images, plus a number of large work files of stacked images. One can quickly accumulate a half-gigabyte of files just to produce a single finished image. Older computers may require a larger hard drive; fortunately, a 250 gigabyte model now costs less than a moderately priced eyepiece and writable CD and DVD are even cheaper. It is not unusual to fill a CD with the imagery needed to create a single master image of a target. Maintaining your files with this directory structure will pay off when you want to return to an image—critical files will have been preserved. As you learn more and more about image processing, you may wish to return to these files to try your new skills.

Processing the large files created by six- and eight-megapixel cameras also calls for large amounts computer RAM. Photoshop routinely uses about four times as much RAM as the image being processed. AIP4Win uses floating point numbers for image calculations, a feature that has advantages for processing astrophotos, but it also needs RAM that is four times larger than the image.

14.7 Color in Astronomical Images

One step in image processing that causes confusion is color correction. This is often aggravated when astrophotographers have the wrong idea about what color a celestial object should be, then try to make their image match that erroneous expectation. Here are colors of various objects:

- Non-emission objects are naturally pale.
- Hydrogen-alpha and H2 emission objects are crimson.

- Galaxy cores are red.
- Starburst regions in galaxies are blue.
- Reflection nebulae are blue.
- Regions of the Milky Way can be either blue or yellow

Color normalization should be first on your list. You do this by first adjusting the image density, and then removing any color bias. A color bias is present if the black points do not match in all color channels. To remove color bias, set the black point in each color (the far left histogram adjustment) just short of the pixel hump in the histogram. If the background is neutral, then the color bias is gone, but if the image target is still off-color, then color balance is the problem. Color balance can be addressed by changing the white point in each color channel (the far right histogram adjustment). It is possible to introduce a color bias by applying improper adjustments to the color balance. It is wise not to be overly aggressive with color balance adjustment because too much can accentuate image noise.

Noise is a limiting factor in how much color correction can be applied to an image. Stacking a number of exposures will increase the signal in each color channel and reduce color noise, but ultimately the image will be limited by noise in the weakest channel. Increased color saturation is better, up to a point. Because digital cameras are weak at recording the red channel, stacking a greater number of images is needed to boost the red signal. But this also proportionally increases the signals in the green and blue channels. To correct this, the increased green and blue signal have to be reduced using the levels control.

Perhaps the most widely used image-sharpening tool is unsharp masking. By changing the filter values, you can improve sharpness to meet the needs of the image. Unsharp masking works by selectively increasing the contrast of the image's higher spatial frequencies while leaving the low frequencies unchanged. *The Handbook of Astronomical Image Processing* provides examples of custom-sharpening filter arrays that can be manually applied to popular image processing programs through their custom filter option. Experimenting with these simple filter arrays will demonstrate how a filter may work with one image but not with another. With a little practice, it is easy to see the relationship between the design of the filter array and the results achieved by it. This will allow the user to "tune" a filter for best results with a particular image.

14.8 Monitor and Printer Output

Ideally, your image processing should be done on cathode ray tube (CRT) style monitors, not LCD screens or laptop screens. The CRT monitors may seem like clunky "old technology," but they still produce better color and wider range of colors and contrast than do the vast majority of LCD monitors.

Your monitor should be turned on for a half hour prior to doing any critical color evaluation with it. If its control panel allows choices, use the 5000K-color temperature setting or one as close to that as possible. Most are shipped from the factory with a much higher color temperature setting and tend to emphasize cooler colors.

Monitor calibration is an important link in the chain of image processing events that is often ignored by astrophotographers. The image is processed until it looks good on the screen, but if the monitor is not calibrated, the image will not look the same when it is printed and, if the calibration is significantly off, the image will display poorly when viewed on other monitors. Not all can be calibrated. A monitor must have a gain control for each color channel in order to allow calibration and inexpensive ones may not have these. Since monitors tend to darken with age, calibration is not something that can be set once and then forgotten. There are calibration devices like the Colorvision Spyder that attaches directly to a CRT or LCD monitor with small suction cups and analyzes its color output. It then installs a custom color calibration input to your monitor's video card. It sells for about $200.

Most prints produced by astrophotographers are printed on inexpensive home-use inkjet printers. The majority of the current crop of printers is capable of producing outstanding prints but they do not necessarily match what you see on the monitor because each produces color in fundamentally different ways. Monitors produce all shades of color with the additive red, green, and blue (RGB) colors while printers produce all shades of color with the subtractive cyan, magenta, and yellow (CMY). To achieve deeper blacks, modern printers also use black (K) as a fourth color in a system known as CMYK. Because of this difference, there is a small range of colors that will display on a monitor but not print from an inkjet printer. This is technically known as the colors being out of gamut. Most people never notice this difference because it is so subtle.

Once your printer is producing pleasing images, you should use the same brand of printing paper and ink to maintain consistency. Printing papers are designed to use a certain type of ink for optimal output. In the search for economy in printing I have tried several aftermarket ink and paper brands and found that the paper and ink recommended by the printer manufacturer consistently produces the best results.

Manufacturers advertise that their device is able to print at very high dots per inch (DPI). But programs often output images to a printer at a far lower DPI; 300 DPI is considered a high-quality output. Setting a printer to 2880 DPI will rarely produce better results than 1440, and sometimes even 720 DPI. I have found that using my printer's maximum resolution takes longer and is a waste of ink. You should experiment with yours to determine what resolution produces the most pleasing results.

However, the normal viewing distance for a print is also a factor to consider when setting the DPI within your processing program. Prints up to 11x14 inches normally look best when printed at 300 DPI because they are usually held in the viewer's hands and examined close up. Larger prints are, by design, usually viewed from greater distances and can use lower DPI settings. Prints 20x30 inches can use 200 DPI while really large prints in the 30x40-inch range can be run as low as 100 DPI.

14.9 A Must-Have Freeware Program

Before I begin our discussion about astrophotography-specific software, you should know about what I consider to be a must-have graphics program that is free. If you have no previous experience with any kind of digital image processing software, I recommend you download the latest version of Irfanview. This is both an image viewer and a powerful image processing program that is simple and intuitive to operate. It supports just about all image formats and can crop, rotate, increase contrast, brightness, color depth, sharpening, file format conversion, and JPG optimization. It is a good choice for beginners to use on single-exposure lunar and planetary images.

I use Irfanview as my default image viewer. By setting the file preferences in Windows to start it any time a JPG or TIF is clicked, the image automatically pops up in it. I think it is superior in functionality and color rendition to the native Windows viewer. It easily allows viewing a series of images that need to be sorted for processing, or for viewing processed images to see the final result. It is also excellent for creating "slide show" presentations.

14.10 An Overview of Popular Image Processing Software

Most digital cameras are sold with a basic image-editing program, such as Photoshop Elements, which is a capable program for terrestrial snapshooting. These programs will allow you to modify the brightness, contrast, and sharpness of images and are fairly intuitive. However, while some of their features can be useful for improving your celestial images, their overall capabilities fall far short of those needed for serious processing. As you have seen throughout this book, most of the digital images reproduced here were made by stacking and averaging multiple images to create a more detailed and denser image. These and other special functions such as dark frame subtraction, contrast stretching, convolution, and deconvolution are beyond the capabilities of these programs.

The calibration of digital images is just as important as is the proper development of the negative or slide in film astrophotography. This is done with dark frame subtraction to eliminate noise and then aligning and stacking many calibrated images into a single master image with a significantly higher signal-to-noise ratio than any one exposure is capable of producing. A large portion of our discussion about image processing will concentrate on these operations because they are central to achieving good results when you do further processing to accentuate fine image detail. Both AIP4Win and ImagesPlus have an automated process that, with a few mouse clicks, begins a sequence of events that completes the calibration, alignment, and stacking process in a few minutes. Previously, it would take an entire evening to accomplish this manually using Photoshop.

Adobe Photoshop is the premier image-editing program with millions of copies in use throughout the world, but it is designed primarily for commercial photography and advertising illustration. While it contains many powerful editing functions that are applicable to astrophotography, it lacks basic capabilities needed by digital astrophotographers. For this reason, Photoshop is often used for final image touch-up after other software has been used for operations that are unique to digital astro-imaging.

Photoshop is expensive—often as much as one or two top-of-the-line eyepieces. But even this cost is relative. Just a decade ago, a new film imager would have invested large amounts of money in darkroom enlarger equipment. Today's digital imager is likely to already own a powerful computer and color printer. In many cases, the cost of specialized software will end up being less than the hardware a film photographer used to purchase in order to produce finished prints.

There are at least a dozen good image processing utility programs that produce excellent results with celestial images and about the same number of exclusive astronomical image processing programs. Many are freeware and one can fruitfully devote many cloudy night hours to their exploration. A partial listing appears in **Appendix B**. However, in this chapter, three commercial programs with which I am most familiar will form the framework for our exploration of digital image processing: Adobe Photoshop, ImagesPlus, and the new updated Version 2 of *Astronomical Image Pro-*

cessing for Windows (AIP4Win), which now extends its features to the three-color RGB images produced by digital cameras.

As you become more familiar with the subject you will discover that each program can do certain important things the others cannot. For instance: AIP4Win gets people started for the least cost and provides the most information on both the theory and practice of image processing; plus you get a full-featured, powerful astronomical image processing program to boot. In my opinion, AIP4Win has the easiest to use interface and the most powerful image calibration, alignment, stacking routines, and color controls on the market today.

ImagesPlus, on the other hand, excels in providing DSLR camera control. The program connects to the Canon 1D Mark II, 10D, 20D, 60D, and 300D, as well as the Nikon D70 through the camera's USB cable. Other features allow analysis of star focus in captured images, conversion of images to 8- or 16-bit TIF format, and an innovative image-grading routine that allows culling those with poor focus or guiding from a series to be aligned and stacked.

Photoshop rounds out my image processing suite by providing easy-to-use functions such as final color and contrast tweaking, mistake fixing with the clone stamp, and powerful composite and mosaic making capability.

In the image processing demonstrations that follow, it is assumed that the camera was set to save images in the RAW file format during the imaging session and that these image files have been converted to either 8-bit or 16-bit TIF files. All camera makers supply software that will convert their version of the RAW format to 8-bit or 16-bit TIF or JPG format. Photoshop 7.0 and earlier versions have limited support for 16-bit images, so users of these versions of Photoshop will want to convert their RAW files to 8-bit TIFs in order to use all its functions.

14.11 Adobe Photoshop

With the exception of 16-bit image support, Photoshop CS offers the user little additional astrophotography capability over older versions. If you already have an earlier version, you would do well to not upgrade that program, but instead invest the money in both AIP4Win and ImagesPlus, and use your older version of Photoshop as a touch-up supplement.

Photoshop's Layers option is used to stack and merge a series of dark frames to produce a master dark frame. This master dark frame is then subtracted from each of a series of images which are then stacked and averaged to create a single master image. However, performing this process with Photoshop is time-consuming and tedious. I will describe the process to demonstrate how it can be done with Photoshop, but after doing this a few times, most people soon move to programs designed specifically for digital astrophotography that automate the calibration and alignment functions.

The first step in any processing sequence is to calibrate the images with a dark frame. The dark frame must, of course, be taken with the same camera using the same exposure and file saving option as with the images that are to be calibrated, and none of these images can be cropped or preprocessed in any way. Ideally, the dark frame itself should be a number of dark frames combined to create a master dark frame. The more dark frames that are used to make the master dark, the better the noise can be averaged out. For simplicity, let us assume that three darks are being combined to create the master dark.

- Open the File window and click on the Open command to open the first dark frame. Immediately save this image with a different file name that explicitly labels it to be the master dark frame. You now have a file titled something like M31_Master_Dark. Next, open the remaining two dark frames. Click on either of these just-opened images to make one the active image and then open the Select menu and click on All. You should see a dashed white-line box surrounding your selected image. Then move to Edit and click on Copy which will copy the selected image into the Windows Clipboard. Close the window containing the selected image and click on the image that you designated to be the master dark file. Move to the Edit menu and select Paste. The selected image is now pasted onto the master dark image. With the remaining dark frame, repeat the process: select, copy the image to the clipboard, close the image and paste the image in the clipboard onto the master dark frame. At this point only the master dark window will be open.

- Where did the "pasted" images go? To find out, move to the Window menu and click on Layers. The layers palate will appear and the original image will be displayed as the background image while the pasted images will be shown as Layer 1 and Layer 2. We are going to average our master dark frame by adjusting the opacity of the two pasted images such that all three images contribute to a final combined one. In Photoshop, images that are layered are seen as if one were looking through a physical stack of film transparencies, so a layer that lies above another obscures those that are below. For averaging, you must be able to "see through" all the layers. To achieve this "visibility" the opacity of each layer must be $1/n$ where n is its place in the stack. This means the opacity of the first (or lowest) layer is 100%, the second 50%, the third 33%, fourth 25%, and so on. If all layers had 100% opacity, the stack would become

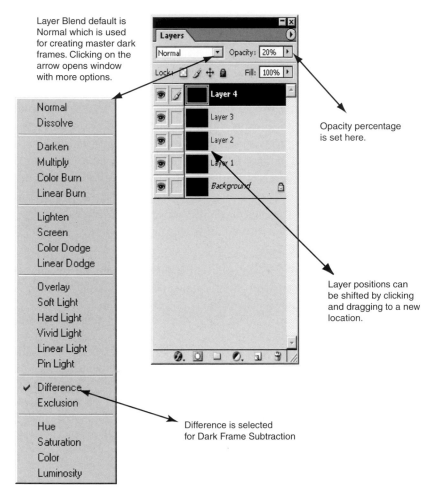

Layer Blend default is Normal which is used for creating master dark frames. Clicking on the arrow opens window with more options.

Opacity percentage is set here.

Layer positions can be shifted by clicking and dragging to a new location.

Difference is selected for Dark Frame Subtraction

Fig. 14.1 *In Photoshop a master dark frame is made by stacking and averaging a number of dark frames using the* Layer *option. Once you have a master dark frame, the* Layer *option is again used to subtract the noise from each of your images to arrive at a calibrated image. You complete the process by stacking and combining your calibrated images.*

opaque and be rendered useless. To set opacity click on Layer 1, and set the Blend Mode to Normal and adjust the Opacity to 50% (see Figure 14.1). Next, click on Layer 2 and set the Blend Mode to Normal and adjust the Opacity to 33%. Finally, click on the Layer menu and select Flatten Image which will merge all three layers into a single master dark frame image. Save this—it is your calibrated master dark frame.

- We are now ready to calibrate our images. Open the first one to be calibrated, and then open the master dark frame. Select and copy the master dark frame. Paste the copied master dark frame onto the image to be calibrated—it will turn black as the dark frame overlays it.

- In the Layers Palate, set the Opacity for both images at 100%, and the Blend Mode to Difference. When OK is clicked, the image will reappear but any sensor noise recorded on the dark frame will have been subtracted from the image. Save this calibrated image.
- The remaining images are then run through this process until all those in the sequence have been calibrated.

The calibrated images can be stacked using Layers and combined in a manner similar to how the master dark frame was created. While a great number of images can be stacked, there is a practical limit to how effective the upper layers will be if they have to be set to very low opacity levels. Because of this, images are usually combined no more than five at a time. If there are more than five images, they are combined in groups, and then the combined result of each group is then combined. For instance, 10 images would be combined in two groups of five each, and then the two resulting images would be combined.

- Open five calibrated images. Next rename one of these as your final "keeper" image. Sequentially select, copy, and paste each of the other four images onto the keeper. In the Layers Palate, place the keeper in the background and the four layers above it (click on the layer and drag it into position). Double click on Layer 1 to open the Layer Style window— set the Opacity to 50% and the Blend Mode to Normal. Then set Layer 2 to 33%, Layer 3 to 25%, and Layer 4 to 20% keeping the Blend Mode set to Normal. Complete the process by opening the Layer menu, and clicking on the Flatten Image command to combine all layers into one image.

While this process is laborious, if done manually, Photoshop has, starting with version 7, scripting options that allow the batch automation of some of these operations.

Once the final calibrated combined image is achieved, click on the Image menu and select Adjustments. Here further options may be executed. These include Brightness and Contrast to set the density and contrast of the image, Levels to correct color bias in the image, and Color Balance for fine-tuning the image color. The Apply Image "add" blending option is also useful for accentuating faint image detail. The Filter menu contains a number of sharpening options and an adjustable-parameter unsharp masking tool to make an image look crisper and better focused.

Wide-field images taken with camera lenses may suffer from blue halos around bright stars. These halos occur because chromatic aberration in the camera lens spreads unfocused blue light around the star. Using a Minus-violet filter reduces, but will not eliminate, the halos in non-apochromatic lenses. Even if images are calibrated and combined in another program, Photoshop can eliminate or greatly reduce these halos. The Replace Color tool is useful for hiding blue halos by turning them the same color as the sky background. If the halos appear around stars that are against both a black sky background and brighter star clouds or nebula, different coloration will have to be applied to hide each set of halos.

- The halos in these different parts of the sky can be separately selected by using the Marquee or Lasso Tools to select areas of the image with similar backgrounds behind the haloed stars. Once the desired areas have been selected, corrections can be applied within these regions.

- From the Image menu, click on Adjustments, then Replace Color. When the Replace Color box is open, the screen cursor turns into an eyedropper icon. Click the eyedropper on one of the blue halos and the Sample preview window in the Replace Color box will turn the same color as the halo. Use the Fuzziness slider to adjust the image preview so only the star halos appear in the preview window. Setting the Fuzziness too high will expand the range of color hues selected and will make undesired changes in other blue areas of the image. Using the Hue, Saturation, and Lightness sliders, you can adjust the color in the Sample window until the halos fade into the sky background. Decrease the saturation to turn the halos gray, and then make them darker by decreasing the lightness. Click OK to complete the color change, and then save the image.

Photoshop has many other image-tweaking options you probably will find useful, but explaining all of them would easily consume another book. A recommended resource is the E-book *Photoshop for Astrophotographers* by Jerry Lodriguss or the forthcoming book *Photoshop Astronomy* by Scott Ireland, which will be available mid-2005.

14.12 Assembling Digital Star Trails with Photoshop

For many people, star trails are an art form. The object of star trails is a pretty picture, often with a scenic terrestrial foreground, and thus the scientific rules that govern lunar and planetary or deep-sky photography do not apply. Like traditional deep-sky digital imaging, star trails are also taken with a series of short exposures that are assembled into a longer one. Since the stars move on the imaging sensor during the exposure, little in the sky besides the star trails themselves is recorded, so the JPG format (which quickly records the image to the memory card) can be used to shorten the interval between successive images.

The first task is to calibrate each exposure segment in the star trail sequence. This is done by subtracting dark frames using the Layers option as described earlier. Some imagers prefer to complete their star trail com-

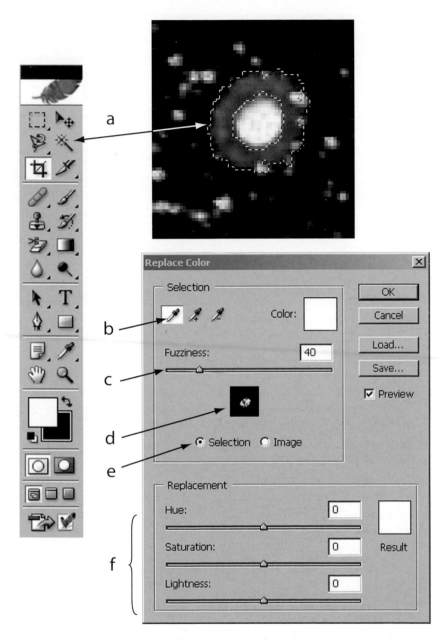

Fig. 14.2 *Photoshop's* Marquee *tool can be used to remove "halos" from stars by selecting similarly-colored pixels. Once the area has been selected (b), the* Replace Color *window (a) is opened and the eyedropper tool is used to sample the colors. The* Fuzziness *adjustment is used to vary the aggressiveness of the color selection within the area defined by the* Lasso Tool *(c). The preview window can be limited to the selection or the entire image (e).* Hue, Saturation, *and* Lightness *in the selected region can be individually adjusted to match the background (f).*

Fig. 14.3 *This final image of M13 was created by using Photoshop 7.0 to combine 5 one-minute exposures with a 200-mm f/4 lens on a Canon 10D set at ISO 400. The camera's RAW format was converted to TIF for processing. The image was processed with dark frame subtraction, aligning and stacking, levels, and unsharp masking. Photo by Robert Reeves.*

posite, and then calibrate the combined completed image. I have found that when the camera tripod is placed on soft ground or grass, the aim of the camera may slowly shift during the image sequence. If the field-of-view shifts more than a pixel or two during the sequence, each off-center image can be moved back into place during the stacking process. This shift will be almost imperceptible in the finished image, but would have rendered the calibration of a combined image less effective. To work around this, I perform the calibration process on each individual frame. This takes longer, but the end result is a cleaner image.

The secret to assembling digital star trails is to use the Lighten mode within Photoshop's Layers palate to composite sequential images. This mode compares the selected image, called the foreground, with the image added in the underlying layer. If the overlaying pixel brightness values are the same for both images, it leaves them unchanged. However, if the pixel values in one image are higher, the lighten mode uses the higher pixel value. When combining sequential images into a single star trail image, the net effect is that the sky brightness never increases beyond that present in the brightest single image; but in the meantime, the combined star trails grow longer as more images are added. Any illuminated foreground object also stays the same. With film, bright foregrounds or bright urban skies would hopelessly overexpose. But by limiting each segment of the digital multiple exposures to an interval that prevents overexposure, the resulting composite of all segments is no brighter than the brightest single image. Light pollution or moonlight glow is thus suppressed to the level recorded on a single short exposure, but the overall image shows the stars combined from all images in the composite.

- Open the first calibrated image in the star trail sequence. Using the Save As option, rename and save this first image with the file name to be used by the completed composite. Open the next image taken in the sequence and click Select and All, then Edit and Copy to move it to the Windows clipboard. The copied image can now be closed. With the first image active, click Edit and Paste to paste the image in the clipboard onto the first image. Click Window then Layers to open the Layers Palate. Click on the Blend mode dropdown menu and select Lighten.

- To check if there has been any shift in camera aim during the sequence of images, magnify the composite by pressing alt and + until the image zooms large enough to see if the same pixels in both images overlap exactly. Examine foreground objects only, because the stars will naturally move between the two exposures. If there is a shift between the two images click Window and Tools to activate the tool bar. Click on the Move Tool and then place the cursor over the composite image. Press the left mouse button and drag the layer into place over the background image until it lines up on a pixel-by-pixel basis.

- Open the rest of the images in the sequence and repeat the above steps to copy and paste them onto the first background image, then apply the lighten mode and align the images if necessary.

The accumulated layered image is a stack of all those added to it, and thus the file size is the total of all the images added to the composite. So, a composite of 10 separate 16-megabyte images totals 160 megabytes. If your computer has limited RAM, Photoshop may have trouble handling images this size since it typically needs about four times as much RAM as the image file size being processed. To work around the file size problem, the image can be flattened which will reduce the file size. Additional layers can then be added to the flattened image. You will probably have to repeat this process as you assemble your image.

To flatten the image, click Layer, then Flatten Image. Once all the exposures have been stacked and flattened into a single image, save your completed star trail composite. Additional processing such as darkening the sky or manipulating any foreground objects is then performed on this composite.

Once the composite is complete, there may be slight gaps between the exposure segments. You may or may not see these depending on how you have oriented your camera. Shooting near the pole will take advantage of the fact that the stars appear to be moving slower so there is less likelihood of gaps. However, near the celestial equator, the stars have faster apparent motion and gaps are more likely. There are several ways you can deal with them. Shooting at less than your camera's highest resolution will bin pixels which will tend to close any tight gaps. Another method is to make a duplicate of the original image, then copy and paste it onto the original. Using the Lighten mode in Layers, slightly shift the pasted image to overlap the gaps or, if gaps are present in a circumpolar composite, the copy can be rotated slightly before pasting.

14.13 ImagesPlus

In lieu of a printed user's manual, my ImagesPlus has a series of instructional AVI videos on CDs that graphically demonstrate its functions. These videos typically last several minutes for each of the several dozen processing options covered. The menu options described here are for ImagesPlus Version 2.0.

14.13.1 Creating a Master Dark Frame

Again, as with other software, the first processing step is the creation of a master dark frame with which to calibrate the actual images.

- From the File drop-down menu, select Image File Operations, then Combine Files. A Windows file selection box will appear. Highlight all the dark frames to be combined, and then click Open. A Combination Methods box will appear. To create a master dark frame, select Average, and then click OK. The combined file is displayed with the default name CombineFilesAvg. The file can be renamed using the File, Save, and Save Copy As commands as needed. See Figure 14.4.

14.13.2 Image Calibration

The next step is to calibrate the images using the master dark frame you just created.

- Click on the Calibrate Images menu, then Calibration Setup.
- In the calibration parameters box, click on the Dark button, then click From Disk to select where the master dark frame is located. Once the dark frame location is entered, click Done. Return to the Calibrate Images and select Calibrate Files. In the Windows file selection box, highlight the images that are to be calibrated. Another window will open asking where

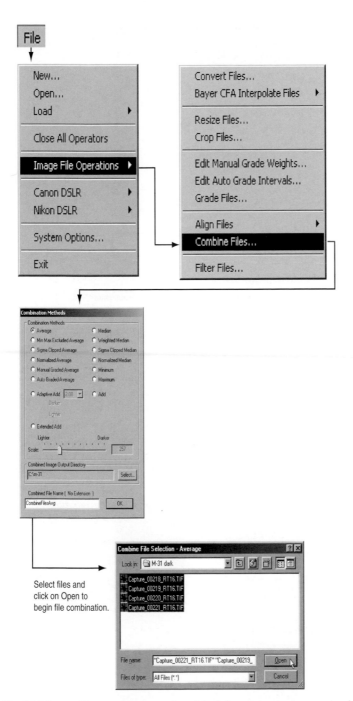

Fig. 14.4 *ImagesPlus combines a series of dark frames to create a master dark frame in the six steps outlined here.*

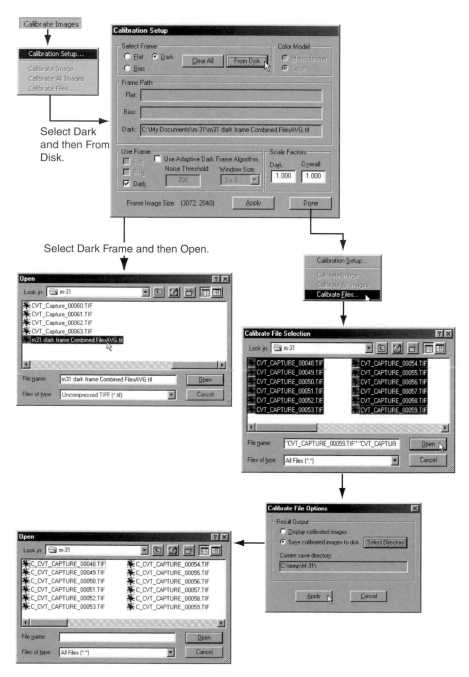

Fig. 14.5 *Once a master dark frame has been created in ImagesPlus the dark frame is loaded along with a series of images that are to be calibrated. When the* Calibrate Images *command is invoked, the process begins and is completed once all the images have had the dark frame subtracted.*

to write the calibrated files. Once this is entered, click Apply and the calibration process begins. The calibrated files are automatically written to the previously selected location and the file names are appended with the prefix C_. The original uncalibrated images remain unchanged.

14.13.3 Grading Images

ImagesPlus has an automated function that allows each image to be graded against either default or custom parameters to determine if it was poorly focused or guided. The grading procedure displays the results in a text file and, by default, assigns a grade of A, B, C, D, or F to each image. When combining files, you should use images that are grade B or better.

- From the File dropdown menu, select Image File Operations, then Grade Files. In the Windows file selection box, highlight the images to be graded and click Open. In the Grade Files box, click on Auto and the grading process will begin. When completed, a text file will display the grade assigned to each image and display the mathematical focus parameters for each color channel of the image used to create the assigned grade.

14.13.4 Alignment and Stacking

You next select the best images for alignment and stacking. As we will see later, the alignment steps to eliminate slight inaccuracies in guiding, field rotation, or flexure in the camera mount are similar to the operation performed in AIP4Win; but unlike the one-time process used in AIP4Win, the alignment stars in ImagesPlus must be selected separately in every image, not just in the first image in the stack.

- From the File menu, select Image File Operations, then Align Files, then Translate, Scale, Rotate.
- In the file selection window, highlight the previously calibrated images that are to be aligned and combined, and then click Open.
- The first image to be aligned will open, followed by a window asking for the image alignment parameters. For the Alignment Method, leave the default Centroid checked on. For the Alignment Type, select Translate+Scale+Rotate. This option will correct any field rotation that occurred between images. In the Alignment Selection area, click on Common Point or Star and leave Reference Image and Auto Advance after Select clicked on. In the Aligned Image Output Directory, select the folder where the aligned images are to be written.
- Now move the cursor so it points directly to a star near the center of the displayed reference image. Press the Shift key while left mouse clicking on the star. The next image in the selection sequence will be loaded and the cursor arrow will be replaced with a box containing a small crosshair.
- On the second image, using either the mouse or arrow keys, adjust the

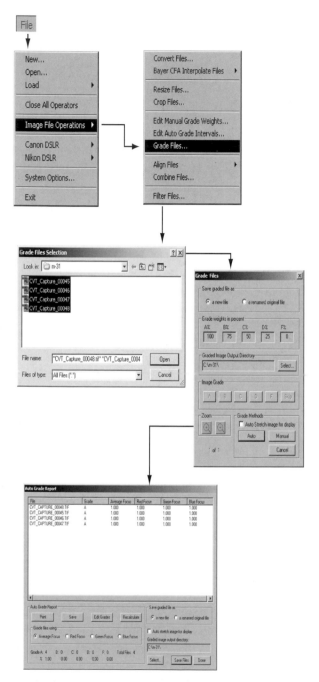

Fig. 14.6 *In this sequence of operations, ImagesPlus evaluates calibrated images and assigns them a grade ranging from A to F. Normally, only A and B images are used in subsequent operations.*

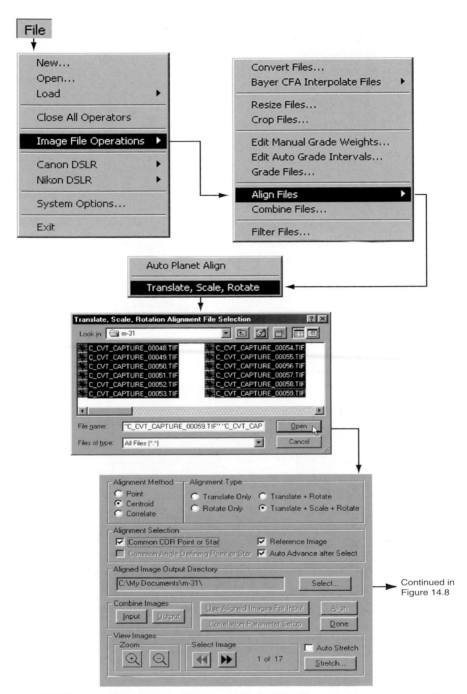

Fig. 14.7 *The ImagesPlus Image Combine operation begins with translation, rotation, and alignment of individual calibrated images. This process is continued in Figure 14.8.*

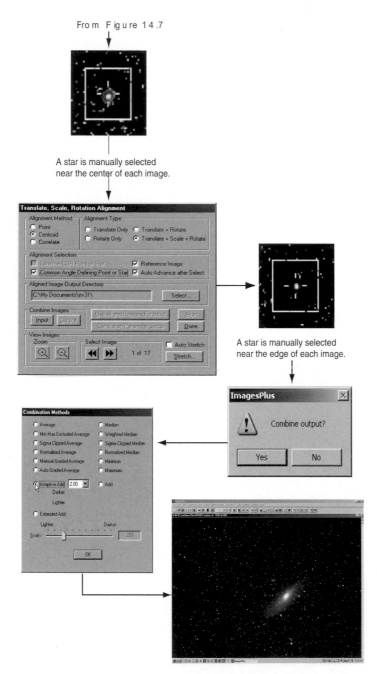

Fig. 14.8 *The image combination process is continued here from Figure 14.7. The images are dimensioned manually by first selecting a star near the center of each. Once the image stack has a designated central star the process is manually repeated with the selection of a star near the edge of the field. Upon completion of the star selection sequence the images are combined to produce a master image suitable for further processing.*

crosshairs so that they are centered on the same alignment reference star as the first image. Once the crosshairs are centered on the star press Shift and left mouse click again. Repeat the process for all the selected images. When the central alignment star in all images has been "tagged," the first reference image will reappear with a red box around the selected star.

- Go back to the alignment parameters selection box and uncheck Common Point or Star and then check Common Angle Defining Point or Star. This will provide the reference to rotate the images to remove any field rotation. Move the cursor arrow so it now points at a star near the edge of the image, then press Shift and right click the mouse. The next image will appear along with the crosshairs. Repeat the Shift and right click selection of the new alignment star until the first reference image reappears with a red box around both stars used for alignment reference.

- Go back to the parameters selection box and click Align to start the automated alignment process. The aligned images will be written with TnRt_ appended to their files names. The original calibrated image files will remain unchanged.

- When the alignment process is complete, a window opens and asks if the aligned images should be combined into a single file. If "Yes" is selected, the same image combination window will appear as seen when creating the master dark frame. This time, select the Adaptive Add option. This will add the images together while suppressing the sky background so a large number of images will not simply add up to a blank white image. When completed, the combined image will be displayed. It can be saved using the File, Save, and Save Copy As options and the file name will be appended with the prefix CombineFilesAdpAdd.

14.13.5 Digital Development

Digital development is a useful tool to bring out the dim details in a combined image. Select the Color menu option, then Brightness, Levels, and Curves, then Digital Development. For initial processing experiments, set the "break point" in the parameter selection box to about 6000 when processing 16-bit images. If the results are not to your liking, reset the slider controls and use the Undo option in the tool bar to return the image to its previous state for another try. Users accustomed to the near-instantaneous action of the Photoshop Undo command will find that ImagesPlus has a noticeable lag in the restoration of the previous image version.

Useful features for early experimentation are the Color Adjustment option under the Color menu that corrects color bias and setting the color balance. Noise reduction and image sharpening can be found under the Local menu, then Smoothing and Noise Reduction. A number of options from this point allow the user to tailor the processing. Under the Restoration

Fig. 14.9 *This final image of M31 was created by using ImagesPlus to combine 17 two-minute exposures through a 200-mm f/4 lens on a Canon 10D set at ISO 400. The camera's RAW format was converted to TIF for processing. The image was created using dark frame subtraction, aligning and stacking, digital development, and adaptive Richardson-Lucy deconvolution. Photo by Robert Reeves.*

menu, select Iterative Restoration, then Adaptive Richardson-Lucy. The default 10 iterations works well to start with, but experimentation will be the best guide in what works best for a particular image.

ImagesPlus has many more processing options, but at this point the user should have a "presentable image" to display and have the basic image data needed to experiment with more advanced processing options.

14.14 AIP4Win

Astronomical Image Processing for Windows (AIP4Win) is widely used by CCD camera imagers. The newly released Version 2.0 now supports the RGB images created by digital cameras. Previously, AIP4Win created color images by merging three separate CCD images taken through red, green, and blue filters. Now the program also applies its feature-rich suite of advanced image processing options to the combined three-color RGB images produced by digital cameras. Indeed, a user scanning the menu options for AIP4Win will find an amazing number of processing options. Do not let the sheer number overwhelm you—image processing is a step-by-step process and the best way to get into a program is to first explore the important options needed to create the basic image. Once a calibrated and stacked image is created, more advanced processing features can be explored at a leisurely pace.

Assuming a master image is going to be created from a sequence of digital exposures, the first menu options to be used will be Calibrate, then Multi-Image. After the master image is created, the View menu allows the usual selection of image viewing and zoom parameters. The Measure menu displays an incredible amount of statistical information about the image. As you work with this tool later, you will find it will help you rapidly evaluate an image at a very basic level. The actual processing of the master image will be performed within the Enhance, Transform, Color, and Multi-Image menus. An explanation of the use and function of each processing tool is displayed from the Help menu by pressing the F1 key while the tool is open.

When AIP4Win is started, a separate window called AIP Data Log is created, which is then automatically minimized. The data log is a text file that records all the processing steps and parameters applied to the image. It can be saved and used as a guide for processing other images. AIP4Win automatically stores the processing history of an image within the FITS file format. Other formats, including TIF, do not store processing history within the image file and the history data log must be saved separately.

Our first task is to apply dark frames to each exposure in order to subtract sensor noise. In my Photoshop discussion, I pointed out that the process of dark frame subtraction and image stacking can be a time-consuming and tedious process. AIP4Win solves this with the automated creation of master dark frames, and then applies the master dark frame to calibrate each image during the one-step process of image alignment and stacking. From the standpoint of ease of use and speed, these automated routines in AIP4Win stand far above other digital camera astronomical image processing programs.

The automatic combination of any number of dark frames to create a master dark frame is performed in the Calibrate menu.

- From the Calibrate drop-down menu, select Setup and the Calibration Setup box will appear. Select Basic as the Calibration Protocol and leave the Average Combine button clicked on. Average Combine is used to create most master dark frames. Median Combine is used if any frames are known to have sharp defects like cosmic ray strikes. It eliminates those images with the greatest and least amount of noise, including those with cosmic ray artifacts, but the resulting master image will be slightly degraded.
- Click on the Select Dark Frame(s) button and a Windows file selection

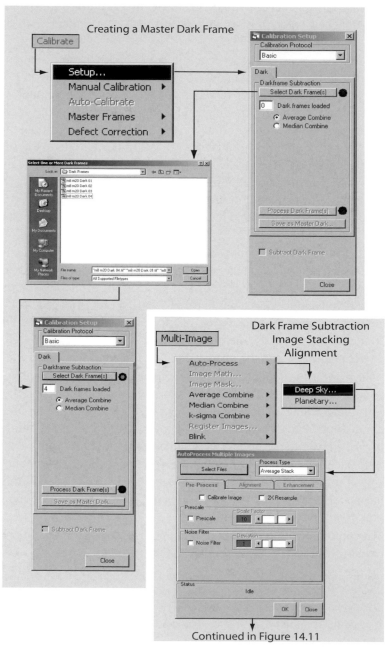

Fig. 14.10 *AIP4Win creates a master dark frame by selecting and merging a series of dark frames. In this sequence, only the Basic Calibration Protocol is demonstrated. Once the master dark frame has been created the dark frame subtraction, image stacking, and alignment process is initiated (and continued in Figure 14.11).*

box opens. Highlight the dark frames to be used, then click Open. The red button next to the selection button will turn green when all the selected dark frames have been loaded. Next, click the Process Dark Frame(s) button. When the Master Dark frame has been created, the red button will turn green and the Save Master Dark button becomes active. The master dark can be saved as a FITS file if it is to be applied to future images, but it is not mandatory to save it to process the current images as long as AIP4Win is not closed. Make sure the Subtract Dark Frame box is selected, and then click OK.

The actual dark frame subtraction, image stacking, and alignment is done in the Multi-Image and Auto-Process options. These are among the most powerful and complex operations in the program and, in my opinion, help make AIP4Win stand out from the others.

- Click on the Multi-Image menu, then Auto-Process, then Deep Sky. The Auto-Process Multiple Images box will appear. Do not use the noise filter in the Pre-Process tab of the Auto-Process Multiple Images box if processing wide-field images that may contain star images only one pixel in size. The noise reduction routine will mistake tiny stars for noise and eliminate them. This option does work well with images taken through a telescope where stars are larger than one pixel.

- Use Average Stack as the process type, and then click the Select Files button. In the Windows file selection box, highlight the files to be calibrated, stacked, and aligned, and then click Open. The image will not open at this time, as one would expect. First, under the Alignment tab, click on the Select Master Image box to select the image to be used as the master reference frame for aligning subsequent images. Once the master image is selected, then the first image will appear on the screen and more options will become active in the Alignment tab.

When automatically aligning and stacking a series of images, always use the first image in the series as the master reference frame. The star drift between succeeding frames will usually move in the same direction, allowing the program to track the star from frame to frame. If an image in another part of the sequence is used as the master, the selected alignment star may initially drift in one direction, and then jump backwards as the earlier images in the sequence are encountered. The program may lose its lock on the alignment star if the jump is too great.

- In the Alignment parameters section of the Alignment tab, click on the Two star button. This will allow a selection of two widely separated stars to align all the images and eliminate any field rotation as well as shifting between images. In the Track Radius section, choose about 15 pixels to set the size of the tracking circle to be placed around two stars. Click on an isolated star on one side of the image to place the tracking circle

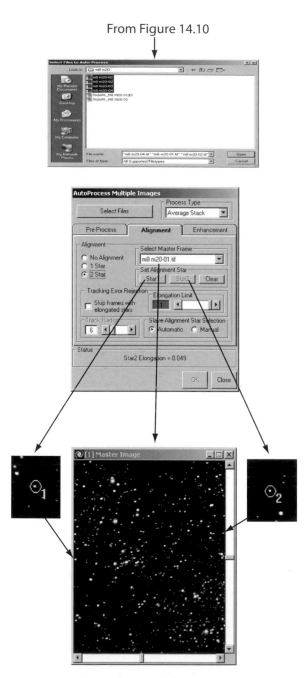

Fig. 14.11 *This is a continuation of the dark frame subtraction, image stacking, and alignment process begun in* **Figure 14.10**. *Here, two stars are selected at opposite outside edges of the first image in the stack. The program then references these stars to the other images in the stack, registers them, and produces a combined image automatically.*

around it. If the circle is offset, re-click the star to move the circle. To lock in the selected alignment star, click on Star 1 in the Set Alignment Star section. Now click on another isolated star on the opposite side of the image to place a tracking circle around it, and then click on Star 2 to lock it in. The Status section at the bottom of the box will display the elongation of the selected star. A number such as 0.15 or 0.20 is usually normal, but if the number is significantly greater, a double star may have been selected.

The alignment circle size is a compromise between being so small the star may not be within the same size circle on the next image, and so big that another star may drift into the circle, creating a misalignment in succeeding images. For this reason, try to select alignment stars that are isolated from others.

If properly aligned, many of today's electronically-controlled telescope mounts are accurate enough to track a target star for a minute or so without guiding corrections. This makes it possible for an automated sequence of one- or two-minute exposures that are tracked, but not guided. A majority of the exposures will exhibit good star images, but statistically, the natural periodic error in any mount will degrade a certain number of them. The Skip frames with elongated stars option commands AIP4Win to eliminate frames with smeared stars from the image stack. The acceptable amount of star elongation can be tailored in the Elongation Limit section.

- Click OK to begin the integrated process of dark frame subtraction, alignment, and stacking. A status bar at the bottom of the box tracks the progress of the operation. When the process is complete, the calibrated and combined image will be displayed. Click Close to close the Auto-Process box, and then save the image using the floppy disk icon at the upper left of the main AIP4Win screen.

At this point, the series of images is calibrated, stacked, and aligned, all in one automated process. Now the actual image processing to bring out faint detail can be performed.

The Image Display Control box normally appears in the upper left of the screen and presents important image data. If multiple images are open, the active image to be displayed can be selected from the Currently Active Image section. Under this are three tabs to select Display, Defaults, and Imager. For now, leave the defaults as they are, but note the Ignore Edge option within Defaults. As we found in Section 3.13, all imaging sensors have a black mask occulting the outer rows of pixels so the camera software can determine what true black is. The Ignore Edge option instructs AIP4Win to ignore the masked pixels so the histogram black pixel count will not be skewed.

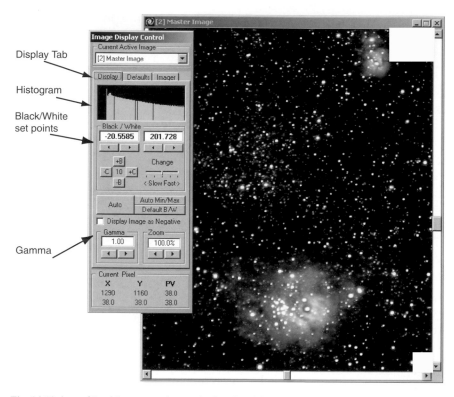

Fig. 14.12 *A combined image can be manipulated with basic image processing operations from AIP4Win's Image Display Control window. Here, the histogram can be inspected, the image black and white points set, and the gamma adjusted. As changes are made, the displayed image automatically updates. Nearly 100 different and more complex operations are available through the various menus.*

The image histogram is shown at the top of the Display tab. Normally, the steeper the left ramp on the histogram, the less background gradient there is in the image. A spike at the right side of the histogram is usually attributable to saturated overexposed stars. In the Black/White section, the pixel values used to display black and white are shown. Either typing in a new value or using the left and right arrow buttons under the value can change these values with a corresponding change in the visual appearance of the displayed image.

The Gamma selection changes the brightness of the image, with numbers higher than 1.0 making the image brighter and numbers less than 1.0 making the image darker. The Zoom window sizes the image for convenient display at your monitor resolution. The default is 50%, but if a different setting is needed, it can be permanently altered in the Defaults tab.

(Pressing the Control + F keys will zoom out to show the full image.)

At the bottom of the Image Display Control box is the Current Pixel values for the pixel directly under the cursor. The X and Y figures are the coordinate location of the selected pixel in the image array. The PV figure is the brightness of the selected pixel with 0 being pure black and 255 pure white. The red, green, and blue numbers at the bottom are the RGB values for the pixel selected. I recommend that you not trust the colors of stars that have saturated the sensor. It is best to work with the histogram to make them white.

An important feature of AIP4Win is that the image displayed will change as histogram pixel values are changed, but the original image is never altered. This allows direct comparison between the new and old images to check the quality of the change. These new images are not automatically written to a file (thus preventing clutter on your hard drive), they are only displayed. If you want to save intermediate images as you process you must explicitly issue a save command.

A good place to start processing the combined image is to work with the Gamma control in the Image Display Control to brighten the non-stellar objects such as nebulae and star clouds. Next, work with the Black/White figures until the left side of the histogram slope is almost at the left side of the histogram display. The displayed image will likely appear pale at this point. Select the Enhance menu, then Histogram Shape. To start with, use the default values and click Apply in the Histogram Shape Input tab. The image will now display a greater contrast, with the dark sky being closer to black and star clouds and nebulae being brighter and more visible.

To start point color processing, select the Color menu, then open the Color Image Tool. Begin with the default Auto Levels setting in the Adjust Color Method, then click Make New Working Image. The resulting image will now have deeper blacks and richer colors. Another color option is to split the image into its color components, then recombine them using a different color method. Under the Color menu, select Split Colors, then Color – LRGB to create a separate color channel for each of the red, green, blue, and luminescence channels. Once each channel is created, use the default settings in the Join Colors Tool option and reselect each channel in its appropriate color selection box; select G2V for Color Join Method and click on the Create Color Image button. Again, a new image is created which displays deeper blacks and brighter colors. If the original image suffered from blue halos surrounding bright stars, they can be shrunk by processing the blue channel to reduce their diameters, which in turn, will reduce the blue halos in the recombined color image.

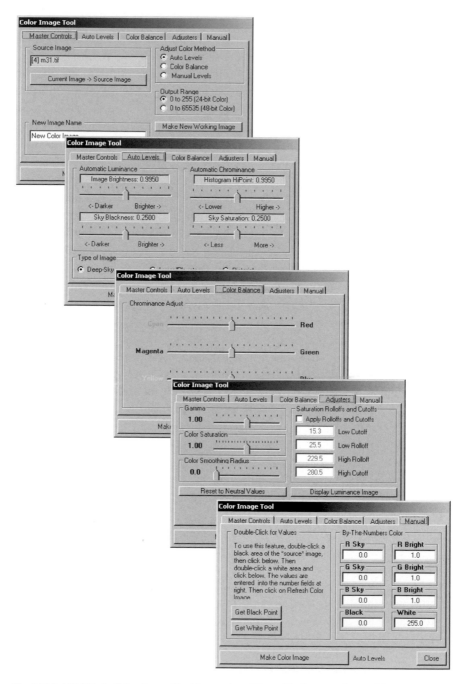

Fig. 14.13 *AIP4Win's Color Image Tool is a series of five tabbed menus that enables a full range of color adjustments.*

Fig. 14.14 *This final image of the M8 and M20 region was created by using AIP4Win to combine 7 two-minute exposures through a 135-mm f/2.8 lens on a Canon 10D set at ISO 400. The camera's RAW format output was converted to TIF for processing. The image was created using dark frame subtraction, aligning and stacking, histogram shaping, and color recombination using the G2V color balance. Photo by Robert Reeves.*

Depending upon the image, the steps we have discussed up to now may be all the processing needed. However, if it is still weak and lacks the desired "snap," it can be processed further using the Color Effects Tool. This tool has many other options including saturation adjustments, color balance, hue, and smoothing.

If the image is slightly mushy, the deconvolution routines will help to sharpen it. For deep-sky images, try Richardson-Lucy deconvolution. When working with planetary images, Van Cittert deconvolution usually works best.

If the stars look slightly dull and the overall image is not as bright and crisp as desired, Brightness Scaling under the Enhance menu will often help with brighten stars and other portions of the image. As with any AIP4Win processing option, if the operation does not work to your satisfaction, just delete the newly created image—your original remains unchanged.

14.15 Conclusion

As seen in the above examples, image processing can evolve into a highly complex sequence of operations, often involving the use of multiple image processing programs. There are seemingly limitless options available to the user who wishes to pursue techniques beyond the basic operations needed for display of the image. For those who desire the best images their digital photographic equipment is capable of taking, this process will consume far more time than that needed to acquire the image under the open sky. Experienced astrophotographers know that every hour spent under the stars will be followed by two or three hours of coaxing the best image possible out of the magical collection of binary numbers the camera stores on the memory card.

It is a fact that digital imaging is the "next big thing" in astrophotography. The astrophotographer who uses a late-model DSLR camera and a modern astronomical image processing program can reflect on the knowledge that they are using cutting edge technology and techniques that were not available, even to the professional, just a decade ago. Digital techniques allow today's astrophotographer to take bold deep steps into the universe that were unimaginable when film photography was the mainstay in astronomy. But remember, this new medium is still technically in its infancy. For those who are amazed at the results achieved so far with digital astrophotography, there is but one simple thought to ponder: You haven't seen anything yet . . . the best is still to come and you have an opportunity to be an active participant!

Appendix A
Equipment Suppliers and Manufacturers

Digital Cameras

B&H Photo
420 Ninth Avenue
New York, NY 10001
800-962-3999
www.bhphotovideo.com/

Canon Cameras
850 Greenbriar Circle
Chesapeak, VA 23320
800-652-2666
www.canonusa.com

Fuji Photo Film USA, Inc.
P.O. Box 7828
Edison, NJ 08818
Attn: Customer Care Dept.
732-857-3487
www.Fujifilm.com

Eastman Kodak Company
343 State Street
Rochester, NY 14650
800-235-6325 (paid support phone number)
www.kodak.com

Konica-Minolta
725 Darlington Avenue
Mahwah, NJ 07430
800-285-6422
kmpi.konicaminolta.us/

Nikon
1300 Walt Whitman Road
Melville, NY 11747
www.nikonusa.com/

Olympus America, Inc.
2 Corporate Center Drive
PO Box 9058
Melville, NY 11747
888-553-4448
www.olympusamerica.com

Pentax Imaging Company
600 12th Street Suite 300
Golden, CO 80401
800-877-0155
www.pentaximaging.com

Sigma Corporation of America
15 Fleetwood Court
Ronkonkoma, NY 11779
800-896-6858
www.sigmaphoto.com/

Miscellaneous equipment and project supplies

Cole-Parmer Instrument Company
625 East Bunker Court
Vernon Hills, IL 60061
800-323-4340
www.coleparmer.com

Eheim Aquarium Pumps
888-343-4662
www.eheim.com

Mosquito Magnet pest control
American Biophysics Corp.
140 Frenchtown Road
N. Kingstown, RI 02852
877-699-8727
www.mosquitomagnet.com/

Peltier coolers
TE Technology, Inc.
1590 Keane Dr.
Traverse City, MI 49686-8257
231-929-3966
www.tetech.com

Batteries and Battery Chargers
Thomas Distributing
128 East Wood
Paris, IL 61944
800-821-2769
www.thomas-distributing.com/

Xantrex
5916 195th Street NE
Arlington, WA 98223
800-670-0707
www.xantrex.com

Voltage Valet Division, Hybrinetics, Inc.
225 Sutton Place
Santa Rosa, California 95407
800-247-6900
www.voltagevalet.com

Scope-Guard Telescope Cases
Holco Company
21102 Timber Ridge Drive
Magnolia, TX 77355
281-259-3305
www.scopeguard.com

Photographic accessories and camera adapters

Alpine Astronomical
208-939-2141
www.alpineas-
tro.com/digital_photography/digital_phot
ography.htm#OPFA

Adirondack Video Astronomy
26 Graves Street
Glens Falls, NY 12801
518-812-0025
www.astrovid.com

Anacortes Telescope and Wild Bird
9973 Padilla Heights Road
Anacortes, WA 98221
800-850-2001
www.BuyTelescopes.com

Baader Planetarium
Zur Sternwarte
D-82291 Mammendorf
Germany
+49-8145-8802
baader-planetarium.de

BC&F Astro Engineering
Metal Planet
The Gatehouse
Straight Furlong
Pymoor, Ely
Cambridgeshire
CB6 2EH
UNITED KINGDOM
www.hfo.org.uk/

CKC Power
23 Graywood Dr.
Orangeburg, NY 10962
845-627-1055
www.ckcpower.com

Foveon, Inc.
2820 San Thomas Expressway
Santa Clara, CA 95051
408-350-5100
www.foveon.com

Hap's Astrocables
2740 Fenimore Drive
Sumter, SC 29150
www.hapg.org/astrocables.htm

Harbortronics
P. O. 2663
Gig Harbor, WA 98335
253-858-7769
www.harbortronics.com

Hoodman Corporation
20445 Gramercy Place Ste. #201
Torrance, CA 90501
800-818-3946
www.hoodmanusa.com/

Hutech Corporation.
23505 Crenshaw Blvd. #225
Torrance, CA 90505
877-289-2674
www.sciencecenter.net/hutech/

Jim's Mobile, Inc.
810 Quail Street, Unit E
Lakewood, CO 80215
303- 233-5353
www.jimsmobile.com

Kendrick Astro Instruments
2920 Dundas St. West
Toronto, Ont M6P 1Y8
416-762-7946
www.kendrick-ai.com/

Kinetronics Corporation
4363 Independence Court
Sarasota, FL 34234
800-624-3204
www.kinetronics.com

Korry Nightshield
Korry Electronics Co.
901 Dexter Avenue North
Seattle, WA 98109
206-281-1300
www.korry.com/products/nightshield/

LensPlus
11969 Livona Lane
Redding, CA 96003
800-659-7770
www.lensadapter.com

Lumicon International
750 Easy Street
Simi Valley, CA 93065
805-520-0047
www.lumicon.com

Newegg.com
132 South 6th Ave.
La Puente, CA 91746
800-390-1119
www.newegg.com/

Orion Telescopes & Binoculars
P.O. Box 1815
Santa Cruz, CA 95061
800-676-1343
www.telescope.com

Photosolve
21272 Chiquita Way
Saratoga, CA 95970
www.photosolve.com

Power-Sonic Corporation
9163 Siempre Viva Road, Suites A-F
San Diego, CA. 92154
619-661-2030
www.power-sonic.com

Radio Shack
200 Taylor Street, Suite 600
Ft. Worth, TX 76102
617-415-3200
www.radioshack.com/

SanDisk
140 Caspian Court
Sunnyvale, CA 94089
408-542-0500
www.sandisk.com/

Santa Barbara Instrument Group
147-A Castilian Drive
Santa Barbara, CA 93117
805-571-7244
www.sbig.com

ScopeTronix
914 SE 14th Place
Cape Coral, FL 33990
866-458-7658
www.scopetronix.com

Sine Patterns LLC
1653 East Main Street
Rochester, NY 14609
585-482-0300
www.sinepatterns.com/

Stellar Technologies International
408 E Bowie
Alamo, TX 78516
800-232-9416
www.stellar-international.com/

Tau Ceti Company
P.O. Box 1101
Conroe, TX 77305
936-539-4073
www.us-astro.com

TeleVue Optics
32 Elkay Drive
Chester, NY 10918
845-469-4551
www.televue.com

Williams Optics
714-209-0388
www.william-optics.com/

Software and Publications

AIP4Win
Willmann-Bell, Inc.
P.O. Box 35025
Richmond, VA 23235
804-320-7016
www.willbell.com

DSLRFocus
www.DSLRFocus.com/

ImagesPlus
Mike Unsold Digital Imaging
193 Tallmadge Road
Rootstown, OH 44272
330-325-0765
www.mlunsold.com/

MegaStar
Willmann-Bell, Inc.
P.O. Box 35025
Richmond, VA 23235
804-320-7016
www.willbell.com

Photoshop CS image processing software
Adobe Systems Incorporated
345 Park Avenue
San Jose, CA 95110-2704
888-724-4508
www.adobe.com/

Photoshop for Astrophotographers
Astropix LLC
P.O. Box 296
Somerdale, NJ 08083
www.astropix.com/PFA/ORDER.HTM

The Sky
Software Bisque
912 12th Street
Golden, CO 80401-1114
800-843-7599
www.bisque.com/

Solar Filters

Baader AstroSolar Safety Film
Marketed in USA by:
Astro-Physics, Inc.
11250 Forest Hills Rd.
Rockford, IL 61115
815-282-1513
www.astro-physics.com

Coronodo Technology Group
1674 South Research Loop Suite 436
Tucson AZ 85710
866-786-9282
www.coronadofilters.com/

DayStar Filters
3857 Schaefer Ave., Suite D
Chino, CA 91710
909-591-4673
www.daystarfilters.com/

Thousand Oaks Optical
Box 4813
Thousand Oaks, CA 91359
805-491-3642
www.thousandoaksoptical.com/

Roger W. Tuthull, Inc.
1255 Toms River Road
Jackson, N J 08527
732-940-0218
www.tuthillscopes.com/

Telescopes and Mounts

Astro-Physics
11250 Forest Hills Road
Rockford, IL 61115.
815-282-1513
www.astro-physics.com/

Celestron International
2835 Columbia Street
Torrance, CA 90503
310-328-9560
www.celestron.com

Losmandy Astronomical Products
Hollywood General Machining, Inc.
1033 N. Sycamore Avenue
Los Angeles, CA 90038
323-462-2855
www.losmandy.com

Meade Instruments Corporation
6001 Oak Canyon
Irvine, CA 92618
800-626-3233
www.meade.com/

Mountain Instruments
1213 So. Auburn St.
Colfax, CA 95713
(530) 346-9113
www.mountaininstruments.com/

Software Bisque
912 12th Street
Golden, CO 80401-1114
800-843-7599
www.bisque.com/

Takahashi Telescopes
Land, Sea & Sky
(Texas Nautical Repair)
3110 S. Shepherd Dr.
Houston, TX 77098
713-529-3551
www.lsstnr.com/main_takahashi.htm

Tele Vue Optics, Inc.
32 Elkay Dr.
Chester, NY 10918
845 469 4551
www.televue.com/

Appendix B
Helpful Web Sites

Digital Terminology and Glossary

www.acdsystems.com/English/Community/Resources/Glossary/index.htm

photography.about.com/library/glossary/blglossary.htm

www.starizona.com/ccd/ccdglossary.htm

support.radioshack.com/support_tutorials/audio_video/digvid-glossary.htm#M

Image Processing Software Suppliers

aberrator.astronomy.net/registax/
RegiStax webcam image stacking and processing program

home.no.net/dmaurer/~dersch/Index.htm
Panorama Tools

Adobe Photoshop Image Editing and Acrobat PDF Reader

www.adobe.com/

Astrostack Webcam Image Stacking and Processing Program

www.astrostack.com/index.html

IRIS, Freeware Image Processing Software From Christian Buil

www.astrosurf.com/buil/us/iris/iris.htm

DSLR Focus and Camera Control Software (PC)

www.dslrfocus.com/

iAstroPhoto Camera Control Software (Mac)

http://homepage.mac.com/stevepur/astrophotography/iAstroPhoto/

Freeware Irfanview Image Viewing And Editing Program

www.irfanview.com/

ImagesPlus Image Processing Software

www.mlunsold.com/

Photoshop Plugin for Sharpening Digital Images

www.ultrasharpen.com/

The Handbook of Astronomical Processing (HAIP) and Astronomical Image Processing for Windows (AIP4Win) Image Processing Software

www.willbell.com/aip/index.htm

Online Calculator for Digiscoping Parameters

www.jayandwanda.com/digiscope/digiscope_calc.html

Heavens-above Satellite Predictions

www.heavens-above.com

Interesting Articles on Astrophotography

www.licha.de/amateur_astronomy_articles.php

CCD and CMOS Sensor Cleaning Tips

photography.about.com/library/weekly/aa120902b.htm
photography.about.com/library/weekly/aa122903d.htm
www.photo.net/equipment/digital/sensorcleaning/
www.pbase.com/copperhill/ccd_cleaning

Listing of Digital Camera Telescope Adapters

www.szykman.com/Astro/Adapters.html

News and Reviews About Digital Photography

www.a-digital-eye.com/
www.dcviews.com/
www.dpreview.com/
imaging-resource.com/DIGCAM01.HTM
photography.about.com
www.robgalbraith.com/
www.steves-digicams.com/

Shopping Price Comparison Guides

www.dealtime.com/
www.pricegrabber.com/
www.pricescan.com/
www.resellerratings.com/
shopper.cnet.com/

Weather and Environment Information

www.spaceweather.com
Space weather

www.cleardarksky.com/csk/
Clear Sky Clock

www.cmc.ec.gc.ca/cmc/htmls/seeing_e.html
Atmospheric seeing forecast

www.inquinamentoluminoso.it/worldatlas/pages/fig1.htm
Light pollution boundary maps

www.intellicast.com/LocalWeather/World/
Intellicast worldwide forecasts

www.ssec.wisc.edu/data/geo.html
Geostationary weather satellite images

weatherimages.org/data/imag192.html
U.S. jet stream forecasts

www.wunderground.com/US/Region/US/Fronts.html
US weather front locations

http://squall.sfsu.edu/
California Regional Weather Server

Web Mailing Lists for Advice from Experienced Users

groups.yahoo.com/group/AstrophotographyCameraLenses/
Astrophotography camera lens mailing list

groups.yahoo.com/group/digital_astro/
Digital astrophotography mailing list with thousands of members

seds.org/mailman/listinfo/astro-photo
Astrophoto Mailing List (APML) film-based astrophotography mailing list

groups.yahoo.com/group/dslrfocus/
DSLRFocus software mailing list

groups.yahoo.com/group/QCUIAG/
Video and webcam mailing list with thousands of members

groups.yahoo.com/group/ToUcam/
Mailing list supporting the astronomical use of Philips ToUcam webcams

Asteroid Databases and Online Ephemeris Services

http://cfa-www.harvard.edu/iau/mpc.html
http://asteroid.lowell.edu/

Optics for Photographers

http://www.vanwalree.com/

Appendix C
Contributors' Personal Web Sites

Antonio Fernandez
http://www.astrosurf.com/afernandez/gallery/widefield/
widefield_ccd_astrophoto.htm

Dale Ireland
http://www.drdale.com/index.htm

Dave Kodama
http://www.eanet.com/kodama/astro/index.htm

Fred Bruenjes
http://www.moonglow.net/

Hap Griffin
http://www.machunter.org/

Jay Ballauer
http://www.allaboutastro.com/

Jeff Ball
http://www.astro-photography.com

Mark Hanson
http://btlguce.digitalastro.net/

Johannes Schedler
http://panther-observatory.com/

José Suro

http://web.tampabay.rr.com/jsuro/index.html

R. A. Greiner

http://www.mapug.com/ragreiner/

Rick Krejci

http://www.ricksastro.com/

Robert Reeves

http://www.robertreeves.com

Paul Hyndman

http://www.astro-nut.com/

Peter Langsford

http://www.users.bigpond.com/lansma/Default.htm

Phil Hart

http://www.philhart.com/gallery/Astrophotography

Terry Lovejoy

http://www.pbase.com/terrylovejoy/

Bibliography

Aguirre, Edwin L. "Astro Imaging with Digital Cameras," *Sky and Telescope* magazine, August 2001, p. 128.

Aguirre, Edwin L. "Deep Sky Imaging with Digital Cameras," *Sky and Telescope* magazine, October 2002, p. 113.

Askey, Phil. "AD Converters," www.dpreview.com/learn/?/Glossary/ Camera_System/AD_Converter_01.htm

Askey, Phil. "Batteries," www.dpreview.com /learn/?/glossary/Camera_System?Baterries_01.htm

Askey, Phil. "Buffer," www.dpreview.com/learn/?/Glossary/Camera_System/Buffer_01.htm

Askey, Phil. "Color Filter Array," www.dpreview.com/learn/?/Glossary/ Camera_System/Colour_Filter_Array_01.htm

Askey, Phil. "Connectivity," www.dpreview.com/learn/?/Glossary/ Camera_System/Connectivity_01.htm

Askey, Phil. "Digital Camera Specifications," www.dpreview.com/reviews/specs.asp

Askey, Phil. "Focal length Multiplier," www.dpreview.com/learn/?/ Glossary/Optical/Focal_length_multiplier_01.htm

Askey, Phil. "RAW Image Format," www.dpreview.com/reviews/canond30/page14.asp

Askey, Phil. "Sensor Sizes," www.dpreview.com/learn/?/ Glossary/Camera_System/Sensor_Sizes_01.htm

Atkins, Bob. "Cleaning the CMOS Sensor of the Canon EOS 10D (and other digital SLRs)," www.photo.net/equipment/digital/sensorcleaning/

Ballauer, Jay. "Digital SLRs versus Astronomical CCDs," www.allaboutastro.com/articles.html

Ballard, James. *Handbook for Star Trackers*, Sky Publishing Corp., Cambridge, MA.

BenDaniel, Matt, "Planning an Astro-Imaging Expedition," *Sky and Telescope* magazine, November, 2002, p. 121.

Berrevoets, Cor. "Processing Webcam Images with RegiStax," *Sky and Telescope* magazine, April, 2004, p. 130

Berry, Richard and Burnell, James. *The Handbook of Astronomical Image Processing*, Willmann-Bell, Inc., Richmond, VA, 2000.

Biggers, Bryan. "Hot Pixels," webpages.charter.net/bbiggers/ DCExperiments/html/hot_pixels.html

Blackwell, John A. "CCD Imaging: Q&A," www.regulusastro.com/regulus/papers/ccdfaq/

Blessing, Jordan. "Digital Camera Imaging With a Telescope," www.scopetronix.com/digicam.htm

Bockaert, Vincent. "Pixels," www.dpreview.com/learn/? /Glossary/Camera_System/Pixels_01.htm

Bockaert, Vincent. "Sensors," www.dpreview.com/learn/'?/ Glossary/Camera_System/Sensors_01.htm

Cannistra, Steve. "Digital Camera, CCD, or Film?", www.starrywonders.com /howtodecide.html

Cidadao, Antonio Jose. "Thoughts on Super-Resolution Planetary Imaging," *Sky and Telescope* magazine, December 2001, p. 127.

Clark, Roger. "Astrophotography Signal-to-Noise with a Canon 10D Camera," clarkvision.com/astro/ canon-10D-signal-to-noise/

Di Cicco, Dennis. "A CCD Camera Buzzword Primer," *Sky and Telescope* magazine, August, 1997, p. 109.

Di Cicco, Dennis. "Afocal What?," *Sky and Telescope* magazine, August, 2001, p. 130.

Di Cicco, Dennis. "Power to Go," *Sky and Telescope* magazine, August, 2000, p. 64.

Douglas, Eric. "How to Predict Seeing," *Sky and Telescope* magazine, January 2000, p. 128.

Dragesco, Jean. *High Resolution Astrophotography*, Cambridge, Great Britain, Press Syndicate of the University of Cambridge, 1995.

Dyer, Alan. "STV: Digital Imaging for the Masses?," *Sky and Telescope* magazine, Jaury 2001, p. 67

Eggers, Roger. "Cashing in Your Chips," www.digitaloutput.net/back%20edit/aug03/feature3.html

Etchells, Dave. "Dave's Definitive Guide to Buying a Digicam," www.imaging-resource.com/ARTS/ BUY/BUY.HTM

Etchells, Dave. "Digital Camera Test Methods," www.imaging-resource.com/TIPS/TESTS/ TESTS.HTM

Fellers, Thomas J. and Davidson, Michael W. "CCD Noise Sources and Signal-to-Noise Ratio" http:/ /micro.magnet.fsu.edu/primer/digitalimaging/concepts/ccdsnr.html

Flam, Faye. "Vapor From Jets Alters Climate," San Antonio Express-News, August 12, 2002.

Fulton, Wayne. "Graphics Interchange Format (GIF)," www.scantips.com/basics9g.html

Fulton, Wayne. "JPEG Joint Photographic Experts Group," www.scantips.com/basics9j.html

Fulton, Wayne. "PNG Portable Network Graphics," www.scantips.com/basics9p.html

Fulton, Wayne. "TIFF Tag Image File Format," www.scantips.com/basics9t.html

Gitto, Paul. "The Arcturus Observatory Mask Page," cometman.com/mask.html

Honis, Gary. "Modifications for Air Cooling," members.tripod.com/~ghonis/c2020zdeep7.htm

Horne, Johnny, and di Cicco, Dennis. "Light Pollution Filters for Cameras," *Sky and Telescope* magazine, June 2001, p. 47.

Howard, Bill. "The Essential Buying Guide Notebooks," *PC* magazine, April 20, 2004, p. 114.

Hyndman, Paul. "Digital Camera Internal Cooling Enhancement," www.astro-nut.com/ice.html

Kamanski, Dave. "Olympus Qs & As on the New $^4/_3$ System," www.a-digital-eye.com/Olympus43Q&A.html

Langsford, Peter. "Projects, Artificial Star," www.users.bigpond.com/ lansma/proj_artificial_star.htm

Langsford, Peter. "Projects ìSerial Cable for Canon EOS 300D," www.users.bigpond.com/ lansma/proj_serialcable.htm

Lehmann, Tracy Hobson. "Buzz Off," *San Antonio Express-News*, August 10, 2002, p. 1E.

MacRobert, Alan. "A New Standard in Solar Filters," *Sky and Telescope* magazine, September 2000, p. 63.

MacRobert, Alan. "Mastering Polar Alignment," *Sky and Telescope* magazine, September 1997, p. 106.

MacRobert, Alan. "Solar Filters: Which Is Best?", *Sky and Telescope* magazine, July 1999, p. 63.

Marshall, Peter. "Digital Dirt and Noise," photography.about.com/library/weekly/aa122903a.htm

Marshall, Peter. "Digital Dirt and Noise," photography.about.com/library/weekly/aa122903b.htm

Marshall, Peter. "Digital Dirt and Noise," photography.about.com/library/weekly/aa122903c.htm

Marshall, Peter. "Digital Dirt and Noise," photography.about.com/library/weekly/aa122903d.htm

Marshall, Peter. "Digital Dirt and Noise," photography.about.com/library/weekly/aa122903e.htm

Marshal, Peter. "Professional Digital Cameras," photography.about.com/library/weekly/aa102802e.htm

Marshall, Peter. "Working With Digital," photography.about.com/library/weekly/aa120902b.htm

Marshall, Peter. "Working With Digital," photography.about.com/library/weekly/aa120902c.htm

Marshall, Peter. "Working With Digital," photography.about.com/library/weekly/aa120902d.htm

Marshall, Peter. "Working With Digital," photography.about.com/library/weekly/aa120902e.htm

McLaughlin, William. "Focus," willmclaughlin.astrodigitals.com/focus1.htm

Michaud, Peter. "Capturing Star trails with a Digital Camera," *Sky and Telescope* magazine, March 2004, p. 126.

Mollise, Rod. "Imaging with Webcams," *Amateur Astronomy* magazine, Fall 2004, p. 32.

Perkins, Philip. "Guiding Techniques for Astrophotography," *Sky and Telescope* magazine, April 2003, p. 130.

Putman, Peter. "Contrast Shmontrast," www.projectorexpert.com/Pages/shmontrast.html

R., Nichola, "CCD & CMOS Cleaning," www.pbase.com/copperhill/ccd_cleaning

Reeves, Robert. "Megapixel Cameras Revive the Art of Lunar Photography," *Amateur Astronomy* magazine #37, Spring 2003, p. 22.

Reeves, Robert. *Wide-Field Astrophotography*, Richmond, Va., Willmann-Bell, Inc., 2000.

Romano, Richard. "Dancing on the Ceiling," www.micropubnews.com/pages/issues/2000/600_foc-digitalcam_mpn.shtml

Rutte, Harrie, and van Venrooij, Martin. *Telescope Optics*, Richmond, Va., Willmann-Bell, Inc. 1988.

Seronik, Gary. "Lunar Mosaics Made Easy," *Sky and Telescope* magazine, Sept. 2004, p. 134.

Smith, Gregory H. *Practical Computer-Aided Lens Design*, Richmond, Va., Willmann-Bell, Inc., 1998.

Vaughn, Chuck. "Knife-Edge Focusing for Astrophotography," www.aa6g.org/Astronomy/Articles/knife-edge.html

Szykamn, Simon. "Setting Digital Cameras for Astronomical Imaging," mysite.verizon.net/~vze4r2c2/Astro/SettingCameras.html

Szykman, Simon. "The Digital Camera Astronomical Imaging FAQ," www.szykman.com/Astro/AstroDigiCamFAQ.html

Tubervile, Jay. "Understanding Vignetting while Digiscoping," www.jayandwanda.com/digiscope/vignette/vignetting.html

Uhlman, Mariam. "DEET Still best at Bugging Mosquitoes," *San Antonio Express-News*, July 4, 2002, p. 2D.

Unk. "Car Battery Frequently Asked Questions (FAQ) Section 1," www.uuhome.de/william.darren/carfaq1.htm

Unk. "CompactFlash Association FAQ," www.compactflash.org/faqs/faq.htm

Unk. "Digital Camera Timeline," www.dpreview.com/reviews/timeline.asp

Unk. "Digital Film Comparison," www.dpreview.com/articles/mediacompare/default.asp

Unk. "Four Thirds System," www.four-thirds.com/index_01.htm

Unk. "Four Thirds System The New Digital SLR Standard," www.four-thirds.com/en/about.html

Unk. "Image Sensors," www.photozone.de/7Digital/digital_3.htm

Unk. "JPEG Image Compression FAQ, part 1/2 ," www.faqs.org/faqs/jpeg-faq/part1/

Unk. "Harbortronics DigiSnap 2000," www.steves-digicams.com/2001_reviews/digisnap2000.html

Unk. "Microdrive Storage Devices," www.steves-digicams.com/microdrive.html

Unk. "NiMH Battery and Charger FAQs," faq.alltekpower.com/faq/index.asp

Unk. "Practical Tips for Using the ST-4 Autoguider," *Sky and Telescope* magazine, April 2003, p. 134.

Unk. "Simple, Accurate Polar Alignment," *Sky and Telescope* magazine, April 2003, p. 133.

Unk. "Sharp Six Megapixel 1/1.8 Type Sensor," www.dpreview.com/news/0312/03121702sharpsixmeg.asp

Unk. "Super Duper," www.micropubnews.com/pages/issues/2000/600_foc-sb-fuji_mpn.shtml

Unk. "T-mounts, Aperture, and Vignetting," www.fli-cam.com/FLIsupport/tmount.htm

Unk. "What is CompactFlash?," www.compactflash.org/faqs/faq.htm

Wisniewski, Joseph S. "The Digital lens FAQ," www.swisarmyfork.com/digital-lens-faq.htm

Index